Almost Everywhere Convergence

Proceedings of the International Conference on
Almost Everywhere Convergence in
Probability and Ergodic Theory
Columbus, Ohio
June 11–14, 1988

Almost Everywhere Convergence

Proceedings of the International Conference on
Almost Everywhere Convergence in
Probability and Ergodic Theory
Columbus, Ohio
June 11–14, 1988

Edited by
Gerald A. Edgar
Louis Sucheston

Department of Mathematics
Ohio State University
Columbus, Ohio

ACADEMIC PRESS, INC.
Harcourt Brace Jovanovich, Publishers
Boston San Diego New York
Berkeley London Sydney
Tokyo Toronto

ACADEMIC PRESS, INC.
1250 Sixth Avenue, San Diego, CA 92101

United Kingdom Edition published by
ACADEMIC PRESS INC. (LONDON) LTD.
24-28 Oval Road, London NW1 7DX

Library of Congress Cataloging-in-Publication Data

International Conference on Almost Everywhere Convergence in Probability and Ergodic Theory (1988 : Columbus, Ohio)
Almost everywhere convergence : proceedings of the International Conference on Almost Everywhere Convergence in Probability and Ergodic Theory, Columbus, Ohio, June 11–14, 1988 / Edited by Gerald A. Edgar, Louis Sucheston.
p. cm.
Includes bibliographical references.
ISBN 0-12-231050-0 (alk. paper)
1. Convergence–Congresses. 2. Inequalities (Mathematics)– –Congresses. 3. Probabilities–Congresses. 4. Ergodic theory– –Congresses. I. Edgar, Gerald A., Date II. Sucheston, Louis. III. Title.
QA295.I56 1988
519.2–dc20 89-17626
CIP

Printed in the United States of America
89 90 91 92 9 8 7 6 5 4 3 2 1

Contents

Contributors ix

Participants xiii

Preface xv

Lower Bounds for Partial Sums of Certain Positive Stationary Processes 1
Jon Aaronson and Manfred Denker

Exact Sequences for Sums of Independent Random Variables 11
André Adler

Unusual Applications of A. E. Convergence 31
E. J. Balder

Filling Scheme Transformations of Coulomb Systems 55
J. R. Baxter and R. V. Chacon

Harmonic Analysis and Ergodic Theory 73
Alexandra Bellow, Roger L. Jones, and Joseph Rosenblatt

Almost Everywhere Convergence of Powers 99
Alexandra Bellow, Roger L. Jones, and Joseph Rosenblatt

Independence Properties of Continuous Flows 121
V. Bergelson

Moving Averages 131
N. H. Bingham

Almost Sure Convergence in Ergodic Theory 145
J. Bourgain

A Pointwise Ergodic Theorem for Positive, Cesaro-Bounded Operators on L_p $(1<p<\infty)$ 153
Antoine Brunel

On the Number of Escapes of a Martingale and Its Geometrical Significance 159
Donald L. Burkholder

Loss of Recurrence in Reinforced Random Walk 179
Burgess Davis

On Pointwise Convergence in Random Walks 189
Yves Derriennic and Michael Lin

Banach Spaces with the Analytic Radon-Nikodým Property and Compact Abelian Groups 195
Gerald A. Edgar

Generalized Measure Preserving Transformations 215
Ulrich Krengel

Tangent Sequences of Random Variables: Basic Inequalities and Their Applications 237
S. Kwapien and W. A. Woyczynski

On a Pointwise Ergodic Theorem for Multiparameter Groups 267
F. J. Martín-Reyes

The Two-Parameter Strong Law for Partially Exchangeable Arrays 281
Terry R. McConnell and Eric Rieders

Multi-Parameter Weighted Ergodic Theorems from Their Single Parameter Version 297
James H. Olsen

Weakly Ergodic Products of (Random) Non-Negative Matrices 305
Steven Orey

The *r*-Quick Version of the Strong Law for Stationary ϕ-Mixing Sequences 335
Magda Peligrad

Almost Everywhere Convergence of Some Nonhomogeneous Averages 349
Karl Petersen

Simple Symmetric Random Walk in Z^d 369
Pal Révész

Almost Superadditive Ergodic Theorems in the Multiparameter Case 393
K. Schürger

On Convergence of Vector-Valued Mils Indexed by a Directed Set 405
Zhen-Peng Wang and Xing-Hong Xue

Contributors

The numbers in parentheses refer to the pages on which the authors' contributions begin.

Jon Aaronson (1), *School of Mathematical Sciences, Tel Aviv University, Ramat Aviv, 69978 Tel Aviv, Israel*

André Adler (11), *Department of Mathematics, Illinois Institute of Technology, Chicago, Illinois 60616*

E. J. Balder (31), *Mathematical Institute, University of Utrecht, Utrecht, The Netherlands*

J. R. Baxter (55), *Department of Mathematics, University of Minnesota, Minneapolis, Minnesota 55455*

Alexandra Bellow (73, 99), *Department of Mathematics, Northwestern University, Evanston, Illinois 60601*

V. Bergelson (121), *Department of Mathematics, The Ohio State University, Columbus, Ohio 43210*

N. H. Bingham (131), *Department of Mathematics, Royal Holloway & Bedford New College, Egham Hill, Egham, Surrey TW20 0EX, England*

J. Bourgain (145), *IHES, 35 route de Chartres, F-91440 Bures-sur-Yvette, France*

Antoine Brunel (153), *Laboratoire des Probabilités, Université Paris VI, 4 place Jussieu, Tour 56, 75230 Paris Cedex 05, France*

Donald Burkholder (159), *Department of Mathematics, University of Illinois, Urbana, Illinois 61501*

R. V. Chacon (55), *University of British Columbia, Department of Mathematics, Vancouver, British Columbia, V6T 1V4 Canada*

Burgess Davis (179), *Department of Statistics, Purdue University, West Lafayette, Indiana 47907*

Manfred Denker (1), *Institut für Math. Stochastik, Universität Göttingen, Lotzestrasse 13, 3400 Göttingen, West Germany*

Yves Derriennic (189), *Université de Bretagne Occidentale, Brest, France*

Gerald A. Edgar (195), *Department of Mathematics, The Ohio State University, Columbus, Ohio 43210*

Roger L. Jones (73, 99), *Department of Mathematics, DePaul University, Chicago, Illinois 60614*

Ulrich Krengel (215), *Department of Mathematical Stochastics, University of Göttingen, Lotzestrasse 13, 3400 Göttingen, West Germany*

S. Kwapien (237), *Institute of Mathematics, Warsaw University, Warsaw 00901, Poland*

Michael Lin (189), *Department of Mathematics, Ben-Gurion University of the Negev, Be'er Sheva, Israel*

F. J. Martín-Reyes (267), *Departamento de Matemáticas, Facultad de Ciencias, Universidad de Málaga, 29071 Málaga, Spain*

Terry R. McConnell (281), *Syracuse University, 200 Carnegie, Syracuse, New York 13210*

James H. Olsen (297), *Department of Mathematics, North Dakota State University, Fargo, North Dakota 58105*

Steven Orey (305), *Department of Mathematics, University of Minnesota, Minneapolis, Minnesota 55455*

Magda Peligrad (335), *Department of Mathematical Sciences, University of Cincinnati, Cincinnati, Ohio 45221*

Karl Petersen (349), *IRMAR, Laboratoire de Probabilités, Université de Rennes I, 35042 Rennes, France, and Department of Mathematics, CB#3250, Phillips Hall, University of North Carolina, Chapel Hill, North Carolina 27599*

Pal Révész (369), *Institute of Statistics, Technical University of Vienna, Wiedner Hauptstrasse 8-10/107, A-1040 Vienna, Austria*

Eric Rieders (281), *State University of New York, College at Brockport, Brockport, New York 14420*

Joseph Rosenblatt (73, 99), *Department of Mathematics, The Ohio State University, Columbus, Ohio 43210*

K. Schürger (393), *Department of Economics, University of Bonn, Bonn, West Germany*

Zhen-Peng Wang (405), *Department of Statistics, East China Normal University, Shanghai, China*

W. A. Woyczynski (237), *Department of Mathematics and Statistics, Case Western Reserve University, Cleveland, Ohio 44106*

Xing-Hong Xue (405), *Department of Statistics, Columbia University, New York, New York 10027*

List of Participants

Jonathan Aaronson
André Adler
Avner Ash
Idriss Assani
Bogdan Baishansky
Erik J. Balder
Rakesh K. Bansal
John Baxter
Alexandra Bellow
Vitaly Bergelson
Erich Berger
N. H. Bingham
Daniel Boivin
Ranko Bojanic
C. Bose
Jean Bourgain
Antoine Brunel
Donald Burkholder
Raphael V. Chacon
Y. S. Chow
Krzysztof Ciesielski
Burgess Davis
W. J. Davis
David Dean
F. W. Carroll
Zita Divis
Lester E. Dubins
A. Dynin
Gerald A. Edgar
Neil Falkner
J. Ferrar
Harvey Friedman
Naresh Jain
William B. Johnson
Roger L. Jones
Michael J. Klass
S. Kwapien
Tze Leung Lai
S. P. Lalley
Michael Lin
F. J. Martín-Reyes
Terry McConnell
Boris Mityagin
S. E. A. Mohammed
James H. Olsen
Steven Orey
Donald Ornstein
Magda Peligrad
Karl E. Peterson
William E. Pruitt
Pal Révész
Joseph Rosenblatt
Klaus Schürger
Robert Stanton
Arkadii Tempelman
Alberto de la Torre
Monique Vuilleumier
Rainer Wittman
Wojbor Woyczynski
Xing-Hong Xue
Radu Zaharopol

Preface

The International Conference on Almost Everywhere Convergence in Probability and Ergodic Theory took place in Columbus, Ohio, on June 11–14, 1988. Financial support was provided by the Institute for Mathematics and Its Application at the University of Minnesota, the National Science Foundation, the United States Army, and The Ohio State University.

There were approximately 60 participants, including mathematicians from the United States, Austria, Canada, the Federal Republic of Germany, France, Great Britain, Hungary, Israel, the Netherlands, the Soviet Union, and Spain.

The organizer of the conference was Louis Sucheston; other members of the organizing committee were Mustafa Akcoglu, Alexandra Bellow, Donald Burkholder, and Gerald Edgar.

There were in recent years important developments in almost everywhere convergence. The lectures at the conference of Alexandra Bellow, Jean Bourgain, and Karl Petersen surveyed progress on the ergodic theorem for subsequences. Bourgain's important results include that for each square integrable function f, the Cesaro averages of some natural subsequences of $T^n f$, e.g. $T^{n^2} f$, converge almost everywhere. There are also results about square roots: Petersen showed that if T_t is a measure preserving flow on a probability space, then the Cesaro average of $f(T_{\sqrt{n}}x)$ converges a.e. Another important new result in a.e. convergence was presented by Antoine Brunel: If T is a positive linear operator on $L_p(1 < p < \infty)$ that is mean bounded, i.e. such that $\sup_n \| (1/n)\sum_{i=1}^{n} T^i \| < \infty$, then (and only then) there is a maximal inequality and pointwise a.e. convergence of Cesaro averages of $T^n f$ for $f \in L_p$. This was proved for contractions in 1974 by M. A. Akcoglu via a dilation theorem, but Brunel's argument is very different: it uses "Brunel's operators" which are contractions associated with mean-bounded operators T. Further interesting almost everywhere results by D. L. Burkholder, S. Orey, and others are included. The proceedings contain some nonlinear theory (U. Krengel), and papers on random walk (B. Davis, P. Révész-P. Erdos). J. R. Baxter and R. V. Chacon prove a basic theorem in physics about stability of mass by the powerful technique developed by Chacon and D. S. Ornstein for almost everywhere convergence (in the Chacon-Ornstein theorem): the filling scheme. The reader will observe that almost everywhere convergence problems and methods are now in full development and of increasing importance.

Lower bounds for partial sums of certain positive stationary processes

Jon Aaronson and Manfred Denker

Jon Aaronson
School of Mathematical Sciences
Tel Aviv University
Ramat Aviv
69978 *Tel Aviv*
Israel

Manfred Denker
Institut für Math. Stochastik
Universität Göttingen
Lotzestr. 13
3400 *Göttingen*
West Germany

ABSTRACT: We find lower bounds for the partial sums of a positive, non–integrable ψ–mixing stationary process satisfying a weak law of large numbers. The method of proof uses infinite measure preserving transformations.

0. Introduction

Recall that a stationary process $(X_k)_{k=1}^{\infty}$ is called ψ–mixing, if

$$\psi(n) := \sup\{\left| \frac{P(B\cap C)}{P(B)P(C)} - 1 \right|\}$$

tends to zero, where the supremum is taken over all $k \geq 1$, and over all sets $B \in \sigma(X_i : i \leq k)$ and $C \in \sigma(X_i : i \geq k+n)$.

We prove

THEOREM 1: Let $(X_k)_{k=1}^{\infty}$ be a positive, ψ–mixing stationary process with $E(X_1) = \infty$. Suppose there are constants $b(n)$, $n \geq 1$ such that

(1) $$\frac{1}{b(n)}\sum_{k=1}^{n} X_k \to 1 \qquad \text{in probability.}$$

ISBN 0-12-231050-0

Then

$$\liminf_{n\to\infty} \frac{1}{b(n/L_2(n))L_2(n)} \sum_{k=1}^{n} X_k \geq 1 \qquad \text{a.e..} \tag{2}$$

If $b(n/L_2(n))L_2(n) \sim b(n)$, then equality holds in (2).

Here, and in the sequel, we denote $L(t) = \log t$ and $L_2(t) = L(L(t))$.

REMARK:

(a) The function $b(t)$ is necessarily regularly varying with index 1 as $t \uparrow \infty$ (see [IL]). Also, if $(X_k)_{k=1}^{\infty}$ is continued fraction mixing (i.e., in addition $\psi(1) < \infty$), then $b(n) \sim nE(X_1 \wedge b(n))$ (see [AD]).

(b) In the case $b(n/L_2(n))L_2(n) \sim b(n)$, the conclusion of theorem 1 was established for i.i.d. random variables in [KT].

The methods we use will prove a similar result when the distribution of $\frac{1}{b(n)} \sum_{k=1}^{n} X_k$ tends to a stable distribution of order $\alpha \in (0,1)$. In this case, when the stable random variable Y_α is normalized so that $E(Y_\alpha^{-\alpha}) = 1$, we have that

$$\liminf_{n\to\infty} \frac{1}{b(n/L_2(n))L_2(n)} \sum_{k=1}^{n} X_k \geq K_\alpha^{-1/\alpha} \qquad \text{a.e.,} \tag{3}$$

where $K_\alpha = \frac{\Gamma(1+\alpha)}{\alpha^\alpha(1-\alpha)^{1-\alpha}}$. We proved in [AD] that equality holds in (3) when $(X_n)_{n=1}^{\infty}$ is continued fraction mixing.

The method of proof uses an estimation of the moments of $N_n = \sup\{k : S_k \leq n\}$, where $S_n = \sum_{k=1}^{n} X_k$. This was done in [AD], using an infinite measure preserving transformation, as described in the next section.

1. Infinite measure preserving transformations

Suppose that $(X, \mathcal{B}, \mu, T)$ is an ergodic measure preserving transformation on the σ–finite measure space $(X, \mathcal{B}, \mu)$. We assume that T is conservative, i.e. that $\sum_{k=1}^{n} f \circ T^k \to \infty$ for some, and hence for all $f \in L_+^1(\mu)$ by Hopf's ergodic theorem ([H]). Recall from [A1] that T is called pointwise dual ergodic, if there is a sequence $a_n = a_n(T)$ such that for all $f \in L^1(\mu)$

$$\lim_{n\to\infty} \frac{1}{a_n} \sum_{k=1}^{n} \hat{T}^k f = \int_X f \, d\mu \quad \text{a.e.,} \tag{4}$$

where $\hat{T}$ denotes the dual operator of $g \mapsto g \circ T$, $g \in L^\infty(\mu)$, restricted to $L^1(\mu)$. The sequence $a_n(T)$ is called a return sequence for T. We state the results in section

2 for pointwise dual ergodic transformations T. However, the results remain valid under the weaker assumption

(5)
$$\limsup_{n\to\infty} \frac{1}{a_n}\sum_{k=1}^{n} \hat{T}^k f \leq \int_X f\,d\mu \quad \text{a.e.,}$$

for all $f \in L^1_+(\mu)$. Under this assumption, for every $\beta > 1$, there is a set $A \in \mathcal{B}$, $\mu(A) = 1$, and $N \in \mathbf{N}$ such that

(6)
$$\sum_{k=1}^{n} \hat{T}^k 1_A \leq \beta a_n \quad \text{a.e. on } A \text{ for every } n \geq N.$$

We assume also (without loss of generality) that the return sequence $a_n(T)$, $n \geq 1$ is of the form

(7)
$$a_n(T) \sim n^\alpha h(n) \quad (n \geq 1)$$

where $0 < \alpha \leq 1$ and h is slowly varying satisfying (w.l.o.g.) $\log h(t) = \eta_0 + \int_0^t \epsilon(s)/s\,ds$ where

(8)
$$\begin{aligned} &(i)\ \ \epsilon(s) = 0 \quad \text{for } 0 < s < 1,\\ &(ii)\ \lim_{s\to\infty} \epsilon(s) = 0\\ &(iii)\ |\, \epsilon(s)\,| \leq \delta < \frac{1}{4}\alpha \quad (\forall s > 0). \end{aligned}$$

<u>THEOREM 2:</u> ([AD]) Let $\beta > 1$ and let $A \in \mathcal{B}$ with $\mu(A) = 1$, such that (6) is satisfied. Then there exists $n_\beta \in \mathbf{N}$ such that for all integers $n \geq n_\beta$ and $n_\beta \leq p \leq L_2(n)^2$,

$$\left(\int_A \left(\sum_{k=1}^{n} 1_A \circ T^k\right)^p d\mu\right)^{1/p} \leq \beta^2 \frac{\Gamma(1+\alpha)}{\alpha^\alpha e^{(1-\alpha)}}\, p^{(1-\alpha)}\, n^\alpha\, h(n/p).$$

A precise proof of this theorem can be extracted from [AD,section 2], all the upper estimations there being valid under the assumptions of theorem 2 here. For the readers convenience, we give a heuristic outline of its ideas.

<u>OUTLINE OF PROOF:</u> Write $S_n = \sum_{k=1}^n 1_A \circ T^k$. It is not hard to see, that for $1 \leq p \leq n$

$$\int_A S_n^p\,d\mu = \int_A \sum_{1\leq k_1,k_2,\ldots,k_p\leq n} \prod_{\nu=1}^{p} 1_A \circ T^{k_\nu}\,d\mu$$

Now

$$\int_A S_n^p\,d\mu \approx (p!)\int_A a(p,n)\,d\mu =: (p!)\overline{a}(p,n), \qquad \text{as } n\to\infty \text{ and } p = O(\mathrm{L}_2(n)^2),$$

where $\approx$ denotes equality to within small errors, and where

$$a(p,n) = \sum_{1\le k_1<k_2<\ldots<k_p\le n} \prod_{\nu=1}^{p} 1_A \circ T^{k_\nu}.$$

Using the assumption on A and changing the order of summation, one easily computes

$$\begin{aligned}\overline{a}(p,n+1) &= \sum_{k=1}^{n} \int_A \sum_{j=1}^{k} \hat{T}^j 1_A \left(a(p,n-k-1) - a(p,n-k)\right) d\mu \\ &\approx \sum_{k=1}^{n} \alpha k^{\alpha-1} h(k)\overline{a}(p,n-k).\end{aligned}$$

It follows that $\overline{a}(p,n)$ can be bounded by its continuous analogue $A(p,n)$, defined recursively by $A(0,n)\equiv 1$ and

$$A(p+1,t) = \int_0^t \alpha\, s^{\alpha-1}\, h(s)\, A(p,t-s)\, ds.$$

Define $M(p,t)$ by

$$A(p,t) =_{def} \frac{\Gamma(1+\alpha)^p}{\Gamma(1+\alpha p)} t^{\alpha p} M(p,t).$$

Let U_p be a random variable with density

$$f_p(x) = \frac{1}{C_p(\alpha)} x^{\alpha-1}(1-x)^{\alpha p} \quad (0<x<1),$$

where

$$C_p(\alpha) = \frac{\Gamma(\alpha)\Gamma(1+\alpha p)}{\Gamma(1+\alpha(p+1))} = \int_0^1 (1-u)^{\alpha-1} u^{\alpha p}\, du.$$

Then, it follows from the definitions of A and M that

$$\begin{aligned}M(p+1,t) &= \frac{1}{C_p(\alpha)} \int_0^1 u^{\alpha-1} M(p,(1-u)t)(1-u)^{\alpha p} h((1-u)t)\, du \\ &= E\big(h(U_p t) M(p,(1-U_p)t)\big).\end{aligned}$$

As $E(U_p)\sim 1/p$,

$$\begin{aligned}M(p+1,t) &\approx h\big(E(U_p)t\big) M\left(p,(1-E(U_p))t\right) \\ &\approx h\Big(\frac{t}{p}\Big) M(p,t). \quad (p = O(L_2(t)^2))\end{aligned}$$

Since $h(t/p) \sim h(t/(p+1))$ as $t/p\to\infty$, we obtain $M(p,t)\approx h(t/p)^p$, $p=O(L_2(t)^2)$. Putting everything together, we finally obtain

$$\int_A S_n^p \, d\mu \approx (p!) \frac{\Gamma(1+\alpha)^p}{\Gamma(1+\alpha p)} n^{\alpha p} h(n/p)^p \qquad (p = O(L_2(n)^2)),$$

and the theorem follows by Stirling's formula. ♡

2. Upper bounds for pointwise dual ergodic transformations

In this section we consider conservative, ergodic, measure preserving transformations T which are pointwise dual ergodic and for which the the return sequence is given by $a_n(T) \sim n^\alpha h(n)$, where $0 < \alpha \leq 1$. The function $h(t) = \eta \exp\bigl(\int_0^t \epsilon(s)/s \, ds\bigr)$ is assumed to be slowly varying, satisfying (8).

Transformations T having Darling–Kac sets (see [A2], [AD] and [DK]), conservative inner functions on the upper half plane $\mathbf{R}^{2+}$, preserving Lebesgue measure (see [A0]) and non–lattice random walks on $\mathbf{R}$ (see [A1]) fall into this class . It is not known in all these examples, whether Darling–Kac sets exist. The following propositions may be considered as a first step towards the identification of the constant $K(T)$ in (9) and (10).

Let

$$K_\alpha = \frac{\Gamma(1+\alpha)}{\alpha^\alpha (1-\alpha)^{1-\alpha}}.$$

We prove

PROPOSITION 1: If $\alpha < 1$, then there exists $K(T) \in [0, K_\alpha]$ such that

$$\limsup_{n\to\infty} \frac{1}{n^\alpha h(n/L_2(n))(L_2(n))^{1-\alpha}} \sum_{k=1}^n f \circ T^k = K(T) \int_X f \, d\mu \tag{9}$$

a.e. for every $f \in L^1_+$.

REMARK: The proof of proposition 1 can be used to show the following: If $\phi(n) \uparrow$ and $\phi(n)/n \downarrow$ as $n \to \infty$, and if $\sum_{n=1}^\infty n^{-1} \exp\bigl[-(\beta\phi(n))\bigr] < \infty$ for all $\beta > 1$, then

$$\limsup_{n\to\infty} \frac{1}{n^\alpha h(n/\phi(n))\phi(n)^{1-\alpha}} \sum_{k=1}^n f \circ T^k < \infty$$

a.e. for every $f \in L^1_+$.

PROPOSITION 2: If $\alpha = 1$, then: (a) There exists $K(T) \in [0,1]$ such that

$$\limsup_{n\to\infty} \frac{1}{nh(n/L_2(n))} \sum_{k=1}^n f \circ T^k = K(T) \int_X f \, d\mu \tag{10}$$

a.e. for every $f \in L^1_+$.

(b) If $h\bigl(h/L_2(n)\bigr) \sim h(n)$ as $n \to \infty$, then

$$\limsup_{n\to\infty} \frac{1}{nh(n)} \sum_{k=1}^n f \circ T^k = \int_X f \, d\mu$$

REMARK: Under additional assumptions on T we have $K(T) = K_\alpha$ (see [AD, section 4]).

PROOF OF PROPOSITION 1: Since T is pointwise dual ergodic, for every $\beta > 1$ there exists a set $A = A_\beta \in \mathcal{B}$ with $\mu(A_\beta) = 1$ satisfying (6) with $\sqrt{\beta}$.

Therefore theorem 2 applies:

$$\int_{A_\beta}\left(\sum_{k=1}^{n} 1_{A_\beta}\circ T^k\right)^p d\mu$$

$$\leq \left(\frac{\beta\Gamma(1+\alpha)}{\alpha^\alpha e^{1-\alpha}}\right)^p p^{(1-\alpha)p}\, n^{\alpha p} h(n/p)^p \tag{11}$$

for all $N \leq p \leq L_2(n)^2$, where N denotes some constant, depending on β. Hence for n large enough, by Markov's inequality and since h is slowly varying,

$$\mu(A_\beta \cap \{\sum_{k=1}^{n} 1_{A_\beta}\circ T^k \geq \beta^3 K_\alpha n^\alpha h(n/t) t^{1-\alpha}\})$$

$$\leq \frac{1}{(\beta^2 K_\alpha n^\alpha h(n/t)t^{1-\alpha})^{at}} \int_{A_\beta}\left(\sum_{k=1}^{n} 1_{A_\beta}\circ T^k\right)^{at} d\mu$$

$$\leq \left(\frac{\Gamma(1+\alpha)a^{1-\alpha}h(n/at)}{\alpha^\alpha \beta K_\alpha h(n/t)e^{1-\alpha}}\right)^{at}$$

$$\leq \exp\Bigg(-(t\beta^{1/(1-\alpha)}\Bigg), \tag{12}$$

where $a = \beta^{1/(1-\alpha)}(1-\alpha)^{-1}$.

Let $\gamma > 1$ and $K_n = K_n(\gamma) = [\gamma^n]$. By (12), setting $t = L_2(n)$, and for some constant M,

$$\sum_{n=1}^{\infty} \mu(A_\beta \cap \{\sum_{k=1}^{K_n} 1_{A_\beta}\circ T^k \geq \beta^2 K_\alpha K_n^\alpha h(K_n/L_2(K_n))(L_2(K_n))^{1-\alpha}\})$$

$$\leq M\sum_{n=1}^{\infty} \exp[-(\beta^{1/(1-\alpha)}L_2(K_n))]$$

$$< \infty,$$

whence

$$\limsup_{n\to\infty} \frac{1}{K_n^\alpha h(K_n/L_2(K_n))(L_2(K_n))^{1-\alpha}} \sum_{j=1}^{K_n} 1_{A_\beta}\circ T^j \leq \beta^2 K_\alpha$$

a.e. on A_β. It follows that

$$\limsup_{n\to\infty} \frac{1}{n^\alpha h(n/L_2(n))(L_2(n))^{1-\alpha}} \sum_{j=1}^{n} 1_{A_\beta}\circ T^j \leq \beta^2\gamma^{1+\alpha} K_\alpha$$

a.e. on A_β. By ergodicity and Hopf's ergodic theorem

$$\limsup_{n\to\infty} \frac{1}{n^\alpha h(n/L_2(n))(L_2(n))^{1-\alpha}} \sum_{j=1}^n f\circ T^j \le \beta^2\gamma^{1+\alpha}K_\alpha \int_X f\,d\mu$$

a.e. for every $f \in L^1_+$ and $\beta, \gamma > 1$.

Since the lim sup is constant, the proposition follows. ♡

PROOF OF PROPOSITION 2: (a) Let $A = A_\beta$ be as in the previous proof for $\beta > 1$. By (11),

$$\int_{A_\beta} \left(\sum_{k=1}^n 1_{A_\beta}\circ T^k\right)^p d\mu \le (\beta n h(n/p)))^p$$

for all $N \le p \le L_2(n)^2$.

By Markov's inequality, for any $y > 0$ and $a \le L_2(n)$,

$$\mu(A_\beta \cap \{\sum_{k=1}^n 1_{A_\beta}\circ T^k \ge ynh(n/L_2(n))\}$$
$$\le \left(\frac{\beta h(n/(aL_2(n)))}{yh(n/L_2(n))}\right)^{aL_2(n)}.$$

and since $K(T) \le 1$, (b) follows. ♡

3. Proof of theorem 1

The proof of the last statement, that $b(n/L_2(n))L_2(n) \sim b(n)$ implies equality in (2), is clear using (1).

First note that $b(n)/n \uparrow \infty$ as $n \uparrow \infty$, and therefore we may assume that X_1 takes values in the natural numbers. We begin constructing a model to apply theorem 2.

Let $(X_k)_{k=1}^\infty$ be defined on the probability space $(A, \mathcal{A}, m)$. We may assume that A is a sequence space equipped with the m–preserving transformation $S(y_k)_{k=1}^\infty = (y_{k+1})_{k=1}^\infty$ so that $X_1(y) = y_1$, whence $X_n(y) = X_1(S^{n-1}y) = y_n$, where $y = (y_k)_{k=1}^\infty$. It is well known that S is ergodic .

Now we let

$$X = \{x = (y,n) : 1 \le n \le X_1(y), y \in A\}$$
$$\mathcal{B} = \bigvee_{n=1}^\infty \mathcal{A} \cap \{X_1 \ge n\} \times \{n\}$$
$$\mu(B \times \{n\}) = m(B) \qquad (B \in \mathcal{A} \cap \{X_1 \ge n\})$$
$$T(y,n) = \begin{cases} (y, n+1), & \text{if } X_1(y) \ge n+1 \\ (Sy, 1), & \text{if } X_1(y) = n. \end{cases}$$

By Kakutani's theorem ([KAK]) T is a conservative, ergodic, measure preserving transformation of the σ–finite measure space $(X, \mathcal{B}, \mu)$. By Kac's formula ([KAC]) $\mu(X) = E(X_1) = \infty$. We identify A with $A \times \{1\}$.

It follows from inequality 1 in [A2, p 1040] (whose proof only uses ψ–mixing and not continued fraction mixing), that

$$(13) \qquad \sum_{k=1}^{n} \hat{T}^k 1_A \leq \beta_n a_n \quad \text{a.e. on } A$$

for all $n \geq 1$, where $\beta_n \downarrow 1$ and $a_n = \int_A \sum_{k=1}^{n} \hat{T}^k 1_A \, d\mu = \int_A N_n \, d\mu$. Here

$$N_n =_{def} \sum_{k=1}^{n} 1_A \circ T^k = \sup\{k : S_k \leq n\}.$$

It is not hard to show using (13) that $\sup_{n\geq 1} \int_A (N_n/a_n)^2 \, d\mu < \infty$, and indeed $N_n/a_n \to 1$ weakly in $L^2(m)$. It also follows from the assumptions that $N_n/a(n) \to 1$ in measure on A, where a is the inverse of b (and therefore regularly varying with index 1).

In view of all this, $a_n \sim a(n)$, and the assumptions of theorem 2 are satisfied.

Thus, for every $\beta > 1$, there exists τ such that for every $\tau \leq p \leq L_2(n)^2$

$$\| N_n \|_p \leq \beta a(n/p) p.$$

Setting $K_n = [\gamma^n]$, $(n \geq 1)$, for $\gamma > 1$, and $p = L(n) = O(L_2(K_n))$, it follows for every $\beta > 1$ that

$$\begin{aligned} & m(N_{K_n} \geq \beta a(K_n/L_2(K_n)) L_2(K_n)) \\ & \leq \frac{1}{(\beta a(K_n/L_2(K_n)) L_2(K_n))^{L(n)}} \| N_n \|_{L(n)}^{L(n)} \\ & \leq \left(\frac{\beta^{1/2} a(K_n/L_2(K_n)) L_2(K_n)}{\beta a(K_n/L_2(K_n)) L_2(K_n)} \right)^{L(n)} \\ & = \beta^{-\frac{1}{2}L(n)}. \end{aligned}$$

We have shown that

$$\limsup_{n\to\infty} \frac{N_n}{a(n/L_2(n)) L_2(n)} \leq 1 \quad \text{a.e.}$$

Write $A(n) = a(n/L_2(n)) L_2(n)$ and $B(n) = A^{-1}(n)$. Both, A and B, are regularly varying with index 1,whence $L_2(B(n)) \sim L_2(n)$, and

$$\liminf_{n\to\infty} \frac{S_n}{B(n)} \geq 1 \quad \text{a.e.}$$

Now

$$n = A(B(n)) = a(B(n)/L_2(B(n)))L_2(B(n))$$
$$\sim a(B(n)/L_2(n))L_2(n),$$

whence $b(n/L_2(n)) \sim B(n)/L_2(n)$ and $B(n) \sim b(n/L_2(n))L_2(n)$. ♡

References

[A0] J. Aaronson: Ergodic theory for inner functions of the upper half plane. Ann. Inst. H. Poincaré (B) **14**, (1978), 233–253

[A1] J. Aaronson: The asymptotic distributional behaviour of transformations preserving infinte measure. J. Analyse Math. **39**, (1981), 203–234

[A2] J. Aaronson: Random f–expansions. Ann. Probab. **14**, (1986),1037–1057

[AD] J. Aaronson, M. Denker: Upper bounds for ergodic sums of infinite measure preserving transformations. Mathematica Gottingensis, preprint (1988)

[DK] D.A. Darling, M. Kac: On occupation times for Markov processes. Trans. Amer. Math. Soc. **84**, (1957), 444–458

[H] E. Hopf: Ergodentheorie. Chelsea Publ. Comp., New York (1948)

[IL] I.B. Ibragimov, Y.V. Linnik: Independent and stationary sequences of random variables. Wolters–Noordhoff, Groningen (1971)

[KAC] M. Kac: Of the notion of recurrence in discrete stochastic processes. Bull. Amer. Math. Soc. **53**, (1947), 1002–1010

[KAK] S. Kakutani: Induced measure preserving transformations. Proc. Imp. Acad. Sci. Tokyo **19**, (1943), 635–641

[KT] M. Klass, H. Teicher: Iterated logarithm laws for asymmetric random variables barely with or without finite mean. Ann. Probab. **5**, (1977), 861–874

EXACT SEQUENCES FOR SUMS OF INDEPENDENT RANDOM VARIABLES

ANDRE ADLER

Illinois Institute of Technology

Chicago, Illinois 60616

For a given sequence of random variables $\{Y_n, n\geq 1\}$ a natural question is, does there exist constants $\{b_n, n\geq 1\}$ such that $\sum_{k=1}^{n} Y_k/b_n \to 1$ almost surely? Theorems are presented showing, under certain constraints, that no such constants can be found. However, through a most interesting class of random variables we see that if these conditions do not hold, then a nontrivial sequence of constants can be obtained. Moreover, by simply changing the parameters associated with this family of random variables it will be shown that the normalized partial sum can converge almost surely to any positive finite constant.

1. Introduction.

Consider a probability space $(\Omega, \mathfrak{F}, P)$ with independent random variables $\{Y_n, n\geq 1\}$. Let $S_n = \sum_{k=1}^{n} Y_k$, $n\geq 1$. A sequence of constants

ISBN 0-12-231050-0

$\{b_n, n\geq 1\}$ is said to be an exact sequence (e.s.) for S_n if $S_n/b_n \to 1$ almost surely (a.s). The term exact sequence is a natural extention of what Rogozin (1968) called an exact upper sequence (e.u.s.). The sequence $\{b_n, n\geq 1\}$ is said to be an e.u.s. if

$$P\{\limsup_{n\to\infty} \frac{|S_n - \mathrm{med}(S_n)|}{b_n} = 1\} = 1.$$

Clearly if $\{Y, Y_n, n\geq 1\}$ are independent and identically distributed (i.i.d.) random variables with $0 < |EY| \leq E|Y| < \infty$, then nEY is an e.s. for S_n.

The search for exact sequences for sums of independent random variables have also gone under the name of the fair games problem. In this context the random variable Y_n represents the winnings from the nth round of some game. Thus S_n is the accumulated winnings from n rounds. If there is an e.s., say $\{b_n, n\geq 1\}$ for S_n, then b_n is a "fair" accumulated entrance fee for this game. Hence the player should pay $b_n - b_{n-1}$, $n\geq 1$, with $b_0 = 0$, per round.

In the early 18th century mathematicians discovered the first Law of Large Numbers for specific random variables, i.e.,

$$\sum_{k=1}^{n} Y_k/n \xrightarrow{P} EY.$$

Thus nEY is a fair solution in the weak sense as long as $0 < |EY| \leq E|Y| < \infty$. If $EY = 0$ they realized that the player need not pay per round, but what should the player pay when $E|Y| = \infty$? This question led to one of the most famous paradoxes in mathematics. The St. Petersburg game was created with all this in mind. The idea was to take a common random variable (the geometric) and transform it so that it no longer had finite expectation.

Feller (1945) and Klass and Teicher (1977) examined solutions in the weak sense. For a weak solution to the St. Petersburg game see Feller (1968, p. 252) or for a weak solution to the generalized St. Petersburg game see Adler and Rosalsky (1987c). We now turn our attention solely to finding exact sequences for sums of independent random variables, or in terms of the fair games problem, solutions in the strong sense.

In a landmark paper, Chow and Robbins (1961) showed that if $\{Y,Y_n,n\geq1\}$ are i.i.d. random variables with $E|Y| = \infty$, then there doesn't exist an e.s. for S_n. Adler and Rosalsky (1987c) improved on this by examining the weighted i.i.d. case (see Theorem 2.4). Likewise, Maller (1978) by using results obtained by Rogozin (1976) proved that if $\{Y,Y_n,n\geq1\}$ are i.i.d. with $EY = 0$, then there doesn't exist an e.s. for S_n under the constraint of $|b_n| \to \infty$. Adler and Rosalsky (1987c) removed the artifical restriction of $|b_n| \to \infty$ (see Theorem 2.5).

In Section 2 we exhibit several strong laws whose common conclusion is that $\{b_n,n\geq1\}$ is not an e.s. for a specific sum of random variables (sometimes weighted, sometimes not). Next, an extraordinary class of strong laws are presented showing that if the hypotheses of the previous theorems are slightly weakened exact sequences do indeed exist. Finally, by modifying two results of Chow and Teicher (1971) we can immediately determine whether or not an e.s. exists in the weighted i.i.d. case.

A few remarks about notation are needed. The symbol C will denote a generic nonzero finite constant which is not necessarily the same in each appearance. Also, as usual, $u_n \sim v_n$ denotes that $u_n/v_n \to 1$ as $n \to \infty$ and $u_n \approx v_n$ means that there is a finite constant M such that $M^{-1} \le u_n/v_n \le M$ as $n \to \infty$, with the license of $u_n \approx \infty$ implying that $u_n = \infty$. Finally, we define $\log x = \log_e \max\{e,x\}$, $x \ge 0$.

2. Some strong Laws of Large Numbers

In this section we state several strong laws. In all, but one, the conclusion under certain constraints, is that there does not exist an exact sequence for the partial sum.

In the weighted i.i.d. case we let $\{a_n, n\ge1\}$ be our weights and $\{b_n, n\ge1\}$ our norming constants. It proves convenient to define $c_n = b_n/|a_n|$.

THEOREM 2.1 [Adler and Rosalsky (1987a)]. Let $\{Y, Y_n, n\ge1\}$ be i.i.d. random variables. If $\{a_n, n\ge1\}$ and $\{b_n, n\ge1\}$ are constants satisfying

$0 < b_n \uparrow \infty$,

$$\max_{1\le k\le n} c_k^2 \sum_{j=n}^{\infty} c_j^{-2} = O(n), \tag{2.1}$$

and

$$\sum_{n=1}^{\infty} P\{|Y| > c_n\} < \infty, \tag{2.2}$$

then

$$\sum_{k=1}^{n} a_k(Y_k - \mu_k)/b_n \to 0 \text{ a.s.},$$

where $\mu_n = EY_n I(|Y_n| \le c_n)$.

As with the previous theorem, the next one is not presented in its full generality. However, we need not assume independence. It owes much to the work of Martikainen and Petrov (1980).

THEOREM 2.2 [Adler and Rosalsky (1987a)]. Let $\{Y, Y_n, n \geq 1\}$ be identically distributed random variables. If $\{a_n, n \geq 1\}$ and $\{b_n, n \geq 1\}$ are constants satisfying $0 < b_n \uparrow \infty$,

$$\max_{1 \leq k \leq n} c_k \sum_{j=n}^{\infty} c_j^{-1} = O(n), \tag{2.3}$$

and (2.2), then

$$\frac{\sum_{k=1}^{n} a_k Y_k}{b_n} \to 0 \text{ a.s.} \tag{2.4}$$

The next theorem examines the mean zero situation. It is obtained by combining Theorem 2.1 and three different constraints. While (2.6) and (2.7) are hypotheses often used in strong laws, (2.5) is more closely associated with the central limit theorem (see Adler and Rosalsky (1988)).

THEOREM 2.3 [Adler and Rosalsky (1987b)]. Let $\{Y, Y_n, n \geq 1\}$ be i.i.d. mean zero random variables and $\{a_n, n \geq 1\}$ and $\{b_n, n \geq 1\}$ be constants satisfying $0 < b_n \uparrow \infty$ and (2.1). Suppose that either

$$P\{|Y| > y\} \text{ is regularly varying with exponent } \rho < -1, \tag{2.5}$$

or

$$c_n \uparrow \text{ and } c_n \sum_{j=1}^{n} c_j^{-1} = O(n), \tag{2.6}$$

or

$$\sum_{j=1}^{n} |a_j| = O(b_n). \tag{2.7}$$

Then (2.4) is equivalent to (2.2).

Theorem 2.4 extends Chow and Robbins (1961) result. It shows, under certain constraints, that there does not exist an e.s. for the sum of independent nonintegrable random variables.

THEOREM 2.4 [Adler and Rosalsky (1987c)]. Let $\{Y, Y_n, n \geq 1\}$ be i.i.d. random variables with $E|Y| = \infty$. If $\{a_n, n \geq 1\}$ are constants satisfying $n|a_n|\uparrow$ and

$$\sum_{j=1}^{n} |a_j| = O(n|a_n|), \tag{2.8}$$

then for every sequence of constants $\{b_n, n \geq 1\}$

$$P\{\lim_{n\to\infty} \sum_{k=1}^{n} a_k Y_k / b_n = 1\} = 0.$$

The last theorem established conditions showing that exact sequences cannot exist when the random variables are without finite mean. Next, we obtain the same conclusion for i.i.d. mean zero random variables.

THEOREM 2.5 [Adler and Rosalsky (1987c)]. If $\{Y, Y_n, n \geq 1\}$ are i.i.d. random variables with $\liminf_{y\to\infty} |EYI(|Y| \leq y)| = 0$, then there does not exist an e.s. for S_n.

Finally, we state a theorem that completely removes the restriction of common distribution. It extends the results of Heyde (1969), Rogozin (1968), and Teicher (1979). In general there are three conditions of interest. Let there exist y_0 and $\lambda > 0$ such that

$$yP\{|Y_n| > y\} \ge \lambda\int_0^y P\{|Y_n| > t\}dt \text{ for all } y \ge y_0 \text{ and } n \ge 1, \tag{2.9}$$

$$y^2P\{|Y_n| > y\} \ge \lambda EY_n^2 I(|Y_n| \le y) \text{ for all } y \ge y_0 \text{ and } n \ge 1, \tag{2.10}$$

and

$$yP\{|Y_n| > y\} \ge \lambda\int_y^\infty P\{|Y_n| > t\}dt \text{ for all } y \ge y_0 \text{ and } n \ge 1. \tag{2.11}$$

THEOREM 2.6 [Adler (1988)]. Let $\{Y_n, n\ge1\}$ be independent random variables and $\{b_n, n\ge1\}$ be positive constants.

(i) Suppose that (2.9) holds when $Y_n \notin L_1$. When $Y_n \in L_1$ let either (2.9) hold or $EY_n = 0$ and (2.11). If $b_n \uparrow \infty$ and

$\sum_{n=1}^\infty P\{|Y_n| > b_n\} < \infty$, then $\lim_{n\to\infty} S_n/b_n = 0$ a.s.

(ii) If (2.10) holds, $b_n \to \infty$, $b_n = O(b_{n+1})$, and $\sum_{n=1}^\infty P\{|Y_n| > b_n\} = \infty$,

then $\limsup_{n\to\infty} |S_n|/b_n = \infty$ a.s.

Now with all these results behind us we will next show examples in which unusual e.s. do exist.

3. Examples

Throughout $\{X, X_n, n\ge1\}$ will be i.i.d. random variables with common density $f(x) = \sigma_\lambda x^{-2}(\log x)^{-\lambda} I_{(e,\infty)}(x)$, $-\infty < x < \infty$. Note that $X \in L_1$ iff $\lambda > 1$ and if $\lambda > 1$, then $EX = \sigma_\lambda/(\lambda-1)$. Also we define

$a'_n = (\log n)^{\alpha}/n$, $b'_n = (\log n)^{\beta}$, $b''_n = (\log_2 n)^{\beta}$, $c'_n = b'_n/a'_n$, and $c''_n = b''_n/a'_n$, $n \ge 1$. Clearly β must be positive, since $\sum_{k=1}^{n} a'_k X_k$ cannot converge almost surely to a finite nonnegative constant. Prior to establishing exact sequences for $\sum_{k=1}^{n} a'_k X_k$ a few lemmas are in order.

LEMMA 3.1. $P\{|X| > x\} \sim \sigma_\lambda x^{-1}(\log x)^{-\lambda}$.

PROOF. This follows immediately from Theorem 1a of Feller (1971, p.281). □

LEMMA 3.2. Equation (2.1) obtains with either $\{c'_n, n\ge1\}$ or $\{c''_n, n\ge1\}$.

PROOF. Since $c'_n = n(\log n)^{\beta-\alpha}$ is eventually increasing and

$$n^2(\log n)^{2(\beta-\alpha)} \sum_{j=n}^{\infty} \frac{1}{j^2(\log j)^{2(\beta-\alpha)}}$$

$$\le n^2(\log n)^{2(\beta-\alpha)} \int_{n-1}^{\infty} \frac{dt}{t^2(\log t)^{2(\beta-\alpha)}}$$

$$= \frac{(1+o(1))n^2(\log n)^{2(\beta-\alpha)}}{(n-1)(\log(n-1))^{2(\beta-\alpha)}}$$

(by Theorem 1a of Feller (1971,p. 281))

$$= O(n),$$

whence (2.1) obtains. Similarly $(c''_n)^2 \sum_{j=n}^{\infty} (c''_j)^{-2} = O(n)$. □

LEMMA 3.3. $\sum_{n=1}^{\infty} P\{|X| > c'_n\} < \infty$ iff $\alpha-\beta-\lambda < -1$. If $\lambda = 1$, then $\sum_{n=1}^{\infty} P\{|X| > c''_n\} < \infty$ iff either

$$\alpha < 0, \text{ or } \alpha = 0 \text{ and } \beta > 1. \tag{3.1}$$

PROOF. Note that $\log c'_n \sim \log n$. Thus

$$\sum_{n=1}^{\infty} P\{|X| > c'_n\} = C \sum_{n=1}^{\infty} \frac{1}{c'_n(\log c'_n)^{\lambda}} \quad \text{(by Lemma 3.1)}$$

$$= C \sum_{n=1}^{\infty} \frac{(\log n)^{\alpha-\beta-\lambda}}{n},$$

which is finite iff $\alpha-\beta-\lambda < -1$. Likewise

$$\sum_{n=1}^{\infty} P\{|X| > c''_n\} = C \sum_{n=1}^{\infty} \frac{(\log n)^{\alpha-1}(\log_2 n)^{\beta}}{n} = O(1)$$

iff (3.1) holds. □

Therefore, via Lemmas 3.2 and 3.3 and Theorem 2.1, $\sum_{k=1}^{n} a'_k(X_k-\mu_k)/b'_n \to 0$ a.s., where $\mu_n = EXI(|X| \le c'_n)$. Note that, via Theorem 2.6(ii), if $\alpha-\beta-\lambda \ge -1$, then $\limsup_{n\to\infty} \sum_{k=1}^{n} a'_kX_k/b'_n = \infty$ a.s. Thus an e.s. for $\sum_{k=1}^{n} a'_kX_k$ only exists when $\alpha-\beta-\lambda < -1$ and $\beta > 0$. To obtain the almost sure limiting behavior of $\sum_{k=1}^{n} a'_kX_k/b'_n$ we need to establish the limiting behavior of $\sum_{k=1}^{n} a'_k\mu_k/b'_n$.

This section is broken down into four subsections. In the first two examples we examine the a.s. limiting behavior of our normed partial sums when $E|X| = \infty$. Then in Example 3.3 we observe the situation when $X \in L_1$. Finally, and most excitingly we obtain an e.s. for the sum of independent mean zero random variables.

EXAMPLE 3.1. If $\alpha-\beta-\lambda < -1$, $\beta > 0$, and $\lambda < 1$, then

$$\lim_{n\to\infty} \frac{\sum_{k=1}^{n} a'_k X_k}{b'_n} \underset{=}{a.s.} \begin{cases} 0 & \text{if } \alpha-\beta-\lambda < -2, \\ \dfrac{\sigma_\lambda}{(1-\lambda)\beta} & \text{if } \alpha-\beta-\lambda = -2, \\ \infty & \text{if } \alpha-\beta-\lambda > -2. \end{cases}$$

PROOF. Since $\log c'_n \sim \log n$

$$\mu_n = EXI(|X| \le c'_n) = \int_e^{c'_n} \frac{\sigma_\lambda \, dx}{x(\log x)^\lambda}$$

$$= \frac{(1+o(1))\sigma_\lambda (\log n)^{1-\lambda}}{1-\lambda} .$$

Thus

$$\lim_{n\to\infty} \frac{\sum_{k=1}^{n} a'_k \mu_k}{b'_n} = \begin{cases} 0 & \text{if } \alpha-\beta-\lambda < -2, \\ \dfrac{\sigma_\lambda}{(1-\lambda)\beta} & \text{if } \alpha-\beta-\lambda = -2, \\ \infty & \text{if } \alpha-\beta-\lambda > -2. \ \square \end{cases}$$

However, if $\lambda = 1$ it can be shown that

$$\lim_{n\to\infty} \frac{\sum_{k=1}^{n} a'_k X_k}{b'_n} \underset{=}{a.s.} \begin{cases} 0 & \text{if } \alpha-\beta < -1, \\ \infty & \text{if } \alpha-\beta \ge -1, \end{cases}$$

when $\alpha-\beta < 0$ and $\beta > 0$. The reason we cannot obtain an e.s. here is that $\{b'_n, n\ge1\}$ is not delicate enough.

EXAMPLE 3.2. Suppose that $\lambda = 1$, $\beta > 0$, and (3.1) hold. If $\alpha \ne -1$, then

$$\lim_{n\to\infty} \frac{\sum_{k=1}^{n} a'_k X_k}{b''_n} \underset{=}{a.s.} \begin{cases} 0 & \text{if } \alpha < -1, \\ \infty & \text{if } -1 < \alpha \le 0, \end{cases}$$

but, if $\alpha = -1$

$$\lim_{n\to\infty} \frac{\sum_{k=1}^{n} a'_k X_k}{b''_n} \underset{=}{a.s.} \begin{cases} 0 & \text{if } \beta > 2, \\ \sigma_1/2 & \text{if } \beta = 2, \\ \infty & \text{if } \beta < 2. \end{cases}$$

PROOF. By Lemmas 3.2 and 3.3 and Theorem 2.1 $\sum_{k=1}^{n} a_k'(X_k-\mu_k)/b_n'' \to 0$ a.s., where

$$\mu_n = EXI(|X| \le c_n'') = \int_e^{c_n'} \frac{\sigma_1 dx}{x\log x} = (1+o(1))\sigma_1 \log_2 n,$$

whence the conclusion obtains. □

We now return to our norming sequence $\{b_n', n\ge1\}$.

EXAMPLE 3.3. If $\alpha-\beta-\lambda < -1$, $\beta > 0$, and $\lambda > 1$, then

$$\lim_{n\to\infty} \frac{\sum_{k=1}^{n} a_k' X_k}{b_n'} \overset{a.s.}{=} \begin{cases} 0 & \text{if } \alpha-\beta < -1, \\ EX/\beta & \text{if } \alpha-\beta = -1, \\ \infty & \text{if } \alpha-\beta > -1. \end{cases}$$

PROOF. Since $\mu_n = EXI(|X| \le c_n') \to EX$ the conclusion follows. □

By changing the values of our parameters (α,β,λ) we can obtain any finite positive almost sure limit for our normed sum of independent random variables. Next, we turn our attention to the mean zero situation.

EXAMPLE 3.4. If $\alpha-\beta-\lambda < -1$, $\beta > 0$, and $\lambda > 1$, then

$$\lim_{n\to\infty} \frac{\sum_{k=1}^{n} a_k'(X_k-EX)}{b_n'} \overset{a.s.}{=} \begin{cases} 0 & \text{if } \alpha-\beta-\lambda < -2, \\ -EX/\beta & \text{if } \alpha-\beta-\lambda = -2, \\ \infty & \text{if } \alpha-\beta-\lambda > -2. \end{cases}$$

PROOF. Let $W = X-EX$ and $W_n = X_n-EX$. Then $P\{|W| > x\} \sim P\{|X| > x\}$, whence $\sum_{n=1}^{\infty} P\{|W_n| > c_n'\} < \infty$ iff $\alpha-\beta-\lambda < -1$. As before we conclude that $\sum_{k=1}^{\infty} a_k'(W_k-\mu_k)/b_n' \to 0$ a.s., where

$$\begin{aligned}\mu_n &= EWI(|W| \le c'_n)\\ &= -EWI(|W| > c'_n)\\ &= -\int_{c'_n}^{\infty} \frac{\sigma_\lambda \, wdw}{(w+EX)^2(\log(w+EX))^\lambda}\\ &= -(1+o(1))\sigma_\lambda \int_{c'_n}^{\infty} \frac{dw}{w(\log w)^\lambda}\\ &= -(1+o(1))EX(\log n)^{1-\lambda},\end{aligned}$$

hence the conclusion follows. □

These examples seem to contradict Theorems 2.2 through 2.6. Next we reexamine the hypotheses of these theorems.

It will be shown that the theorems previously mentioned are quite optimal, for if the hypotheses are slightly weakened the conclusions fail. Before we analyze these conditions a few lemmas are needed, their proofs are omitted. Note that c'_n and c''_n are eventually increasing.

LEMMA 3.4.

$$c'_n \sum_{j=n}^{\infty} \frac{1}{c'_j} \approx \begin{cases} n(\log n) & \text{if } \alpha-\beta < -1,\\ \infty & \text{if } \alpha-\beta \ge -1.\end{cases}$$

LEMMA 3.5.

$$c''_n \sum_{j=n}^{\infty} \frac{1}{c''_j} \approx \begin{cases} n(\log n) & \text{if } \alpha < -1,\\ n(\log n)(\log_2 n) & \text{if } \alpha = -1 \text{ and } \beta > 1,\\ \infty & \text{otherwise.}\end{cases}$$

LEMMA 3.6.

$$c'_n \sum_{j=1}^{n} \frac{1}{c'_j} \approx \begin{cases} n(\log n)^{\beta-\alpha} & \text{if } \alpha-\beta < -1,\\ n(\log n)(\log_2 n) & \text{if } \alpha-\beta = -1,\\ n(\log n) & \text{if } \alpha-\beta > -1.\end{cases}$$

LEMMA 3.7.

$$\sum_{j=1}^{n} a_j' \approx \begin{cases} 1 & \text{if } \alpha < -1, \\ \log_2 n & \text{if } \alpha = -1, \\ (\log n)^{\alpha+1} & \text{if } \alpha > -1. \end{cases}$$

Clearly, all the hypotheses of Theorem 2.1 hold. Hence the only hypothesis of Theorem 2.2 that can fail is (2.3). Noting Lemmas 3.4 and 3.5 we see that in all possible cases (2.3) doesn't hold, failing ever so barely in some situations.

Using Lemma 3.1 we see that $P\{|X| > x\}$ is indeed regularly varying, but with exponent $\rho = -1$. Hence (2.5) cannot be extended. Just as delicate, we see that, via Lemma 3.6, (2.6) fails by only a factor of a slowly varying function. Finally, for (2.7) to hold we need $\alpha-\beta \leq -1$ and since $X \in L_1$, we also must have $\lambda > 1$. Therefore $\alpha-\beta-\lambda < -2$, but recalling Example 3.4, an exact sequence exists only when $\alpha-\beta-\lambda = -2$. Thus, as expected Theorem 2.3 is not applicable.

Observing Theorem 2.4 we see that na_n' is nondecreasing when $\alpha \geq 0$, but (2.8) fails by only a factor of a slowly varying function no matter our choice of α (see Lemma 3.7). Example 3.4 shows the necessity of the hypothesis of common distribution in Theorem 2.5. Conditions (2.9) and (2.11) fail, hence Theorem 2.6(i) is not violated. However (2.10) does hold. This is an interesting fact in itself, since (2.9) implies (2.10). Thus, when $\sum_{n=1}^{\infty} P\{|X_n| > b_n\} = \infty$ it turns out that $\{b_n, n\geq 1\}$ is not an e.s. for S_n via Theorem 2.6(ii).

4. New Results

We conclude with two theorems that generalize those of Chow and Teicher (1971). Conditions are given showing when and when not there exists constants $\{C_n, n\geq 1\}$ such that $\sum_{k=1}^{n} a_k Y_k / b_n - C_n \to 0$ a.s. Our goal has been to establish similar strong laws with C_n converging to some finite nonzero constant.

THEOREM 4.1. Let $\{Y_n, n\ge 1\}$ be i.i.d. random variables with

$$\liminf_{y\to\infty} yP\{|Y_1| > y\} > 0. \tag{4.1}$$

If $\{a_n, n\ge 1\}$ and $\{b_n, n\ge 1\}$ are constants with $b_n = O(|b_{n+1}|)$, $c_n \to \infty$, and $\sum_{n=1}^{\infty} c_n^{-1} = \infty$, then there doesn't exist constants $\{C_n, n\ge 1\}$ with $\sum_{k=1}^{n} a_k Y_k / b_n - C_n \to 0$ a.s.

PROOF. Let $\{Y_n^*, n\ge 1\}$ be a symmetrized version of $\{Y_n, n\ge 1\}$. Suppose the conclusion fails. Then $\sum_{k=1}^{n} a_k Y_k^* / b_n \to 0$ a.s. Hence

$$\left|\frac{a_n Y_n^*}{b_n}\right| \le \left|\frac{\sum_{k=1}^{n} a_k Y_k^*}{b_n}\right| + \left|\frac{b_{n-1}}{b_n}\right| \cdot \left|\frac{\sum_{k=1}^{n-1} a_k Y_k^*}{b_{n-1}}\right| \to 0 \quad \text{a.s.}$$

Thus, via the Borel-Cantelli lemma $\sum_{n=1}^{\infty} P\{|Y_n^*| > c_n\} < \infty$. However for all large y

$$\begin{aligned} 4yP\{|Y_n^*| > y\} &= 4yP\{|Y_n - Y_n'| > y\} \\ &\ge 4yP\{|Y_n| > 2y\}P\{|Y_n'| < y\} \\ &\ge 2yP\{|Y_1| > 2y\} \ge \epsilon, \end{aligned}$$

for some $\epsilon > 0$. Therefore, for a suitably chosen n_0

$$\sum_{n=1}^{\infty} P\{|Y_n^*| > c_n\} \ge C \sum_{n=n_0}^{\infty} c_n^{-1} = \infty,$$

a contradiction. □

Returning to our random variables $\{X_n, n\geq1\}$ we see that (4.1) obtains iff $\lambda \leq 0$ (Lemma 3.1). On the other hand $\sum_{n=1}^{\infty} (c_n')^{-1} = \infty$ iff $\alpha-\beta \geq -1$. Therefore $\alpha-\beta-\lambda \geq -1$, but we've already established the fact that $\{b_n', n\geq1\}$ is not an e.s. for $\sum_{k=1}^{n} a_k' X_k$ when $\alpha-\beta-\lambda \geq -1$.

Finally, we present necessary and sufficient conditions for determining whether or not the centering sequence $\{C_n, n\geq1\}$ exists.

THEOREM 4.2 Let $\{Y, Y_n, n\geq1\}$ be i.i.d. random variables and $L(y)$ and $a(y)$ be positive functions defined on $[0,\infty)$. Let $a_n = a(n)$, and $\{b_n, n\geq1\}$ be positive constants with $b_n = O(b_{n+1})$ and c_n nondecreasing. Suppose $ya(L(y))$ is nondecreasing and that either $a(L(y))$ is nonincreasing or $a(L(c_{[y]}))$ is slowly varying at infinity, as $y \to \infty$. Furthermore, let

$$0 < \liminf_{n\to\infty} \frac{c_n}{n} a(L(c_n)) \leq \limsup_{n\to\infty} \frac{c_n}{n} a(L(c_n)) < \infty. \qquad (4.2)$$

Then

$$\frac{\sum_{k=1}^{n} a_k Y_k}{b_n} - C_n \to 0 \text{ a.s.} \qquad (4.3)$$

for some constants $\{C_n, n\geq1\}$ iff

$$E|Y|a(L(|Y|)) < \infty. \qquad (4.4)$$

PROOF. In view of (4.2) there exists positive finite constants M_1 and M_2 such that if n is sufficiently large

$$M_1 \leq \frac{c_n}{n} a(L(c_n)) \leq M_2.$$

Then if n is large

$$\max_{1\le k\le n} c_k^2 \sum_{j=n}^{\infty} c_j^{-2} = c_n^2 \sum_{j=n}^{\infty} c_j^{-2}$$

$$\le \frac{M_2^2 n^2}{[a(L(c_n))]^2} \sum_{j=n}^{\infty} \frac{[a(L(c_j))]^2}{M_1^2 j^2} = O(n),$$

if either $a(L(y))$ is nonincreasing or $a(L(c_{[y]}))$ is slowly varying.

For a suitably choosen n_0

$$\sum_{n=n_0}^{\infty} P\{|Y| > c_n\} = \sum_{n=n_0}^{\infty} P\{|Y|a(L(|Y|)) > c_n a(L(c_n))\}$$

$$\le \sum_{n=n_0}^{\infty} P\{|Y|a(L(|Y|)) > M_1 n\} < \infty \quad \text{(by (4.4))}.$$

Using Theorem 2.1 (4.3) obtains, where C_n may be chosen to be

$\sum_{k=1}^{n} a_k EYI(|Y| \le c_k)/b_n$.

Next suppose that (4.3) holds. Let $\{Y_n^*, n\ge 1\}$ be a symmetrized version of $\{Y_n, n\ge 1\}$. Then $\sum_{k=1}^{n} a_k Y_k^*/b_n \to 0$ a.s., hence $Y_n^*/c_n \to 0$ a.s. Thus

$$\sum_{n=1}^{\infty} P\{|Y_1^*| > c_n\} < \infty.$$

Let $c_0 = 0$ and note that

$$E|Y_1^*|a(L(|Y_1^*|)) = \sum_{n=1}^{\infty} E|Y_1^*|a(L(|Y_1^*|))I(c_{n-1} < |Y_1^*| \le c_n)$$

$$\le \sum_{n=1}^{\infty} c_n a(L(c_n))P\{c_{n-1} < |Y_1^*| \le c_n\}$$

$$\leq C \sum_{n=1}^{\infty} nP\{c_{n-1} < |Y_1^*| \leq c_n\}$$

$$= C \sum_{n=1}^{\infty} \sum_{k=1}^{n} P\{c_{n-1} < |Y_1^*| \leq c_n\}$$

$$= C \sum_{k=1}^{\infty} P\{|Y_1^*| > c_{k-1}\} < \infty,$$

which is tantamount to (4.4). □

Note that for our sequence $c_n' = n(\log n)^{\beta-\alpha}$, all the hypotheses of Theorem 4.2 are satisfied with $a(L(y)) = (\log y)^{\alpha-\beta}$. Therefore $\{C_n, n\geq 1\}$ exists iff $EX(\log X)^{\alpha-\beta} < \infty$, which agrees with our previous constraint, $\alpha-\beta-\lambda < -1$. Choosing $a(L(y)) = (\log y)^{\alpha}(\log_2 y)^{-\beta}$ we see that Theorem 4.2 is applicable when $c_n = n(\log n)^{-\alpha}(\log_2 n)^{\beta}$. Thus, necessary and sufficient conditions for the existence of the sequence $\{C_n, n\geq 1\}$ is $EX(\log X)^{\alpha}(\log_2 X)^{-\beta} < \infty$, which occurs iff (3.1) holds.

If $\sum_{k=1}^{n} Y_k/b_n - C_n \rightarrow 0$ a.s. we say that $\sum_{k=1}^{n} Y_k/b_n$ is stable when centered at C_n. For a further discussion on stability see Chapter Four of Stout (1974).

REFERENCES

ADLER, A. (1988). On the law of the iterated logarithm for nonidentically distributed random variables. **Stochastic Anal. Appl. 6** 117-127.

ADLER, A. and ROSALSKY, A. (1987a). Some general strong laws for weighted sums of stochastically dominated random variables. **Stochastic Anal. Appl.** 5 1-16.

ADLER, A. and ROSALSKY, A. (1987b). On the strong law of large numbers for normed weighted sums of i.i.d. random variables. **Stochastic Anal. Appl. 5** 467-483.

ADLER, A. and ROSALSKY, A. (1987c). On the Chow-Robbins "fair" games problem. Dept. of Statistics, Univ. of Florida, Tech. Report No. 288.

ADLER, A. and ROSALSKY, A. (1988). Some generalized central limit theorems for weighted sums with infinite variance. Dept. of Statistics, Univ. of Florida, Tech. Report No. 294.

CHOW, Y.S. and ROBBINS, H. (1961). On the sums of independent random variables with infinite moments and "fair" games. **Proc. Nat. Acad. Sci. U.S.A. 47** 330-335.

CHOW, Y.S. and TEICHER, H. (1971). Almost certain summability of independent, identically distributed random variables. **Ann. Probab. 42** 401-404.

FELLER, W. (1945). Note on the law of large numbers and "fair games." **Ann. Math. Statist. 16** 301-304.

FELLER, W. (1968). **An Introduction to Probability Theory and Its Applications**, Vol I, 3rd ed. Wiley, New York.

FELLER, W. (1971). **An Introduction to Probability Theory and Its Applications,** Vol II, 2nd ed. Wiley, New York.

HEYDE, C.C. (1969). A note concerning behaviour of iterated logarithm type. **Proc. Amer. Math. Soc. 23** 85-90.

KLASS, M. and TEICHER, H. (1977). Iterated logarithm laws for asymmetric random variables barely with or without finite means. **Ann. Probab. 5** 861-874.

MALLER, R.A. (1978). Relative stability and the strong law of large numbers. **Z. Warsch. verw. Gebiete 43** 141-148.

MARTIKAINEN, A.I. and PETROV, V.V. (1980). On a theorem of Feller. **Theory Probab. Appl. 25** 191-193.

ROGOZIN, B.A. (1968). On the existence of exact upper sequences. **Theory Probab. Appl. 13** 667-672.

ROGOZIN, B.A. (1976). Relatively stable walks. **Theory Probab. Appl. 21** 375-379.

STOUT, W.F. (1974). **Almost Sure Convergence.** Academic Press, New York.

TEICHER, H. (1979. Rapidly growing random walks and an associated stopping time. **Ann. Probab. 7** 1078-1081.

UNUSUAL APPLICATIONS OF A.E. CONVERGENCE

E.J. Balder
Mathematical Institute, University of Utrecht
Utrecht, Netherlands

INTRODUCTION

In [2b] the present author obtained an abstract extension of Komlós' theorem (see also [2a] for "ordinary" extensions), and used this to give applications to various subjects (e.g., weak compactness criteria in L_1-spaces, Prohorov's theorem for transition probabilities, compactness criteria in generalized Köthe spaces). In all these cases classical (sequential) convergence - of a weak type, involving integral expressions - is strengthened into the kind of a.e. convergence for arithmetic averages that figures in Komlós' original result (by its subsequence character this type of convergence is a topological concept in the sense of e.g. [9]). Incidentally, it is interesting to recall that Komlós' result [10] was originally proven by means of a delicate truncation argument, involving martingale convergence, and that it

ISBN 0-12-231050-0

has been seminal for the development of the so-called "subsequence principle" ([7],[1],[3]).

In this paper we shall illustrate the usefulness of the abstract version of Komlós' theorem by more, new applications (to wit, an existence result in L_1-spaces, a weak relative compactness result for multifunctions, an approximation result for multifunctions and a lower semicontinuity result for integral functionals). We shall also show that the abstract version of Komlós' theorem (and its proof) have natural counterparts for martingales.

1 ABSTRACT VERSION OF KOMLÓS' THEOREM

Here we recapitulate the main result of [2b]. Let $(T, \mathcal{T}, \mu)$ be an arbitrary measure space, and E a convex subset of a Hausdorff topological vector space (actually, it is enough to require this space to be closed for nonnegative – or even convex – linear combinations). As usual, $\mathcal{B}(E)$ denotes the Borel σ-algebra on E. Let $h : T \times E \rightarrow [0, +\infty]$ be a given function such that for every $t \in T$

$$h(t, \cdot) \text{ is convex and sequentially inf-compact on } E \tag{1.1}$$

(the latter means that for every sequence $\{x^k\}$ in E with $\beta' := \sup_k h(t, x^k) < +\infty$ there exist $x_* \in E$ and a subsequence $\{x_{k_i}\}$

of $\{x_k\}$ such that $\{x_{k_i}\}$ converges to x_* and $h(t, x_*) \leq \beta'$). Let $\mathcal{A} := \{a_j\}$ be a given collection of functions $a_j : T \times X \to \mathbb{R}, j \in \mathbb{N}$, such that for every $t \in T$

$$a_j(t, \cdot) \text{ is affine and sequentially continuous on } E. \tag{1.2}$$

We suppose that these satisfy the following growth condition with respect to h : for every $j \in \mathbb{N}$ there exist $C_j > 0$ and $\phi_j \in \mathcal{L}^1_{\mathbb{R}}$ (the space of integrable real-valued functions on T) such that

$$| a_j(t, x) | \leq C_j h(t, x) + \phi_j(t) \text{ on } T \times E. \tag{1.3}$$

A subset D of E is said to be *countably separated* by the collection $\mathcal{A}$, if for every $t \in T$ the functions $a_j(t, \cdot) : E \to \mathbb{R}, j \in \mathbb{N}$, separate the points of D (i.e., for any $x, y \in D, x = y$ if and only if $a_j(t, x) = a_j(t, y)$ for all $j \in \mathbb{N}$). Finally, let us agree to call a function $f : T \to E$ *$\mathcal{A}$-scalarly measurable* if the function $a_j(\cdot, f(\cdot))$ is $\mathcal{T}$-measurable for every $j \in \mathbb{N}$.

Theorem 1.1 *Let $\{f_k\}$ be a sequence of $\mathcal{A}$-scalarly measurable functions $f_k : T \to E$ such that the following* tightness *condition holds:*

$$\beta := \sup_{k \in \mathbb{N}} \int_T^* h(t, f_k(t)) \mu(dt) < +\infty, \tag{1.4}$$

where $\int_T^*$ denotes outer integration. Suppose also that for some null set $N \in \mathcal{T}$

$$cl\ co \cup_k f_k(T\backslash N) \text{ is countably separated by } \mathcal{A}. \tag{1.5}$$

Then there exist an $\mathcal{A}$-scalarly measurable function $f_* : T \to E$ and a subsequence $\{f_m\}$ of $\{f_k\}$ such that for every further subsequence $\{f_{m_i}\}$ of $\{f_m\}$

$$\lim_{n\to\infty} \frac{1}{n}\sum_{i=1}^{n} f_{m_i}(t) = f_*(t) \text{ a.e. in } T, \tag{1.6}$$

$$\int_T^* h(t, f_*(t))\mu(dt) \leq \beta. \tag{1.7}$$

As shown in [2b], this result subsumes Komlós' theorem as well as its extension given in [2a]. (It should be pointed out that Komlós' theorem is essential for the proofs in [2a], [2b]). Note that the a.e. convergence in Theorem 1.1 takes place in the original topology on E (so to satisfy the tightness condition, this topology will have to be sufficiently coarse). An immediate consequence of Theorem 1.1 is as follows. A function $g : T \times E \to (-\infty, +\infty]$ is said to have growth property (G') with respect to h if for every $\epsilon > 0$ there exists $\phi \in L^1_{\mathbb{R}}$ such that

$$g^-(t,x) \leq \epsilon h(t,x) + \phi(t) \text{ on } T \times E.$$

Corollary 1.2 *In Theorem 1.1 the inequality*

$$\liminf_{m\to\infty} \int_T^* g(t, f_m(t))\mu(dt) \geq \int_T^* g(t, f_*(t))\mu(dt). \tag{1.8}$$

holds for every function $g : T \times E \to (\infty, +\infty]$ having growth property (G') with respect to h, and such that $g(t, \cdot)$ is convex and sequentially lower semicontinuous on E for every $t \in T$.

2 APPLICATIONS

In addition to the applications of Theorem 1.1 given in [2b], mentioned in the previous section, we shall discuss here some more instances where Theorem 1.1 can be of use. As a very easy introductory application, consider the following slight extension of [5, Prop. 3.7] (in [5] this result is attributed to C. Hess). We suppose here that μ is a finite measure, and that V is a Banach space with separable dual space V'. (The norm on V and the dual norm on V' will be denoted by $\|\cdot\|$ and $\|\cdot\|'$ respectively.)

Proposition 2.1 *Suppose that $\{f_n\}$ is a sequence in $\mathcal{L}^1_V$ such that*

$$\liminf_{n\to\infty} \int_T \|f_n\| d\mu < +\infty, \tag{2.1}$$

$$f_n(t) \text{ is relatively compact for } \sigma(V, V') \text{ a.e. in } T, \tag{2.2}$$

and suppose that there exists a countable dense subset D of V' such that for every $v' \in D$

$$\lim_{n\to\infty} < f_n(t), v' > \ \textit{exists a.e. in } T \tag{2.3}$$

Then there exists $f_* \in \mathcal{L}^1_V$ *such that*

$$\lim_{n\to\infty} f_n(t) = f_*(t) \ \textit{in } \sigma(V, V') \ \textit{a.e. in } T \tag{2.4}$$

Proof. By hypothesis (2.1), there exists a subsequence f_k of f_n such that

$$\sup_{k\in\mathbb{N}} \int_T \|f_k\| d\mu < +\infty.$$

Now apply Theorem 1.1 to this sequence, equipping $E := V$ with the topology $\sigma(V, V')$ and defining

$$h(t, v) := \begin{array}{l} \|v\| \text{ if } v \in \text{ cl co } f_k(t) \\ +\infty \text{ otherwise} \end{array}$$

$$a^{v'}(t, v) := < v, v' >, v' \in D.$$

Then by Krein's theorem and the Eberlein-Smulian theorem (1.1) holds. Also, (1.2)-(1.4) are obvious, and (1.5) holds for the countable collection $\mathcal{A} := \{a^{v'} : v' \in D\}$. So by Theorem 1.1 there exist an $\mathcal{A}$-scalarly measurable function $f_* : T \to V$ and a subsequence $\{f_m\}$ of $\{f_k\}$ for which (1.6)-(1.7) hold. By (2.3) it follows then that for every $v' \in D$

$$\lim_{n\to\infty} < f_n(t), v' > = \lim_{m\to\infty} < f_m(t), v' > =$$

$$= \lim_{n\to\infty} \frac{1}{n} \sum_{m=1}^{n} < f_m(t), v' > \text{ a.e. in } T,$$

so we conclude from (1.6), denseness of D and boundedness of $\{f_n(t)\}$ (by (2.2)) that for every $v' \in V'$

$$\lim_{n\to\infty} < f_n(t), v' > = < f_*(t), v' > \text{ a.e. in } T.$$

This amounts to (2.4). Finally, by (1.6)–(1.7) and separability of V it follows that f_* is strongly measurable and integrable. □

Our second application is as follows. Let μ and V be as above. The collection of all convex $\sigma(V, V')$-compact subsets of V will be denoted as $\mathcal{P}_{ck}(V)$. The Hausdorff metric on $\mathcal{P}_{ck}(V)$ (corresponding to $\|\cdot\|$ on V) is denoted by H, and the diameter of a set in $\mathcal{P}_{ck}(V)$ by $|\cdot|$ (thus $|A| = H(A, \{0\})$) [6, II]. Recall that a function $f : T \to \mathcal{P}_{ck}(V)$ is also called multifunction, and that f is defined to be *scalarly measurable* if for every $v' \in V'$ the *support function* $s(v' \mid f(\cdot))$ of f, given by

$$s(v' \mid f(t)) := \sup_{v \in f(t)} < v, v' >,$$

is $\mathcal{T}$-measurable. We have the following new result:

Theorem 2.2 *Suppose that $\{f_k\}$ is a sequence of scalarly measurable multifunctions $f_k : T \to \mathcal{P}_{ck}(V)$ such that for some $g : T \to \mathcal{P}_{ck}(V)$*

$$f_k(t) \subset g(t) \text{ a.e. in } T \tag{2.5}$$

$$\int_T^* \mid g(t) \mid \mu(dt) < +\infty. \tag{2.6}$$

Then there exist a scalarly measurable multifunction $f_* : T \to \mathcal{P}_{ck}(V)$ and subsequence $\{f_m\}$ of $\{f_k\}$ such that

$$\lim_{n\to\infty} s(v' \mid \frac{1}{n}\sum_{m=1}^{n} f_m(t)) = s(v' \mid f_*(t)) \text{ for all } v' \in V' \text{ a.e.}$$

Proof. Consider the convex set $E := \mathcal{P}_{ck}(V)$, equipped with the initial topology with respect to all functionals $s(v' \mid \cdot), v' \in V'$. Define

$$h(t,x) := \begin{matrix} \mid x \mid \text{ if } x \subset g(t) \\ +\infty \text{ otherwise} \end{matrix},$$

$$a_j(t,x) := s(v'_j \mid x) := \sup_{v\in x} < v, v'_j >,$$

where $\{v'_j\}$ is the subset of V' that figures in the proof of [6, III.35]. From this proof it follows that (1.5) holds. Note that for every $v' \in V'$, $s(v' \mid \cdot)$ is additive and positively homogeneous on E (a consequence of [6, II.17]). Hence (1.2) holds. By [6, II.18]

$$| x |= \sup_{w \in U} | s(w \mid x) |,$$

so $| \cdot |= H(\cdot, \{0\})$ is convex and lower semicontinuous on E. The sequential inf-compactness in (1.1) follows from the fact that for every $t \in T$ the relative $\sigma(V, V')$-topology on the compact set $g(t)$ is metrizable (note that $(V, \sigma(V, V'))$ is Suslin), so by [6, II.4, Remark] the set $\{x \in E : x \subset g(t)\}$ is (sequentially) compact for the *corresponding* Hausdorff metric. It is then straightforward to show that this set is also sequentially compact for the topology on E, so (1.1) has been validated. By

$$| s(v' \mid x) | \leq \sup_{v \in x} \|v\| \|v'\|' =| x \mid \|v'\|'$$

the inequality (1.3) holds for all functions a_j. Of course, $\mathcal{A}$-scalar measurability of the functions f_k holds by definition. Finally, (1.4) holds by (2.5)–(2.6). We may now apply Theorem 1.1. This gives immediately the desired result. □

Remark 2.3 As noted in section 1, the results in [2b] remain valid if the space E is a convex subset of a space which is not a vector space, but is endowed with addition and multiplication for *nonnegative* scalars. The space $E = \mathcal{P}_{ck}(V)$ serves as an example of such a case : note that in $\mathcal{P}_{ck}(V)$ we do not have $x + (-x) = \{0\}$, for instance. (Nevertheless, it would be possible here to embed $\mathcal{P}_{ck}(V)$ in a vector space; see [6, II.19].)

It is possible to rephrase Theorem 2.2 in case V is a separable dual Banach space. Rather than doing so directly, we shall discuss a slightly different application, in which this variant of Theorem 2.2 figures implicitly. This result is a generalization of a best approximation result due to Lu'u [11]; the proof is of interest because it shows how easily one can manipulate with the a.e. convergence obtained in Theorem 1.1. As before, let μ be finite and let V be this time the separable dual of a Banach space W. V will be equipped with the weak star topology $\sigma(V, W)$. Let f_0 be a fixed element in the space $\mathcal{L}^1_{\mathcal{P}_{ck}(V)}(T, \mathcal{T})$ of all scalarly $\mathcal{T}$-measurable multifunctions $f : T \to \mathcal{P}_{ck}(V)$ which are integrably bounded, i.e., with integrable diameter $| f(\cdot) |$. (By [6, II.18] and Lipschitz continuity of $s(\cdot \mid x)$ for every $x \in \mathcal{P}_{ck}(V)$ we have

$$| x | = \sup_{w \in U} | s(w \mid x) | = \sup_{w \in D} | s(w \mid x) |,$$

where D denotes any countable dense subset in the unit ball U of the separable space W; hence the diameter $| f(\cdot) |$ is measurable.)

Let $\mathcal{S}$ be a sub-σ-algebra of $\mathcal{T}$, and consider the approximation problem (P) of minimizing the functional

$$\Delta(f, f_0) := \int_T H(f(t), f_0(t))\mu(dt)$$

over the space $\mathcal{L}^1_{\mathcal{P}_{ck}(V)}(T, \mathcal{S})$ of all scalarly $\mathcal{S}$-measurable and integrably bounded multifunctions having values in $\mathcal{P}_{ck}(V)$. Recall that H denotes the Hausdorff distance on $\mathcal{P}_{ck}(V)$ corresponding to the norm $\|\cdot\|$ on V.

Theorem 2.4 *There exists a best approximant in problem (P): i.e., there exists $f_* \in \mathcal{L}^1_{\mathcal{P}_{ck}(V)}(T, \mathcal{S})$ such that*

$$\Delta(f_*, f_0) = \alpha := \inf\{\Delta(f, f_0) : f \in \mathcal{L}^1_{\mathcal{P}_{ck}(V)}(T, \mathcal{S})\},$$

provided that $\mathcal{L}^1_{\mathcal{P}_{ck}(V)}(T, \mathcal{S})$ is nonempty.

Proof. Of course, the underlying measure space for this application is $(T, \mathcal{S}, \mu)$. Let $E := \mathcal{P}_{ck}(V)$ be equipped with the same topology as in Theorem 2.2. Let $\{f_k\}$ be a minimizing sequence for (P). By the triangle inequality for H we get

$$\int_T | f_k | \, d\mu \leq \Delta(f_k, f_0) + \int_T | f_0 | \, d\mu.$$

The first term on the right converges to α; hence (1.1) holds with

$$h(t, x) := | x | \, .$$

As noted in the proof of Theorem 2.2, the function $| \cdot |$ is convex and lower semicontinuous on E. Now it is also sequentially inf-compact: for any sequence $\{x^k\} \in E$ with $\beta' := sup_k \mid x^k \mid < +\infty$ all sets are contained in the ball B around 0 with radius β'. By the Alaoglu-Bourbaki theorem the ball B is compact in V and hence metrizable (V is Suslin). Thus, the sequence $\{x^k\}$ contains a subsequence converging to some $x^* \subset B$ in the corresponding Hausdorff metric [6, II.4, Remark]. This implies the convergence of this subsequence to x^* in the sense of E. It will be noted that this argument is almost the same as the corresponding one used to prove Theorem 2.2. Of course, this is no coincidence : we can refer to that proof for the entire remainder of the derivation of the following consequence of Theorem 1.1: there exist a scalarly $\mathcal{S}$-measurable multifunction $f_* : T \rightarrow E$ and a subsequence $\{f_m\}$ of $\{f_k\}$ with

$$\lim_{n\to\infty} s(v' \mid \frac{1}{n} \sum_{m=1}^{n} f_m(t)) = s(v' \mid f_*(t)) \text{ for all } v' \in V' \text{ a.e.,} \quad (2.7)$$

$$\int_T \mid f_*(t) \mid \mu(dt) < +\infty. \quad (2.8)$$

From this it follows that f_* belongs to $\mathcal{L}^1_{\mathcal{P}_{ck}(V)}(T, \mathcal{S})$. Finally, by [6, II.18] we have that

$$H(x, f_0(t)) = \sup_{w \in U} [s(w \mid x) - s(w \mid f_0(t))]. \tag{2.9}$$

Hence, for every $t \in T$ the function $H(\cdot, f_0(t))$ is convex and lower semicontinuous on E. Now

$$\alpha = \lim_{m \to \infty} \Delta(f_m, f_0) = \lim_{n \to \infty} \frac{1}{n} \sum_{m=1}^{n} \Delta(f_m, f_0) \geq$$

$$\geq \int_T \liminf_{n \to \infty} \frac{1}{n} \sum_{m=1}^{n} H(f_m(t), f_0(t)) \mu(dt),$$

where the two identities are elementary, and the inequality holds by Fatou's lemma. By convexity of $H(\cdot, f_0(t))$ this gives

$$\alpha \geq \int_T \liminf_{n \to \infty} H(\frac{1}{n} \sum_{m=1}^{n} f_m(t), f_0(t)) \mu(dt).$$

Finally, it follows from this, by (2.7) and lower semicontinuity of $H(\cdot, f_0(t))$, that $\alpha \geq \Delta(f_*, f_0)$. □

Next, let us give an application to a more complicated lower semicontinuity problem than the one encountered in the above proof. We shall prove a very general lower semicontinuity result for integral functionals that plays an important role in the theory of optimal control and the theory of differential inclusions [4]. Again we suppose μ to be finite. Let Y be a metric space, and let V be a separable reflexive Banach space.

Theorem 2.5 *Suppose that* $\{y_k\}$ *and* y_∞ *are measurable functions* $y_k : T \to Y$ *such that*

$$\lim_{k\to\infty} y_k(t) = y_\infty(t) \text{ in } Y \text{ a.e. }, \tag{2.10}$$

Suppose also that $\{v_k\}$ *is a sequence in* $\mathcal{L}^1_V$ *such that*

$$\int_T \|v_k(t)\|\mu(dt) < +\infty. \tag{2.11}$$

Then there exist a function $v_* \in \mathcal{L}^1_V$ *and a subsequence* $\{v_m\}$ *of* $\{v_k\}$ *such that*

$$v_*(t) \in \cap_{p=1}^{\infty} cl\ co\ \{v_m(t) : m \geq p\} \text{ a.e. in } T \tag{2.12}$$

$$\liminf_{m\to\infty} I_l(y_m, v_m) \geq I_l(y_\infty, v_*) \tag{2.13}$$

for every function $l : T \times Y \times V \to [0, +\infty]$ *such that* $l(t, y_\infty(t), \cdot)$ *is convex on* V *for every* $t \in T$, *and* $l(t, \cdot, \cdot)$ *is lower semicontinuous in every point of the set* $\{y_\infty(t)\} \times V$. *Here*

$$I_l(y_m, v_m) := \int_T^* l(t, y_m(t), v_m(t))\mu(dt), \text{ etc.}$$

Proof. Let $E := M^1_+(\hat{\mathbb{N}} \times V)$, the space of all probability measures on $\hat{\mathbb{N}} \times V$, where $\hat{\mathbb{N}} := \mathbb{N} \cup \{\infty\}$ is the (Alexandrov) compactification of the set of natural numbers. We shall equip E with the usual narrow (or weak) topology. We define $h : T \times E \to [0, +\infty]$ by

$$h(t,x) := \int_{\hat{\mathbb{N}}\times V} \|v\| x(d(k,v)).$$

Then $h(t,\cdot)$ is obviously convex. Also, if $\beta' := \sup_k h(t,x^k) < +\infty$ holds for a sequence $\{x^k\}$ in E, then for every $\alpha > 0$ we have elementarily that $x^k(\hat{\mathbb{N}} \times \{v : \|v\| > \alpha\}) \leq \beta'/\alpha$, which shows that $\{x^k\}$ is tight, and hence relatively compact by Prohorov's theorem (recall that $\sigma(V,V')$ is the default topology on V). Note that $\hat{\mathbb{N}} \times V$ is a Suslin space. Then $M_+^1(\hat{\mathbb{N}} \times V)$ is also Suslin [8, III.60], which implies that any of its compact subsets is metrizable [8, III.66], hence sequentially compact. Thus, $h(t,\cdot)$ is sequentially inf-compact on E. This shows that (1.1) holds. Let $\mathcal{C}_b(\hat{\mathbb{N}} \times V)$ be the set of all bounded continuous real-valued functions on $\hat{\mathbb{N}} \times V$. $M_+^1(\hat{\mathbb{N}} \times V)$ was seen to be Suslin for the narrow topology, so by [6, III.31] there exists a countable subset $\{c_j\}$ of $\mathcal{C}_b(\hat{\mathbb{N}} \times V)$ separating the points of E. Hence, the functions a_j, defined by

$$a_j(t,x) := \int_{\hat{\mathbb{N}}\times V} c_j dx,$$

satisfy (1.5) and also (1.2)-(1.3). Let $f_k : T \to E$ be defined by

$$f_k(t) := \epsilon_{(k,v_k(t))} := \text{ Dirac probability measure at } (k, v_k(t)).$$

Then $a_j(\cdot, f_k(\cdot)) = c_j(k, v_k(\cdot))$ is evidently $\mathcal{T}$-measurable, so f_k is $\mathcal{A}$-scalarly measurable. Also, $h(t, f_k(t)) = \|v_k(t))\|$ by our definition of h, so condition (1.4) also holds. We may therefore apply Theorem 1.1: there exist an $\mathcal{A}$-scalarly measurable $f_* : T \to E$ and a subsequence $\{f_m\}$ of $\{f_k\}$ for which (1.6)-(1.7) hold. This gives marginally for every subsequence $\{f_{m_i}\}$ of $\{f_m\}$

$$\delta_*(t) := f_*(t)(\hat{\mathbb{N}} \times \cdot) = \lim_{n\to\infty} \frac{1}{n} \sum_{i=1}^{n} \epsilon_{v_{m_i}(t)} \text{ a.e .} \tag{2.14}$$

For the other marginal $f_*(t)(\cdot \times V)$ we immediately find

$$f_*(t)(\cdot \times V) = \lim_{n\to\infty} \frac{1}{n} \sum_{m=1}^{n} \epsilon_{\{k\}} = \epsilon_{\{\infty\}},$$

so we can conclude that

$$f_*(t) = \epsilon_{\{\infty\}} \times \delta_*(t) \text{ a.e .} \tag{2.15}$$

Note that by (1.7) the barycenter $v_*(t)$ of the probability measure $\delta_*(t)$, given by

$$v_*(t) := \int_T v\delta_*(t)(dv),$$

is well-defined a.e. Setting v_* equal to zero on the exceptional set, we conclude that $v_* \in \mathcal{L}^1_V$, by Jensen's inequality. From narrow convergence in (2.14) it also follows that for every $p \in \mathbb{N}$

$$\delta_*(t)(C_p(t)) \geq \limsup_{n\to\infty}(n-p+1)/n = 1 \text{ a.e. in } T,$$

where $C_p(t) :=$ cl $\{v_m(t) : m \geq p\}$. Thus, the probability measure $\delta_*(t)$ is carried by $C_p(t)$ a.e., which implies that its barycenter $v_*(t)$ must belong to cl co $C_p(t) =$ cl co $\{v_m(t) : m \geq p\}$ a.e. This proves (2.12). Finally, let l be as in (2.13), and denote the left side of (2.13) by $\gamma \in [0, +\infty]$. There exists a subsequence $\{m_i\}$ of $\{m\}$ such that $\gamma = \lim_i I_l(y_{m_i}, v_{m_i}) = \lim_n \frac{1}{n}\sum_{i\leq n} I_l(y_{m_i}, v_{m_i})$. By Fatou's lemma - which continues to hold for outer integrals - this implies

$$\gamma \geq \int_T^* \liminf_{n\to\infty} \frac{1}{n}\sum_{i=1}^{n} l(\cdot, y_{m_i}, v_{m_i})d\mu. \tag{2.16}$$

Let $g : T \times E \to [0, +\infty]$ be defined by

$$g(t, x) := \int_{\hat{\mathbb{N}}\times V} l(t, k, v)x(d(k, v));$$

Then for every $t \in T, g(t, \cdot)$ is affine and lower semicontinuous on E, as $l(t, y_\cdot, \cdot)$ is seen to be lower semicontinuous on $\hat{\mathbb{N}} \times V$. Thus, (2.13)-(2.14) give

$$\gamma \geq \int_T^* \liminf_{n\to\infty} g(t, \frac{1}{n}\sum_{i=1}^{n} \epsilon_{v_{m_i}(t)}) \geq$$

$$\geq \int_T^* g(t, \epsilon_{\{\infty\}} \times \delta_*(t))\mu(dt) =$$

$$= \int_T^* [\int_V l(t, y_\infty(t), v)\delta_*(t)(dv)]\mu(dt).$$

Lastly, by Jensen's inequality and the above it follows from the convexity hypothesis for $l(t, y_\infty(t), \cdot)$ for every $t \in T$ that

$$l(t, y_\infty(t), v)\delta_*(t)(dv) \geq l(t, y_\infty(t), v_*(t)) \text{ a.e. },$$

and so $\gamma \geq I_l(y_\infty, v_*)$. □

Remark 2.6 In particular, if $\{v_k\}$ converges weakly to some $v_\infty \in \mathcal{L}^1_V$ in the weak topology $\sigma(\mathcal{L}^1_V, \mathcal{L}^\infty_{V'}[V])$, then $v_*(t) = v_\infty(t)$ a.e., as is easy to show using Corollary 1.2. Actually, only in this form the result of Theorem 2.5 deserves the name "lower semicontinuity result": Note that v_* no longer depends on the subsequence $\{f_m\}$ (for $v_* = v_0$ a.e.). For this reason the inequality

$$\liminf_{k \to \infty} I_l(y_k, v_k) \geq I_l(y_\infty, v_\infty)$$

holds for any function l having the stated properties. Another observation is that (2.12) can be strenghtened into a (1.6)-like property. (To do this one has to use tools that fall outside the scope of this paper; viz.. Chacon's biting lemma and seminormality results.)

3 MARTINGALE ANALOG

We shall now give a martingale analog of Theorem 1.1. Let $(T, \mathcal{T}, \mu)$ be a probability space, and let E be some topological space. Let $h : T \times E \to [0, +\infty]$ be a given function such that for every $t \in T$

$$h(t, \cdot) \text{ is sequentially inf-compact on } E. \tag{3.1}$$

Let $\mathcal{A} := \{a_j\}$ be a given collection of functions $a_j : T \times E \to \mathbb{R}, j \in \mathbb{N}$, that for every $t \in T$

$$a_j(t, \cdot) \text{ is sequentially continuous on } E. \tag{3.2}$$

We suppose that these satisfy the following growth condition with respect to h: there exist $C > 0$ and $\phi \in \mathcal{L}^1_{\mathbb{R}}$ such that for all $j \in \mathbb{N}$

$$| a_j(t, x) | \leq Ch(t, x) + \phi(t) \text{ on } T \times E \tag{3.3}$$

So in comparison with section 1 the convexity-affinity conditions have been dropped, and C, ϕ in (3.3) have become universal. Finally, the arbitrary nature of $\{f_k\}$ has to be replaced by a martingale property. Let $\{\mathcal{T}_n\}$ be an increasing sequence of sub-σ-algebras of $\mathcal{T}$. Let us agree to call a sequence $\{f_n\}$ of functions $f_n : T \to E$ $\mathcal{A}$-*scalarly adapted* if for every $n \in \mathbb{N}$ the functions $a_j(\cdot, f_n(\cdot)), j \in \mathbb{N}$, and $h(\cdot, f_n(\cdot))$ are $\mathcal{T}_n$-measurable.

Theorem 3.1 *Let $\{f_n\}$ be a $\mathcal{A}$-scalarly adapted sequence of functions $f_n : T \to E$ such that the following tightness condition holds:*

$$\beta := \sup_{k \in \mathbb{N}} \int_T h(t, f_n(t))\mu(dt) < +\infty, \tag{3.4}$$

and such that

$$\{h(\cdot, f_n(\cdot))\} \text{ is a submartingale with respect to } \{\mathcal{T}_n\}. \tag{3.5}$$

Suppose also that for some null set $N \in \mathcal{T}$ the countable separation condition (1.5) holds and that for every $j \in \mathbb{N}$

$$\{a_j(\cdot, f_n(\cdot))\} \text{ is a martingale with respect to } \{\mathcal{T}_n\}. \tag{3.6}$$

Then there exists an $\mathcal{A}$-scalarly measurable function $f_* : T \to E$ such that

$$\lim_{n\to\infty} f_n(t) = f_*(t) \text{ a.e. in } T, \tag{3.7}$$

$$\int_T h(t, f_*(t))\mu(dt) \leq \beta. \tag{3.8}$$

Moreover,

$$\lim_{n\to\infty} \sup_{j\in\mathbb{N}} | a_j(t, f_n(t)) - a_j(t, f_*(t)) |= 0. \tag{3.9}$$

Proof. The proof of (3.7)-(3.8) is almost completely identical to the proof of Theorem 1.1 as given in [2b]. Instead of applying Komlós' theorem to the sequences $\{a_j(\cdot, f_n(\cdot))\}$ and $\{h(\cdot, f_n(\cdot))\}$, we now apply the classical convergence results for integrable (sub)martingales (the respective integrability conditions hold because of (3.1),(3.3)). Thus, we find the existence of functions $\psi_j \in \mathcal{L}^1_{\mathbb{R}}, j \in \mathbb{N} \cup \{0\}$, such that

$$\lim_{n\to\infty} a_j(t, f_n(t)) = \psi_j(t) \text{ a.e. in } T, \tag{3.10}$$

for every $j \in \mathbb{N}$, and such that

$$\lim_{n\to\infty} h(t, f_n(t)) = \psi_0(t) \text{ a.e.} \tag{3.11}$$

(These identities now take the place of (2.5) in [2b].) From (3.1), (3.11) it follows that $\{f_n(t)\}$ has at least one limit point a.e. Just as in [2b] the countable separation property, in conjunction with (3.10), is used to argue that a.e. all such pointwise limit points must coincide. For the details we refer to [2b]. Finally, (3.9) follows from a well-known lemma [13, V.2.9] on the convergence of countable suprema of (sub)martingales, by virtue of conditions (3.3), (3.4) and (3.6). □

Of course, the analog of Corollary 1.2 in the martingale setting is now an utterly obvious application of Fatou's lemma. More interesting are the applications of Theorem 3.1. We shall leave it to the reader to apply Theorem 3.1 to the cases considered in section 2. He will then discover that the analog of (2.7) is actually a well-known convergence result of Neveu [12] (whose proof differs from ours – e.g. no compactness is used to obtain the limit f_*), which generalizes the well-known convergence theorem for vector-valued martingales [13, V.2.8].

REFERENCES

1 D.J. Aldous, Limit theorems for subsequences of arbitrarily-dependent sequences of random variables, Z. Wahrscheinlichkeitstheorie Verw. Gebiete 40 (1977), 59-82.

2a E.J. Balder, Infinite-dimensional extension of a theorem of Komlós, Probab. Th. Rel. Fields, to appear.

2b E.J. Balder, New sequential compactness results for spaces of scalarly integrable functions, J. Math. Anal. Appl., to appear.

3 I. Berkes and E. Peter, Exchangeable random variables and the subsequence principle, Probab. Th. Rel. Fields 73 (1986) 395-413.

4 G. Buttazzo, Semicontinuity, Relaxation and Integral Representation problems in the Calculus of Variations, Pitman Lecture Notes, to appear.

5 C. Castaing, Quelques résultats de convergence des suites adaptées, Séminaire d'Analyse Convexe Montpellier 17 (1987) 2.1-2.24.

6 C. Castaing and M. Valadier, Convex Analysis and measurable multifunctions, Lecture Notes in Mathematics 580, Springer-Verlag, 1977.

7 S.D. Chatterji, A principle of subsequences in probability theory, Adv. Math. 13 (1974), 31-54.

8 C. Dellacherie and P.-A. Meyer, Probabilités et Potentiel, Hermann, Paris, 1975.

9 M. Dolcher, Topologie e strutture di convergenza, Ann. Scuola Norm. Sup. Pisa Cl. Sci. 14 (1960), 63-92.

10 J. Komlós, A generalisation of a problem of Steinhaus, Acta Math. Acad. Sci. Hungar. 18 (1967), 217-229.

11 D.Q. Lu'u, Best approximations in the space of closed convex valued multi-functions, Séminaire d'Analyse Convexe Montpellier 12 (1982), 19.1-19.21.

12 J. Neveu, Convergence presque sûre de martingales multivoques, Ann. Inst. Henri Poincaré B-8 (1972) 1-7.

13 J. Neveu, Martingales à Temps Discret, Masson, Paris, 1972.

FILLING SCHEME TRANSFORMATIONS OF COULOMB SYSTEMS

J.R. Baxter and R.V. Chacon

University of Minnesota, Department of Mathematics, Minneapolis, Minnesota, USA

University of British Columbia, Department of Mathematics, Vancouver, B.C., Canada

1. Introduction

The filling scheme technique originated in ergodic theory [2], as a tool to prove the Maximal Ergodic Lemma. In [14] this procedure was extended to continuous time, by a rather difficult construction. Consideration of space-time reduced functions [13] led to considerable technical simplifications in the continuous-time case, providing a good example of the usefulness of potential theory techniques in the theory of Markov processes. In present paper, conversely, the filling scheme is applied to classical potential theory, to obtain estimates for the electrostatic energy of a system of charges. Using these inequalities an elementary proof of the Stability of Matter Theorem is given, in Sections 3 and 4. Our basic method is a continuation of the approach used in [1]. We note that very different approaches to estimating electrostatic energy are used in [5] and [12]. The filling scheme method does not seem to give very precise estimates, at least in its present form, but it is general, and easy to apply. It is valid for the potential associated with any reasonable Markov process, although we will restrict our rigorous discussion to the case of Brownian motion in $\mathbb{R}^3$.

It is convenient to work with general distributions of charge, which we represent by

ISBN 0-12-231050-0

signed measures on $\mathbb{R}^3$. Of course, in applications the only measures that need be considered either have well-behaved volume or surface densities, or are point measures. We will use δ_x to denote the unit point measure concentrated at x. For any finite signed measure, we define the classical potential field Pot μ by

$$\text{Pot}\,\mu\,(y) = \int 1/|y-x|\ \mu(dx).$$

Pot μ is defined everywhere when μ is nonnegative, and is finite quasi-everywhere (q.e.), meaning up to a polar set (i.e. a set of capacity 0). Thus Pot μ is defined quasi-everywhere when μ is a signed measure. In particular, Pot μ is defined a.e. with respect to Lebesgue measure on $\mathbb{R}^3$. We recall that the potential of a probability measure μ is the **occupation time density** of Brownian motion started with initial distribution μ, as it moves to ∞.

For any nonnegative measures μ and ν, we define the mutual energy $<\mu,\nu>$ by

$$<\mu,\nu> = \iint 1/|y-x|\ \ \mu(dx)\nu(dy) = \int(\text{Pot}\ \mu)d\nu = \int(\text{Pot}\ \nu)d\mu.$$

We define $<\mu,\nu>$ in the same way when μ and ν are signed measures, provided that $<|\mu|,|\nu|> <\infty$, where $|\mu|,|\nu|$ denote the total variation measures associated with μ and ν respectively. As the notation suggests, $<\mu,\mu> \geq 0$ for all signed measures μ such that $<|\mu|,|\mu|> <\infty$.

For any signed measures $\mu_1,\ldots,\mu_n$ we define

$$V(\mu_1,\ldots,\mu_n) = \Sigma_{1\leq i<j\leq n}<\mu_i,\mu_j>, \tag{1.1}$$

assuming that the inner products are finite. As observed in [1] it is also useful to work with a quantity Γ that includes the self-energies of some measures. Thus if $\mu_1,...,\mu_n$, ν are nonnegative measures such that $<\mu_i,\nu> <\infty$ for all i, let

$$\Gamma(\mu_1,...,\mu_n;\nu) = \Sigma_{1\leq i<j\leq n}<\mu_i,\mu_j> - \Sigma_{1\leq i\leq n}<\mu_i,\nu> + (1/2)<\nu,\nu>. \qquad (1.2)$$

We may think of the μ_i as positive charge, ν as negative charge (or vice versa),but we use postive measures, and let the sign be part of the formula for Γ.

Theorem 1 Let $\mu_1,...,\mu_n$, ν be nonnegative measures, Pot ν finite ν-a.e. Then there exist nonnegative measures $\gamma_1,...,\gamma_{n+1}$, such that (a) $\nu=\gamma_1+...+\gamma_{n+1}$, (b) $\gamma_i(\mathbb{R}^3)\leq\mu_i(\mathbb{R}^3)$, (c) Pot $\gamma_i \leq$ Pot μ_i everywhere, (d) Pot γ_i = Pot μ_i $(\gamma_{i+1}+...+\gamma_{n+1})$-a.e), for i=1,..,n.

Theorem 1 is proved in [1]. It follows easily from the potential-theory construction of the filling scheme. The measures γ_i are obtained by applying the filling scheme construction sequentially to Brownian motion starting from an initial distribution μ_i, with target measure $\nu-(\gamma_1+...+\gamma_{i-1})$, i=1,...,n (it makes little difference in the result if instead we fill all measures simultaneously). We think of the μ_i-particles as "cancelling" with any ν-particles they encounter. γ_i is simply the mass of ν-particles which are cancelled by μ_i-particles, i=1,...,n.

Condition (b) of Theorem 1 follows from (c). If $<\mu_i,\nu> <\infty$ for all i, (a),(c), and (d) imply that for all k=1,...,n,

$$\Gamma(\mu_1,...,\mu_n;\nu) = \Gamma(\mu_1,...,\mu_k;\gamma_1+...+\gamma_k) + \Gamma(\mu_{k+1},...,\mu_n;\gamma_{k+1}+...+\gamma_{n+1}) + \Omega, \qquad (1.3)$$

where $\Omega = \Sigma_{1\leq i\leq k}\Sigma_{k+1\leq j\leq n}<\mu_i-\gamma_i,\mu_j> \geq 0$. By induction we have at once that for all k=1,...,n,

$$\Gamma(\mu_1,\ldots,\mu_n;\nu) \geq \Sigma_{1\leq i\leq k}\Gamma(\mu_i,\gamma_i) + \Gamma(\mu_{k+1},\ldots,\mu_n;\gamma_{n+1}). \qquad (1.4)$$

Because of [13], the filling scheme may be regarded as topic in potential theory, and for this reason probabilistic language may be avoided, at least in the classical potential theory case we consider in the present paper. However, the filling scheme picture seems more intuitive than the reduced function approach, at least for probabilists. We will say that when a system of charges is regarded as the initial distribution for Brownian motion B_t, and a stopping time τ is defined (which may or may not be the filling scheme) that the distribution of B_t on the set $\{\tau>t\}$ is the "moving" charge. Because of the interpretation of potential as occupation time density, we have as a general heuristic:

$$\partial/\partial t \text{ Pot (all charge)} = -\text{ C(density of moving charge)}. \qquad (*)$$

We take C=1, by an appropriate change of time scale. There may be other charge distributions present which do not take part at all in the Brownian motion, of course. If we denote the energy of the entire system of charges at time t by V(t), then (∗) easily gives the rule for finding dV(t)/dt:

$$dV/dt = -\Sigma_i \int \text{(density of ith moving charge) d(all charges that interact with the ith moving charge)}. \qquad (**)$$

(If we allow self energies, a factor of 1/2 in self-energy terms is assumed here.) However, if we wish to perform the filling scheme, and use **cancellation** of charge, then we must add correction terms to our previous expression for dV/dt, to take account of the fact that not all charges interact with each other, so that cancellation of positive and negative charges may increase V. To illustrate this way of regarding the process, we will sketch a proof of (1.3) with n=2, k=1. Let μ_2 be held fixed, while filling ν from μ_1. We will denote the moving charge at time t by $\mu_1(t)$, the unfilled part of ν at time t by $\nu(t)$, and the filled part of ν at time t by $\gamma(t)$, and let $\Gamma(t)= \Gamma(\mu_1(t),\mu_2;\nu(t))$. $\gamma_1=\gamma(\infty)$, $\gamma_2+\gamma_3=\nu(\infty)$. Thus $\Gamma(\mu_2;\gamma_2+\gamma_3)-\Gamma(\mu_1,\mu_2;\nu)=\Gamma(\infty)-\Gamma(0)$, and we may verify (1.3) by computing the integral of $d\Gamma/dt$. In computing the contribution to $d\Gamma/dt$ from (**), we note that because of the nature of the filling scheme, positive μ_1-charge never moves over negative ν-charge, because it would be cancelled! And of course μ_1-charge does not interact with itself. Thus the only contribution to $d\Gamma/dt$ from (**) is $-\int(\text{density of } \mu_1(t))d\mu_2$. By (*), the integral of this quantity over time is simply $\int(\text{Pot } \gamma_1 - \text{Pot } \mu_1)d\mu_2=-\langle\mu_1-\gamma_1,\mu_2\rangle$, which is the quantity Ω of (1.3). However, there is another contribution to $d\Gamma/dt$, due to cancellation. Positive charge is cancelled at the same rate, $d\gamma(t)/dt$, as negative charge, but μ_1-charge does not interact with μ_1-charge, so the cancellation rate of positive energy is only $\langle d\gamma(t)/dt,\mu_2\rangle$, while the cancellation rate of negative energy is $\langle d\gamma(t)/dt,\mu_1(t)+\mu_2\rangle$. Thus cancellation contributes a term $\langle d\gamma(t)/dt,\mu_1(t)\rangle$ to $d\Gamma/dt$. Since cancellation necessarily takes place on a set where no charge has moved previously, we see by (*) that $\text{Pot } (\mu_1(t)+\gamma(t))= \text{Pot } (\mu_1(0)+\gamma(0))=\text{Pot } \mu_1$ on this set. Hence $\langle d\gamma(t)/dt,\mu_1(t)\rangle= \langle d\gamma(t)/dt,\mu_1\rangle-\langle d\gamma(t)/dt,\gamma(t)\rangle$. Integrating this term over time gives $-\Gamma(\mu_1;\gamma_1)$. (1.3) follows, and we are done.

For our future purposes it is more convenient to express Theorem 1 in terms of V rather than Γ.

Corollary Let $\mu_1,\ldots,\mu_n$, $\nu_1,\ldots,\nu_m$ be nonnegative measures, such that $\langle\mu_i,\nu_j\rangle <\infty$ for all i and j, and Pot ν_j is finite ν_j-a.e. for all j. Then there exist nonnegative measures γ_i,ζ_j, such that $\gamma_1+\ldots+\gamma_n=\zeta_1+\ldots+\zeta_m$, $\zeta_j\leq\nu_j$ for each j, $\gamma_i(\mathbb{R}^3)\leq\mu_i(\mathbb{R}^3)$, and

$$V(-\mu_1,\ldots,-\mu_n,\nu_1,\ldots,\nu_m) \geq -\Sigma_{1\leq i\leq n}<\gamma_i,\mu_i-\gamma_i/2> -\Sigma_{1\leq j\leq m}<\zeta_j,\nu_j-\zeta_j/2>, \qquad (1.5)$$

To prove the corollary, we simply let $\nu=\nu_1+\ldots+\nu_m$ and apply (1.4) with k=n. Since $\gamma_1+\ldots+\gamma_n\leq\nu_1+\ldots+\nu_m$, we can find $\zeta_1,\ldots,\zeta_m$ with $\zeta_j\leq\nu_j$ such that $\gamma_1+\ldots+\gamma_n=\zeta_1+\ldots+\zeta_m$. (We may take ζ_j to be the part of ν_j which is filled by $\mu_1+\ldots+\mu_n$.) The corollary follows by a straightforward computation.

It is sometimes useful to note the obvious fact that

$$<\zeta_j,\nu_j-\zeta_j/2> \leq (1/2)<\nu_j,\nu_j>. \qquad (1.6)$$

A system of charges $q_1,\ldots,q_k$ at positions $x_1,\ldots,x_k$ in $\mathbb{R}^3$ is considered to have an electrostatic potential energy $V = \Sigma_{1\leq i<j\leq k}\, q_iq_j/|x_i-x_j|$, so that V is equal to $V(q_1\delta_{x_1},\ldots,q_k\delta_{x_k})$ in our present notation. In Section 2 we will apply (1.5) when the μ_i are point charges and the ν_j are uniform surface distributions on spheres. This gives a lower bound for the energy of a system of point charges, which we then apply in Section 3.

2. An Inequality for Point Charges

In what follows we will often work with balls and spheres, so we write

$$B(a,R)=\{x: x\in\mathbb{R}^3, |x-a|<R\}, \quad \partial B(a,R)=\{x: x\in\mathbb{R}^3, |x-a|=R\}.$$

Let $x,y\in\mathbb{R}^3$, R>0. If we imagine a unit positive charge at y and a unit negative charge at x, the work which must be done to expand the positive charge to a uniform distribution on the surface of the sphere $\partial B(y,R)$ is clearly $(1/|x-y|-1/R)\vee 0$. We will denote this quantity by W(y,R,x).

For any closed set A in $\mathbb{R}^3$, and any finite measure μ on $\mathbb{R}^3$, we define the **harmonic measure** μ' for μ on A to be the result of sweeping μ onto A using the classical balayage operation. Of course μ' is just the hitting distribution of Brownian motion on A from initial distribution μ. Then Pot μ – Pot μ' is nonnegative on $\mathbb{R}^3$, is 0 quasi-everywhere on A, and is the classical Green's potential of μ on A^c.

In the next proposition we consider positive charges z_j at positions y_j and unit negative charges at x_i. Intuitively, we will expand each positive charge to a sphere $\partial B(y_j,R_j)$. This requires U-energy to overcome the attractive forces but also releases repulsive F-energy, where U and F are defined below. We use (1.5) of Section 1 to estimate the V of the resulting configuration. To facilitate this estimate, we move each δ_{x_i} to a harmonic measure on $\partial B(y_j,L_{ij})$ for each j. This requires extra U-energy.

Proposition 1 Let $y_1,...y_M,x_1,...,x_N \in \mathbb{R}^3$, $z_1,...,z_M$ real, $z_j \geq 0$. Let $R_1,...,R_M$ be real, $R_j>0$. Let ν_j denote the uniform measure on $\partial B(y_j,R_j)$ with total charge z_j. Finally let L_{ij} be real, $R_j \leq L_{ij} \leq \infty$ for i=1,..,N, j=1,..,M and let ρ_{ij} denote the harmonic measure for δ_{x_i} on $\partial B(y_j,L_{ij})$ (if $L_{ij}=\infty$ then $\rho_{ij}=0$). Then there exist nonnegative measures λ_{ij},ζ_j such that

$$\zeta_j=\Sigma_{1\leq i\leq N}\lambda_{ij}\leq \nu_j \text{ for all j}, \quad \Sigma_{1\leq j\leq M}\lambda_{ij}(\mathbb{R}^3)\leq 1 \text{ for all i}, \tag{2.1}$$

and

$$V(-\delta_{x_1},...,-\delta_{x_N},z_1\delta_{y_1},...,z_M\delta_{y_M}) + U \geq -G_1 - G_2 + F, \tag{2.2}$$

where

$$U = \Sigma_{1\leq i\leq N,1\leq j\leq M} z_j W(y_j,L_{ij},x_i), \tag{2.3}$$

$$G_1=\Sigma_{1\le i\le N,1\le j\le M}<\lambda_{ij},\rho_{ij}>,\quad G_2=\Sigma_{1\le j\le M}<\zeta_j,\nu_j-\zeta_j/2>, \tag{2.4}$$

$$F = (1/2)\Sigma_{1\le j,k\le M,j\ne k}z_jz_k(1/|y_j-y_k|-1/R_j)\vee 0. \tag{2.5}$$

F is easily seen to be an underestimate for the repulsive work done by the positive charges z_j when expanded to the distributions ν_j.

Proof of Proposition 1. We let $\mu_i=\delta_{x_i}$, and then use the notation of (1.5). Since $\gamma_1+...+\gamma_n=\zeta_1+...+\zeta_m$, we can find λ_{ij} such that $\Sigma_{1\le i\le N}\lambda_{ij}=\zeta_j$ and $\Sigma_{1\le j\le M}\lambda_{ij}=\gamma_i$, and by (1.5)

$$V(-\delta_{x_1},...,-\delta_{x_N},\nu_1,...,\nu_M) \ge -\Sigma_{1\le i\le n}<\gamma_i,\delta_{x_i}> - G_2. \tag{2.6}$$

Let $U_1 = \Sigma_{1\le i\le N,1\le j\le M}z_jW(y_j,R_j,x_i)$. Let η_{ij} denote the uniform measure on $\partial B(y_j,L_{ij})$ with total charge z_j. Then $<\eta_{ij},\delta_{x_i}>= <\eta_{ij},\rho_{ij}>=<\nu_j,\rho_{ij}>$. Hence
$z_j(W(y_j,L_{ij},x_i)-W(y_j,R_j,x_i)) = <\nu_j,\delta_{x_i}>-<\eta_{ij},\delta_{x_i}> = <\nu_j,\delta_{x_i}>-<\nu_j,\rho_{ij}> =$
$\int(\text{Pot }\delta_{x_i}-\text{Pot }\rho_{ij})d\nu_j \ge \int(\text{Pot }\delta_{x_i}-\text{Pot }\rho_{ij})d\lambda_{ij}$, so

$$U-U_1 \ge \Sigma_{1\le i\le N,1\le j\le M}(<\lambda_{ij},\delta_{x_i}>-<\lambda_{ij},\rho_{ij}>). \tag{2.7}$$

Since F underestimates the repulsive work, we have

$$V(-\delta_{x_1},...,-\delta_{x_N},z_1\delta_{y_1},...,z_M\delta_{y_M}) + U_1 \ge$$
$$V(-\delta_{x_1},...,-\delta_{x_N},\nu_1,...,\nu_M) + F. \tag{2.8}$$

(2.2) follows from (2.6),(2.7) and (2.8).

3. The Stability of Matter.

We consider the nonrelativistic Schroedinger model. Suppose there are M nuclei, with fixed positions y_j and charges z_j, j=1,...,M, and N electrons, with variable positions x_i, i=1,...,N and charge −1. Let the wave function ψ for the electrons be defined on $(\mathbb{R}^3)^N$, and let H be defined by

$$H\psi = -\Sigma_{1\leq i\leq N}\Delta_{x_i}\psi + (V(-\delta_{x_1},\ldots,-\delta_{x_N},z_1\delta_{y_1},\ldots,z_M\delta_{y_M}))\psi. \tag{3.1}$$

We always assume ψ is smooth, say continuous second derivatives, and $\int|\psi|^2=1$. Electrons are Fermions, which means that for finding a lower bound for H we may restrict our attention to wave functions ψ for which there is a partition of {1,...,N} into two sets I_1 and I_2, such that ψ is antisymmetric with respect to any exchange of coordinates within I_1 or I_2. Finally, we suppose that $1\leq z_j\leq z$ for all j. Under these assumptions,

Theorem 2 There exists a constant C, depending only on z, such that for any ψ as described above,

$$(\psi,H\psi) \geq - C\ \mathrm{Min}(M,N). \tag{3.2}$$

This lower bound was first proved by Dyson and Lenard in their papers on the stability of matter [4], [7], in the form

$$(\psi,H\psi) \geq - CN. \tag{3.3}$$

Later Lieb and Thirring [11] gave a new and physically motivated proof, obtaining

$$(\psi,H\psi) \geq -C(z_1{}^{7/3}+...+z_M{}^{7/3}+N), \tag{3.4}$$

with a much better bound, both in terms of the size of the constant C and the dependence on the nuclear charge sizes z_j. Their approach was based on a comparison of H with the Thomas-Fermi approximation (cf. [9],[10]).

In [5], Fefferman gave an estimate of the form

$$(\psi,H\psi) \geq -CM, \tag{3.5}$$

for systems with $z_j=z$ for all j and $z\geq 40$, with a good value for C. His method of proof gives the same bound for all values of z, although the value of C is not as sharp in the general case [6].

In this section we will give an elementary proof of Theorem 2 based on the inequality of Section 2. Our method does not give a good estimate of the value of the constant C, though it may be possible to improve this approach in the future. To indicate the sort of argument used, we first note that the kinetic energy, T, is defined by

$$T = (\psi,-\Sigma_{1\leq i\leq N}\,\Delta_i\psi) = \Sigma_{1\leq i\leq N}\int_{(\mathbb{R}^3)^N} |\nabla_i\psi|^2.$$

T becomes large whenever electrons are likely to be in a small region, or are close together. The methods in [4] and [11] are based on inequalities which relate the kinetic energy to properties of the electrons which are defined independently of the particular nuclear configuration $y_1,...,y_M$. In the present paper we do not prove an inequality of this sort, but simply show that the kinetic energy can be used to pay for the cost in energy terms of expanding the nuclear charges to spherical shells, and then apply the classical estimate of Section 2.

Proof of Theorem 2

We denote the minimum of the nuclear charges z_j by z_o, the maximum by z. Let p be an integer, $p \geq 1$, α, β, S real, $\alpha, \beta, S > 0$. The values of α and β are arbitrary but fixed, while p and S are subject to restrictions which will be given later.

Let A_j denote the distance from y_j to its pth-nearest neighbour (y_j is the 0th-nearest neighbour of y_j). Let $S_j = A_j \wedge S$, and let $R_j = (1+\alpha)S_j$. Let J denote the set of indices j such that $S_j = A_j$. $S_j = S$ if $j \in J^c$. Let $L_{ij} = L_j = (1+\alpha+\beta)S_j$ for all i and j.

All the notations and conclusions of Proposition 1 of Section 1 now apply. We note that

$$F \geq (1/2)\Sigma_{j\in J} z_j z_o p\alpha/R_j. \tag{3.6}$$

Clearly $<\lambda_{ij},\rho_{ij}> \leq \|\lambda_{ij}\|/\beta S_j$ for all i and j. Using (2.1),

$$G_1 \leq \Sigma_{j\in J} z_j/\beta S_j + \mathrm{Min}(zM,N)/\beta S. \tag{3.7}$$

In estimating G_2, we recall (1.6): $<\zeta_j, v_j - \zeta_j/2> \leq (1/2)<v_j, v_j>$. Summing over $j \in J$ gives at most $\Sigma_{j\in J} z_j^2/2R_j$. The sum over the rest is no more than $z^2M/2(1+\alpha)S$. Since $<\zeta_j, v_j - \zeta_j/2> \leq <\zeta_j, v_j> \leq \|\zeta_j\| z/R_j$, the sum over the rest is also no more than $zN/(1+\alpha)S$, where we use (2.1). Hence

$$G_2 \leq \Sigma_{j\in J} z_j^2/2Rj + z\mathrm{Min}(zM/2,N)/(1+\alpha)S. \tag{3.8}$$

We will require p to be large enough that

$$(1/2)z_o p\alpha \geq (1+\alpha)/\beta + z/2 + C_0, \tag{3.9}$$

where C_0 is a constant (depending on α and β) to be determined later.

Combining (2.2),(3.6),(3.7), (3.8), (3.9),

$$V(-\delta_{x_1},\ldots,-\delta_{x_N},z_1\delta_{y_1},\ldots,z_M\delta_{y_M}) + U \geq$$
$$-\mathrm{Min}(C_1M,C_2N)/S + \Sigma_{j\in J}C_0/R_j. \qquad (3.10)$$

We note that (3.10) is still a classical bound, valid for arbitrary electron configurations, relating the total energy V to U. We now consider the wave function ψ. For any function f on $(\mathbb{R}^3)^N$, let Av(f) denote $\int_{(\mathbb{R}^3)^N} |\psi|^2 f\, dx_1\ldots dx_N$. Since $(\psi,H\psi)=T+Av(V(-\delta_{x_1},\ldots,-\delta_{x_N},z_1\delta_{y_1},\ldots,z_M\delta_{y_M}))$, we see that if (3.9) holds,

$$(\psi,H\psi) \geq T - Av(U) - \mathrm{Min}(C_1M,C_2N)/S + \Sigma_{j\in J}C_0/Rj. \qquad (3.11)$$

We will show in Section 4 that for S sufficiently small (depending on p) we have

$$T - Av(U) \geq -C_3\mathrm{Min}(M,N)/S - C_4\Sigma_{j\in J}1/R_j, \qquad (3.12)$$

where C_4 does not depend on p or S, so Theorem 2 follows by setting $C_0=C_4$, choosing p, and finally choosing S.

4. Expansion Costs

In this section we shall obtain estimates for the quantity Av(U) of Section 3. First we note the obvious fact that

$$\Sigma_{1\leq j\leq M}\mathbf{1}_{B(y_j,S_j/2)} \leq p. \qquad (4.1)$$

As a consequence of (4.1) we can deduce a simple geometrical fact:

$$\Sigma_{1\le j\le M} S_j \mathbf{1}_{B(y_j,\gamma S_j)} \le 24pS\gamma^2(\gamma+1/2) \text{ for } \gamma\ge 1. \tag{4.2}$$

Proof of (4.2) It is clearly enough to prove (4.2) at 0. Let J′ be the set of j such that $|y_j|<\gamma S_j$. Let $d\mu=(1/|x|^2)dx$. Since $1/|x|$ is harmonic on $\{x\neq 0\}$, and the square function is convex, $1/|x|^2$ is subharmonic on $\{x\neq 0\}$. Hence if $S_j/2\le|y_j|<\gamma S_j$, $\mu(B(y_j,S_j/2))\ge(1/|y_j|^2)\pi S_j^3/6\ge\pi S_j/6\gamma^2$. If $|y_j|<S_j/2$, then clearly $\mu(B(y_j,S_j/2))\ge(1/|S_j|^2)\pi S_j^3/6=\pi S_j/6$. Using (4.1), $\Sigma_{1\le j\le M} S_j \mathbf{1}_{B(y_j,\gamma S_j)}(0)\le \Sigma_{j\in J'}(6/\pi)\gamma^2\mu(B(y_j,S_j/2))\le(6/\pi)\gamma^2 p\mu(B(0,(\gamma+1/2)S)$, and (4.2) follows.

We will obtain our estimate for the expansion cost U by standard techniques. Although the intermediate results we obtain are well known, it seems hard to find a precise reference. Thus for the convenience of the reader we will give a fairly complete proof of the analytical facts that we use. We will use a variational approach to the existence of eigenfunctions for the operators involved, since it avoids questions of domains of operators and existence of boundary derivatives. We do need to consider the space $H^1(D)$, where D is an open set in $\mathbb{R}^k$. $H^1(D)$ is the space of real L^2-functions on D whose weak first derivatives are also in L^2. That is, $\psi\in H^1(D)$ means: $\psi\in L^2(D)$, and for i=1,..,k there exists $\lambda_i\in L^2(D)$ such that for any smooth function ϕ with compact support in D, $\int_D\psi\partial_i\phi dx=-\int_D\lambda_i\phi dx$.

Let D be a bounded open set in $\mathbb{R}^3$, and let w,v be real Borel functions on D, with w>0 bounded away from 0 and bounded on D, and v bounded outside a compact subset of D. For any functions $f,g\in H^1(D)$, define $t_w(f,g)$, $q(f,g)$ by

$$t_w(f,g) = \int_D w\nabla f\cdot\nabla g dx, \quad q(f,g) = t_w(f,g) + \int_D vfg dx. \tag{4.3}$$

Lemma 1 Suppose that for any $\varepsilon>0$ there exists a constant $C(\varepsilon)$ such that for every smooth function f on D with compact support in D,

$$\int_D |v|\, f^2 dx \leq \varepsilon \int_D w|\nabla f|^2 dx + C(\varepsilon)\int_D f^2 dx. \tag{4.4}$$

Then (4.4) holds for all $f\in H^1(D)$, possibly with a diffferent $C(\varepsilon)$. There exist functions $\phi_j\in H^1(D)$, $j=0,1,\ldots$, and real numbers $\lambda_j\to\infty$, such that for any function $g\in H^1(D)$,

$$q(\phi_j,g) = \lambda_j \int_D \phi_j g\, dx. \tag{4.5}$$

and such that $\{\phi_j\}$ is orthonormal and complete in $L^2(D)$. We may choose ϕ_j so that for each j,

$$\lambda_j = q(\phi_j,\phi_j) = \inf\{q(g,g): g\in H^1(D), \int_D g^2 dx=1, \int_D \phi_i g\, dx=0 \text{ for all } i<j\}. \tag{4.6}$$

Proof If f has compact support in D we can approximate f by a convolution with a smooth function, then apply (4.4) to the smoothed version of f, and use Fatou's Lemma. Any $f\in H^1(D)$ can be written as the sum of an f_1 with compact support in D plus an f_2 with support only where v is bounded, so (4.4) holds, with a change in $C(\varepsilon)$. The existence of a minimizing g for the inf in (4.6) is an easy consequence of (4.4) and Rellich's Lemma. (4.5) is a necessary condition for ϕ_j to be an extreme value, the condition $\lambda_j\to\infty$ follows from Rellich's Lemma again, and completeness of $\{\phi_j\}$ then follows (cf. [3], Chapter VI, Sections 2 and 3), so the lemma is proved.

In our application, the set D will be be an open ball centered at y, and we will have

$v(x) \le c/|x-y|$. Thus (4.4) will hold by, for example, [3], Chapter VI, Section 5.

Now, using the notation of Lemma 1, and assuming (4.4),(4.5) and (4.6), we will extend all the functions on D (including w and v) to $\mathbb{R}^3$ by defining them equal 0 on D^c. We then extend $\{\phi_j\}_{j=0,1,\ldots}$ to an orthonormal basis $\{\phi_j\}_{j\in\mathbb{Z}}$ for $L^2(\mathbb{R}^3)$ by choosing $\{\phi_j\}_{j=-1,-2,\ldots}$ to be an orthonormal basis for $L^2(D^c)$. We let $\lambda_j=0$ for $j<0$, noting that (4.5) now holds for all $j\in\mathbb{Z}$. Let Π_m=the span of $\{f_1\otimes\ldots\otimes f_N\}$, where $f_i\in\{\phi_{-m},\ldots,\phi_m\}$ for each i. Let Λ_N be the set of $\psi\in L^2((\mathbb{R}^3)^N)$ such that: (a) for each i and any fixed values of x_k, $k\neq i$, as a function of x_i on D, ψ is in $H^1(D)$; and (b) for each i, $\int_{(\mathbb{R}^3)^N} \mathbf{1}_D(x_i)|\nabla_{x_i}\psi|^2 dx_1\ldots dx_N<\infty$. We note that $\Lambda_N\supset\Pi_m$ for each m. Let T_w, Q be the bilinear forms on Λ_N defined by

$$T_w(\psi,\eta) = \int_{(\mathbb{R}^3)^N}\Sigma_{1\le i\le N}\, w(x_i)\nabla_{x_i}\psi\cdot\nabla_{x_i}\eta\, dx_1\ldots dx_N,$$

$$Q(\psi,\eta) = T_w(\psi,\eta) + \int_{(\mathbb{R}^3)^N}\Sigma_{1\le i\le N}\, v(x_i)\psi\eta\, dx_1\ldots dx_N. \qquad (4.7)$$

Then by (4.6) for any $\psi\in\Lambda_N$,

$$Q(\psi,\psi) \ge \lambda_0\int_{(\mathbb{R}^3)^N}\Sigma_{1\le i\le N}\mathbf{1}_D(x_i)\psi^2 dx_1\ldots dx_N. \qquad (4.8)$$

(4.5) shows at once that if $\psi\in\Lambda_N$, $\eta(x_1,\ldots,x_N)=\phi_{j_1}(x_1)\ldots\phi_{j_N}(x_N)$, then

$$Q(\psi,\eta)=\int_{(\mathbb{R}^3)^N}\Sigma_{1\le i\le N}\,\lambda_{j_i}\psi(x_1,\ldots,x_N)\phi_{j_1}(x_1)\ldots\phi_{j_N}(x_N)dx_1\ldots dx_N \qquad (4.9)$$

Lemma 2 For $\psi\in\Lambda_N$, $Q(\psi,\psi)\ge\limsup_{m\to\infty}Q(\psi_m,\psi_m)$, where ψ_m=proj ψ on Π_m.
Proof Claim 1: $T_w(\psi_m,\psi_m)$ is bounded in m. Proof of Claim 1: By (4.9), $Q(\psi_m,\psi_m)=Q(\psi_m,\psi)$. By (4.4), for any $\varepsilon>0$, $Q(\psi_m,\psi_m)\ge T_w(\psi_m,\psi_m)-\varepsilon T_w(\psi_m,\psi_m)-C(\varepsilon)N\int_{(\mathbb{R}^3)^N}\psi_m^2 dx_1\ldots dx_N$. Also $2Q(\psi_m,\psi)\le T_w(\psi_m,\psi_m)+T_w(\psi,\psi)+\varepsilon T_w(\psi_m,\psi_m)+\varepsilon T_w(\psi,\psi)+$

$C(\varepsilon)N\int_{(\mathbb{R}^3)^N} \psi_m{}^2 dx_1...dx_N + C(\varepsilon)N\int_{(\mathbb{R}^3)^N} \psi^2 dx_1...dx_N$, and Claim 1 follows.

Claim 2: $\int_{(\mathbb{R}^3)^N}\Sigma_{1\le i\le N} |v|(x_i)(\psi-\psi_m)^2 dx_1...dx_N \to 0$ as $m\to\infty$. Proof of Claim 2: For any $\varepsilon>0$, $\int_{(\mathbb{R}^3)^N}\Sigma_{1\le i\le N} |v|(x_i)(\psi-\psi_m)^2 dx_1...dx_N \le$ $\varepsilon T_w(\psi-\psi_m,\psi-\psi_m)+C(\varepsilon)N\int_{(\mathbb{R}^3)^N} (\psi-\psi_m)^2 dx_1...dx_N$. $T_w(\psi-\psi_m,\psi-\psi_m)\le 2T_w(\psi,\psi)+2T_w(\psi_m,\psi_m)$, so Claim 2 follows from Claim 1.

To finish the proof of the lemma, we note that $2Q(\psi_m,\psi_m)=2Q(\psi_m,\psi)\le$ $T_w(\psi_m,\psi_m)+T_w(\psi,\psi)+\int_{(\mathbb{R}^3)^N}\Sigma_{1\le i\le N} v(x_i)(\psi^2+\psi_m{}^2)dx_1...dx_N+$ $\int_{(\mathbb{R}^3)^N}\Sigma_{1\le i\le N} |v|(x_i)(\psi-\psi_m)^2 dx_1...dx_N = Q(\psi_m,\psi_m)+Q(\psi,\psi)+$ $\int_{(\mathbb{R}^3)^N}\Sigma_{1\le i\le N} |v|(x_i)(\psi-\psi_m)^2 dx_1...dx_N$. Hence $Q(\psi_m,\psi_m)\le$ $Q(\psi,\psi)+\int_{(\mathbb{R}^3)^N}\Sigma_{1\le i\le N} |v|(x_i)(\psi-\psi_m)^2 dx_1...dx_N$, and the lemma follows from Claim 2.

The argument in the next result occurs somewhere in every proof of the stability of matter.

Corollary to Lemma 2 Let C denote the sum of the absolute values of those λ_j which are negative. Then if $\psi\in\Lambda_N$ satisfies the Fermion antisymmetry condition of Section 3,

$$Q(\psi,\psi) \ge -2C\int_{(\mathbb{R}^3)^N} \psi^2 dx_1...dx_N. \tag{4.10}$$

Proof Lemma 2 shows that we may restrict our attention to $\psi\in\Pi_m$, and hence to finite linear combinations of functions of the form $\psi_{j_1...j_N}=\phi_{j_1}(x_1)...\phi_{j_N}(x_N)$. The $\psi_{j_1...j_N}$ are orthogonal both with respect to the standard inner product and with respect to Q, by (4.9). Antisymmetry shows that we need only consider linear combinations of functions $\psi_{j_1...j_N}$ such that for any k, $j_i=k$ occurs for at most two values of i, and the result follows.

We now can derive the estimates needed in Section 3. Fix $\gamma=1+\alpha+\beta$. Let $D_j=B(y_j,\gamma S_j)$, $w_j(x)=S_j$ on $B(y_j,\gamma S_j)$, $v_j(x)=-z_j(1/|x-y_j|-1/\gamma S_j)\vee 0$. Let Q_j be the corresponding form defined by (4.7). (4.10) together with a simple scaling argument shows that for some constant C_5, if $\psi \in \Lambda_N$ satisfies the Fermion antisymmetry condition,

$$Q_j(\psi,\psi) \geq -(\psi,\psi)C_5/S_j. \tag{4.11}$$

Hence for our wave function ψ with $(\psi,\psi)=1$, we have

$$\Sigma_{1\leq j\leq M}\, Q_j(\psi,\psi) \geq -C_5\Sigma_{j\in J}1/S_j - C_5M/S. \tag{4.12}$$

We now define the **charge density** associated with an arbitrary wave function ψ on $(\mathbb{R}^3)^N$, with $\int_{(\mathbb{R}^3)^N} |\psi|^2=1$ as usual. We define the charge density ρ_i for the i-th electron by $\rho_i(x)=\int_{(\mathbb{R}^3)^{N-1}} |\psi|^2(x_1,\ldots,x_{i-1},x,x_{i+1},\ldots,x_N)dx_1\ldots dx_{i-1}dx_{i+1}\ldots x_N$. The total charge density ρ is defined by $\rho=\rho_1+\ldots+\rho_N$. (If we consider $(\mathbb{R}^3)^N$ as a sample space with probability density $|\psi|^2$, and consider $x_1,\ldots,x_N$ as random variables defined on $(\mathbb{R}^3)^N$, then ρ_i is the density of the distribution of x_i.)

(4.8) together with scaling shows that for some constant C_6,

$$Q_j(\psi,\psi) \geq -\, C_6 \int_{D_j} \rho(x)dx/S_j. \tag{4.13}$$

Using (4.13) and (4.2) for $j\in J^c$, and (4.11) for $j\in J$, we have

$$\Sigma_{1\leq j\leq M}\, Q_j(\psi,\psi) \geq -C_5\Sigma_{j\in J}1/S_j - C_6 24p\gamma^2(\gamma+1/2)N/S. \tag{4.14}$$

For $S>0$ such that $24pS\gamma^2(\gamma+1/2)\leq 1$, we have $T \geq \Sigma_{1\leq j\leq M}\, T_{w_j}(\psi,\psi)$, so that $T - Av(U) \geq \Sigma_{1\leq j\leq M}\, Q_j(\psi,\psi)$, and hence (3.12) holds as claimed.

References

1. J.R. Baxter, Inequalities for Potentials of Particle Systems, Illinois J. Math. 24 (1980), 645-652.

2. R.V. Chacon, D.S. Ornstein, A general ergodic theorem, Illinois J. Math 4 (1960), 153-160.

3. R. Courant, D. Hilbert, Methods of Mathematical Physics vol. 1, Wiley, New York 1953.

4. F.J. Dyson, A. Lenard, Stability of matter I, J. Math. Phys. 8 (1967), 423-434.

5. C. Fefferman, The N-body problem in quantum mechanics, Comm. Pure and Appl. Math. 39 (1986), S67-S109.

6. C. Fefferman, private communication.

7. A. Lenard, F.J. Dyson, Stability of matter II, J. Math. Phys. 8 (1968), 698-711.

8. A. Lenard, Lectures on the Coulomb Stability Theorem, in Statistical Mechanics and Mathematics Problems, ed. A. Lenard, Lecture Notes in Physics 20, Springer-Verlag, New York 1973.

9. E.H. Lieb, The stability of matter, Rev. Modern Phys. 48 (1976), 553-569.

10. E.H. Lieb, B. Simon, The Thomas-Fermi theory of atoms, molecules, and solids, Adv. in Math. 23 (1977), 22-116.

11. E.H. Lieb, W.E. Thirring, A bound for the kinetic energy of fermions which proves the stability of matter, Phys. Rev. Lett. 35 (1975), 687-689.

12. E.H. Lieb, H.-T. Yau, The stability and instability of relativistic matter, preprint.

13. P.-A. Meyer, Le schéma de remplissage en temps continu, Séminaire de Probabilities IV, Lecture Notes in Mathematics no. 124, Springer-Verlag, Berlin 1972, 170-194.

14. H. Rost, The stopping distributions of a Markov process, Invent. Math. 14 (1971), 1-16.

HARMONIC ANALYSIS AND ERGODIC THEORY

Alexandra Bellow*

Northwestern University, Department of Mathematics

Evanston, IL 60601

Roger L. Jones

DePaul University, Department of Mathematics

Chicago, IL 60614

Joseph Rosenblatt*

Ohio State University, Department of Mathematics

Columbus, OH 43210

INTRODUCTION

There has always been a very close connection between the fields of harmonic analysis and ergodic theory. A number of techniques from harmonic analysis have played a crucial role in obtaining certain ergodic theory results. This is especially striking in the recent very important work of J. Bourgain. In some cases ergodic theory results have suggested parallel results in harmonic analysis. The connection between these two fields has often not been fully exploited. This paper examines some of these connections, and uses the relationship between

* Partially supported by NSF

ISBN 0-12-231050-0

harmonic analysis and ergodic theory to obtain some new results.

Throughout the paper, (X,Σ,m) will denote a non–atomic probability space, and T an invertible ergodic measure preserving point transformation of X onto itself. If A is a subset of real numbers then $|A|$ will denote the Lebesgue measure of the set A. If A is considered only as a subset of Z, then $|A|$ will be used to denote the number of elements from Z contained in A.

SECTION 1

If $f \in L^p(R^1)$, for some p, $1 \le p < \infty$, then we can form the Poisson integral of f,

$$u(x,y) = P_y \star f(x) = \int_{-\infty}^{\infty} f(x-t) \frac{y}{\pi(t^2+y^2)} dt. \tag{1}$$

In 1906 Fatou [13] proved that if we consider the function $u(x_0+x,y)$, and let (x,y) approach $(0,0)$, then $u(x_0+x, y)$ will converge to $f(x_0)$ for a.e. choice of x_0, provided (x,y) approaches the point $(0,0)$ from within a cone which has its vertex at $(0,0)$. Convergence in such a region is called non–tangential convergence.

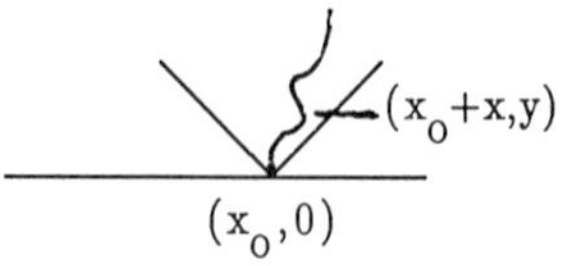

In 1927 Littlewood [17] showed that the conclusion of Fatou's Theorem is false if the point (x,y) is allowed to approach the boundary along any tangential curve. In 1949 Zygmund [23] gave a second proof of Littlewood's result.

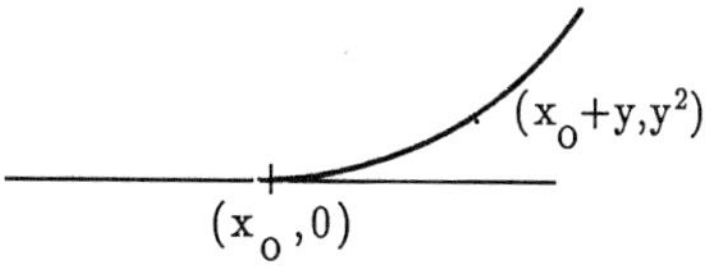

In 1975 Akcoglu and del Junco [1] showed that for T an ergodic measure preserving point transformation, averages of the form

$$\frac{1}{\sqrt{n}} \sum_{k=0}^{\sqrt{n}} f(T^{n+k}x) \tag{2}$$

can diverge a.e. for f the characteristic function of a measurable set. Subsequent to their work, other authors (including A. Bellow and V. Losert [9], J. Rosenblatt [19], and M . Schwartz [20]) have shown that similar averages may also fail to converge a.e. In fact del Junco and Rosenblatt [12] showed that for averages of the form (2) convergence may fail in a dramatic way. They showed that given any $\epsilon>0$, there exists a set E with $|E|<\epsilon$ and such that

$$\limsup_{n\to\infty} \frac{1}{\sqrt{n}} \sum_{k=0}^{\sqrt{n}} f(T^{n+k}x) = 1 \text{ a.e.}$$

while

$$\liminf_{n\to\infty} \frac{1}{\sqrt{n}} \sum_{k=0}^{\sqrt{n}} f(T^{n+k}x) = 0 \text{ a.e.}$$

(In [4] Bellow demonstrated that this dramatic failure of convergence also occurs for ergodic averages along certain subsequences, and in [19] Rosenblatt showed that it occurs for averages of the form $\frac{1}{2^n} \sum_{k=0}^{n} \binom{n}{k} f(T^k x)$.)

It was only recently realized [6] that the negative results in ergodic theory mentioned above correspond to the much earlier negative results of Littlewood (1927) and Zygmund (1949). Having noticed this, the stronger negative results in ergodic theory contained in [6] and [19] suggested a stronger negative result on harmonic functions. This negative result will be the content of Theorem 2′ below. Moreover, positive results in harmonic analysis regarding a form of convergence more general than non–tangential convergence suggested (and led to a proof of) a positive convergence result for subsequences of averages such as (2).

The question of how one could approach the boundary and expect a harmonic function to have a limit seemed settled, until 1984, when Nagel and Stein [18] showed that it was possible to approach the boundary along a path that could not be contained in any cone, and still have a limit exist for each f in $L^p(R)$, $1 \le p < \infty$. Their proof was simplified by Sueiro, [21].

Let Ω be a bounded subset of the upper half plane, R^2_+ $(= \{(x,y)\colon x \in R,\ y > 0\})$, and assume $(0,0)$ is its only limit point on the real axis. Nagel and Stein give necessary and sufficient conditions for this set Ω to be a good approach region in the sense that

$$\lim_{\substack{(x,y) \to (0,0) \\ (x,y) \in \Omega}} u(x_0 + x, y) = f(x_0) \text{ a.e.}$$

For bad regions, they show that the associated maximal operator

$$M_\Omega^P f(x_0) = \sup_{(x,y) \in \Omega} |u(x_0 + x, y)| \tag{3}$$

fails to be weak type (p,p) for every p, $1 \le p < \infty$.

(The reader should think of Ω as a curve which starts at $(0,0)$ and moves into the upper half plane.)

Fix an aperture α and define

$$\Omega_\alpha = \{(z,s)\in R_+^2 : s\le 1 \text{ and } |z-x| < \alpha(s-r) \text{ for some } (x,r)\in\Omega\},$$

and

$$\Omega_\alpha(s) = \{x : (x,s)\in\Omega_\alpha\}.$$

Nagel and Stein then prove the following theorem:

Theorem 1. *a) Assume there exist constants* $\alpha>0$ *and* $A<\infty$ *so that* $|\Omega_\alpha(\lambda)|\le A\lambda$, *then the maximal operator* M_Ω^P *is weak type (1,1) and strong type (p,p) for* $1<p\le\infty$.

b) Conversely, if the maximal function M_Ω^P *is weak type (p,p) for some* $p\ge 1$, *then for every* $\alpha>0$, *there is a constant* $A_\alpha<+\infty$ *such that* $|\Omega_\alpha(\lambda)| \le A_\alpha\lambda$.

The conclusion of a) quickly leads to the convergence result by standard arguments, and the failure of b) quickly implies that a limit cannot exist.

For many purposes, one approximate identity is as good as any other. Consequently Nagel and Stein are also able to prove that the symmetric maximal operator

$$M_\Omega^S f(x_0) = \sup_{(x,y)\in\Omega} \left|\frac{1}{2y}\int_{-y}^{y} f(x_0+x+t)\,dt\right|, \tag{4}$$

or the one sided maximal operator

$$M_{\Omega}^{r}f(x_0) = \sup_{(x,y)\in\Omega} \left|\frac{1}{y}\int_0^y f(x_0+x+t)\,dt\right|, \tag{5}$$

both of which are defined for locally integrable functions, can replace the Poisson maximal operator (3), in the above Theorem.

Their theorem has the following useful corollary:

Corollary 1. *The maximal operators M_{Ω}^{P} and $M_{\Omega_\alpha}^{P}$ are either both weak type (p,p) or neither is. The same is true for the associated pair of symmetrical maximal operators $(M_{\Omega}^{s}, M_{\Omega_\alpha}^{s})$ (4), and the associated pair of one sided maximal operators $(M_{\Omega}^{r}, M_{\Omega_\alpha}^{r})$ (5).*

Proof: Note that the union of cones with aperture α placed at each point in Ω_α is exactly the set Ω_α. Therefore the region Ω satisfies the condition of Theorem 1, if and only if the region Ω_α does.

Exploiting the analogy between harmonic analysis and ergodic theory, Bellow, Jones, and Rosenblatt [6] established that the analogous theorem holds in the ergodic theory setting.

Let Ω be an infinite collection of lattice points with positive second coordinate. Define

$$\Omega_\alpha = \{(z,s) \mid |z-y| \le \alpha(s-r) \text{ for some } (y,r) \text{ in } \Omega,\ (z,s) \text{ a lattice point}\}.$$

The cross section of Ω_α at integer height $s>0$ is denoted by $\Omega_\alpha(s)$ and defined by

$$\Omega_\alpha(s) = \{k \mid (k,s) \in \Omega_\alpha\}.$$

Define the ergodic maximal function associated to the set Ω by

$$M_\Omega f(x) = \sup_{(k,n)\in\Omega} \left|\frac{1}{n}\sum_{j=0}^{n-1} f(T^{k+j}x)\right|. \tag{6}$$

We then have:

Theorem 1'. *a) Assume there exist constants $A<\infty$ and $\alpha>0$ such that $|\Omega_\alpha(\lambda)| \le A\lambda$ for all integer $\lambda>0$; then M_Ω is weak type (1,1) and strong type (p,p) for $1<p\le\infty$.*

b) If M_Ω is weak type (p,p) for some finite $p>0$ then for every $\alpha>0$ there exists $A_\alpha<\infty$ such that for all integer $\lambda>0$ we have $|\Omega_\alpha(\lambda)| \le A_\alpha\lambda$.

With this result we can give a very fast proof of the negative result of Akcoglu and del Junco.

Let $\Omega = \{(n,\sqrt{n}) : n>0\}$. Note that the cross section $\Omega_1(\lambda)$ has measure on the order of λ^2. For any fixed $A<+\infty$, the equation $\lambda^2<A\lambda$ cannot be true for all λ. Consequently, part b of Theorem 1' implies we cannot have a weak maximal inequality. Hence we cannot have convergence.

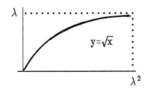

In [6] the following positive result is also proved.

Corollary 1′. *If* $\{n_k\}$ *and* $\{\ell_k\}$ *satisfy the growth conditions* $n_{k+1} > n_k+\ell_k$ *and for some fixed* $j \geq 1$, $\ell_k > c\cdot n_{k-j}$, *then* $\lim_{k\to\infty} \frac{1}{\ell_k} \sum_{i=0}^{\ell_k-1} f(T^{n_k+i}x)$ *exists a.e. for all* $f \in L^1$.

Note that a special case of this result is that if $n_k = 2^{2^k}$ and $\ell_k = \sqrt{n_k}$ then we have convergence. This should be contrasted with the negative result of Akcoglu and del Junco for averages of the form (2), which says we have divergence if we use the full sequence.

There are several further corollaries to Theorem 1′.

Corollary 2′. [Bellow and Losert, unpublished] *Let* $n_k = r^k$ *and* $\ell_k = as^k$ *with* $1<s<r$, *a and s positive integers. Then for each choice of* $p \geq 1$,

$$\lim_{k\to\infty} \frac{1}{\ell_k} \sum_{j=0}^{\ell_k-1} f(T^{n_k+j}x)$$

fails to exist a.e. for some $f \in L^p$.

Corollary 3′. *Assume* $r>1$, *and let* $n_k = \left[r^k\right]$ *and* $\ell_k = o(n_k)$, *then for each choice of* $p \geq 1$,

$$\lim_{k\to\infty} \frac{1}{\ell_k} \sum_{j=0}^{\ell_k - 1} f(T^{n_k+j}x)$$

fails to exist a.e. for some f *in* L^p.

Corollary 4′. [20] Let $L \geq 1$ be given, let $n_k = k^L$ *and* $\ell_k = o(n_k)$, *then for each choice of* $p \geq 1$,

$$\lim_{k\to\infty} \frac{1}{\ell_k} \sum_{j=0}^{\ell_k - 1} f(T^{n_k+j}x)$$

fails to exist a.e. for some f *in* L^p.

Actually, a stronger negative result is true in the ergodic theory setting. In [6] the following theorem is proved:

Theorem 2. *Let* $\Omega = \{(n_k, l_k) \mid n_k \nearrow, l_k \nearrow \infty\}$. *If the linear growth condition on* $|\Omega_\alpha(\lambda)|$ *fails, then we have the "strong sweeping out property", i.e. given* $\epsilon > 0$, *there exists a set* E *with* $m(E) < \epsilon$, *such that*

$$\lim\sup \frac{1}{l_k} \sum_{j=0}^{l_k - 1} \chi_E(T^{n_k+j}x) = 1 \ \text{a.e.}$$

and

$$\lim\inf \frac{1}{l_k} \sum_{j=0}^{l_k - 1} \chi_E(T^{n_k+j}x) = 0 \ \text{a.e.}$$

This suggests an extension of the negative results of Littlewood and Zygmund such that we have "the strong sweeping out property" along the entire real line when we take a tangential approach to the boundary:

Theorem 2′. *Let Ω be a bounded subset of R^2_+ which has (0,0) as its only limit point on the real axis. Assume the linear growth condition $|\Omega_\alpha(\lambda)| < A\lambda$ fails. Then we have "the strong sweeping out property", i.e. given $\epsilon>0$ there exists a set E such that $|E| \le \epsilon$,*

$$\limsup_{\substack{(x,y)\to(0,0)\\(x,y)\in\Omega}} P_y * \chi_E(x_0+x) = 1 \text{ for a.e. } x_0 \tag{7}$$

and

$$\liminf_{\substack{(x,y)\to(0,0)\\(x,y)\in\Omega}} P_y * \chi_E(x_0+x) = 0 \text{ for a.e. } x_0. \tag{8}$$

We first prove a lemma which is similar to Theorem 2′, but uses convolution with dilates of $\frac{1}{2}\chi_{[-1,1)}(x)$ rather than the Poisson kernel. Let $\varphi_y(x) = \frac{1}{2y}\chi_{[-1,1)}(x/y)$. We then have:

Lemma 1. *Let Ω be a bounded subset of R^2_+ which has (0,0) as its only limit point on the real axis. Assume the linear growth condition $|\Omega_\alpha(\lambda)| < A\lambda$ fails. Then we have "the strong sweeping out property" for convolution with dilates of φ, i.e. given $\epsilon>0$ there exists a set E such that $|E| \le \epsilon$,*

$$\limsup_{\substack{(x,y)\to(0,0)\\(x,y)\in\Omega}} \varphi_y * \chi_E(x_0+x) = 1 \text{ for a.e. } x_0$$

and

$$\liminf_{\substack{(x,y)\to(0,0)\\(x,y)\in\Omega}} \varphi_y * \chi_E(x_0+x) = 0 \text{ for a.e. } x_0.$$

Before proving this lemma, we will prove a slightly weaker version, which we can then use to prove Lemma 1. Consider our space to be the probability space $[-1/2, 1/2)$, with all arithmetic mod 1.

Lemma 2. *Assume that the set* Ω *consists of a sequence of points* $\{(x_k, y_k)\}_{k=1}^{\infty}$ *converging to* $(0,0)$, *and that the linear growth condition fails. Then given* $\epsilon>0$, *there exists a set* B *such that* χ_B *has period 1,* $|B\cap[-1/2,1/2)|<\epsilon$, *and*

$$\limsup_{\substack{(x,y)\to(0,0)\\(x,y)\in\Omega}} \frac{1}{2y}\int_{-y}^{y}\chi_B(x_0+x-t)\,dt = 1 \tag{9}$$

while

$$\liminf_{\substack{(x,y)\to(0,0)\\(x,y)\in\Omega}} \frac{1}{2y}\int_{-y}^{y}\chi_B(x_0+x-t)\,dt = 0 \tag{10}$$

for almost every x_0 *in* $[-1/2,1/2)$.

Proof. Define on $[-1/2,1/2)$ the linear operators

$$T_k f(x) = \int f(x+x_k-t)\varphi_{y_k}(t)dt$$

where f is defined on $[-1/2, 1/2)$ and extended periodically to R. The operators T_k meet conditions i) and ii) of Theorem 3 in [6]. In addition they commute with the mixing family of measure preserving transformations $\{S_\theta\}$ where $S_\theta(x) = x+\theta$ mod 1. Thus condition iii) of that theorem is satisfied. It remains to show that for each $\epsilon>0$ and integer n, there exist sets A_p such that if $E_p = \{\sup_{k\geq n} |T_k\chi_{A_p}| \geq 1-\epsilon\}$ then $\sup_p \dfrac{|E_p|}{|A_p\cap[-1/2,1/2)|} = \infty$.

Since the growth condition fails for Ω it also fails if we remove the first n points from Ω. Let Ω' denote this smaller region. Further, we can choose n so large and λ_0 so small that $\Omega_1'(\lambda_0) \subset [-1/2,1/2)$. Note that if $\lambda'<\lambda''$ then $\Omega_1'(\lambda') \subset \Omega_1'(\lambda'')$.

By our assumption that the linear growth condition fails, we can assume that there is a sequence $\{\lambda_p\}$, $\lambda_p \searrow 0$ and $\lambda_p<\lambda_0$, such that $|\Omega_1'(\lambda_p)| > p\lambda_p$. Let A_p be a set such that χ_{A_p} is 1 if $|x| < 2\lambda_p$, zero elsewhere in $[-1/2,1/2)$, and is periodic with period 1. Note if $z \in \Omega_1'(\lambda_p)$ (i.e. $(z,\lambda_p) \in \Omega_1'$) then $|z-x| < (\lambda_p - y)$ for some $(x,y) \in \Omega'$. Thus we have $|z-x| < \lambda_p$ and $\lambda_p > y$. From this we get

$$\sup_{k \geq n} T_k \chi_{A_p}(-z) \geq \frac{1}{2y} \int_{-y}^{y} \chi_{A_p}(-z+x-t)\, dt.$$

The domain of integration starts in the set $A_p \cap [-1/2,1/2)$ and ends in $A_p \cap [-1/2,1/2)$, and hence $\sup_{k \geq n} T_k \chi_{A_p}(-z) = 1$. The fact that this works for each $z \in \Omega_1'(\lambda_p)$ implies that

$$|E_p| \geq |\{\sup_{k \geq n} T_k \chi_{A_p}(-z) = 1\}| \geq |\Omega_1'(\lambda_p)| \geq p\lambda_p = \tfrac{p}{4} |A_p \cap [-1/2,1/2)|.$$

Thus all the conditions of Theorem 3 of [6] are satisfied, and we have "the strong sweeping out property" on our probability space $[-1/2, 1/2)$.

Proof of Lemma 1. Assume first that Ω is a sequence as in Lemma 2. Consider the set B constructed in Lemma 2. Note that for each $x_0 \in (-1/2,1/2)$ such that (9) and (10) hold, (9) and (10) continue to hold independent of how B is extended outside $[-1/2,1/2)$. To see this simply observe that for each $x_0 \in (-1/2,1/2)$ the averages considered do not extend beyond $(-1/2,1/2)$ for (x,y) sufficiently close to $(0,0)$.

By translation, we can replace the interval $[-1/2, 1/2)$ by any interval of unit length. Let $I_{2k} = [k,k+1)$ and $I_{2k+1} = [-k-1,-k)$, $k = 0,1,2,\cdots$. Place a set B_k of measure $\epsilon/2^{k+1}$ in the kth such interval, where B_k has the strong sweeping out property in the interval I_k, and χ_{B_k} is zero outside I_k. Let E be the union of the sets B_k, $k=0,1,2,\cdots$. We thus have a set E on the entire line such that $|E| < \epsilon$, and with the property that the lim sup $\chi_E \star \varphi_y(x_o+x) = 1$ and lim inf $\chi_E \star \varphi_y(x_o+x) = 0$ for a.e. x_o on the entire line, as $(x,y) \to (0,0)$ through the set Ω.

It remains to show that more general Ω can be allowed. By the assumption that the growth condition fails, we can find a sequence $\{\lambda_n\}$ converging to zero such that $|\Omega_1(\lambda_n)| \geq 8n\lambda_n$. Thus there exist $\{z_1^n,\cdots,z_n^n\} \subset \Omega_1(\lambda_n)$ such that $|z_i^n - z_j^n| > 4\lambda_n$ for $i \neq j$. Let $(x_k^n,y_k^n) \in \Omega$ such that (z_k^n,λ_n) is in the cone based at (x_k^n,y_k^n). If Ω^n consists of the points (x_k^n,y_k^n) then from geometric considerations we have that $|\Omega_1^n(2\lambda_n)| \geq 2n\lambda_n$. The collection

$$\Omega^* = \{(x_k^n,y_k^n)\colon k=1,2,\cdots,n \text{ and } n=1,2,\cdots\}$$

can be ordered to give the sequence required in the above argument, completing the proof in the general case.

Proof of Theorem 2′. Note that as in the proof of Lemma 1, we can assume that Ω consists of a sequence of points converging to $(0,0)$. Let the sequence be denoted by $\{(x_k,y_k)\}_{k=1}^{\infty}$. Further, we can first assume that we are working on the probability space $[-1/2, 1/2)$. Define the sequence of operators $\{T_k\}$ by $T_kf(x) = \int_{-\infty}^{\infty} f(x+x_k-t)P_{y_k}(t)\,dt$ where f is defined on $[-1/2, 1/2)$ and extended periodically.

Given $\epsilon>0$, there exists a positive number $b = b(\epsilon)$ such that $\frac{1}{\pi}\int_{-b}^{b}\frac{1}{1+t^2}\,dt > 1-\epsilon/2$. If $\Omega^b = \{(x,by)\colon (x,y)\in\Omega\}$ then it is easy to see that if Ω does not satisfy the growth condition, then neither does Ω^b. Thus by Lemma 2, there exists a set E such that χ_E has period 1, $|E\cap[-1/2,1/2)|<\epsilon$ and $\limsup_{(x,y)\in\Omega}\frac{1}{2by}\int_{-by}^{by}\chi_E(x_0+x-t)\,dt = 1$ for almost every choice of x_0. Hence given ϵ', for almost every point x_0 we can find an infinite set of $(x,y)\in\Omega$ such that $\frac{1}{2by}\int_{-by}^{by}\chi_E(x_0+x-t)\,dt > 1-\epsilon'$. Consequently, $|\{t\in(-yb,yb)\colon\chi_E(x_0+x-t)=1\}| \geq 2yb - 2yb\epsilon'$. Note that the Poisson kernel is radial decreasing, and the largest value of $P_y(t) = \frac{1}{\pi y}$. Thus we have

$$\begin{aligned}
P_y\star\chi_E(x_0+x) &= \int_{-\infty}^{\infty}P_y(t)\chi_E(x_0+x-t)\,dt \\
&\geq \int_{-by}^{by}P_y(t)\chi_E(x_0+x-t)\,dt \\
&= \int_{-by}^{by}P_y(t)\,dt - \int_{-by}^{by}P_y(t)\chi_{E^c}(x_0+x-t)\,dt \\
&\geq 1-\epsilon/2 - \frac{1}{\pi y}\int_{-by}^{by}\chi_{E^c}(x_0+x-t)\,dt \\
&\geq 1-\epsilon/2 - \frac{1}{\pi y}2yb\epsilon' \geq 1-\epsilon
\end{aligned}$$

for a suitable choice of ϵ'.

We can now apply Theorem 1.3 of [12] to conclude that given $\epsilon>0$ there exists a set E, $|E\cap[-1/2,1/2)| < \epsilon$, such that χ_E has period 1, and we have

$$\limsup_{(x,y)\in\Omega} P_y\star\chi_E(x_0+x) = 1 \tag{11}$$

while

$$\liminf_{(x,y)\in\Omega} P_y\star\chi_E(x_0+x) = 0, \tag{12}$$

for almost every choice of x_0 in $[-1/2, 1/2)$. By translation, this interval can be replaced by any interval of unit length.

Assume we are now on $[0,1)$. Note that the lim sup and lim inf do not change if the set E is modified outside the interval $[0,1)$. To see this, let F be a set which agrees with E in the interval $[0,1)$. Given $\delta>0$ there exists a $Y = Y(\delta)$ such that $\int_{|t|>\delta} P_y(t)\,dt < \delta$ for all $y<Y$. Let x_0 be a point in $(0,1)$ such that (11) and (12) hold. Pick δ such that $0<\delta<x_0<1-\delta$. Then pick $(x,y) \in \Omega$ so small that $\delta < x_0+x < 1-\delta$, and $y<Y(\delta)$. We have

$$\begin{aligned}
P_y{\star}\chi_F(x_0+x) &= \int_{-\infty}^{\infty} P_y(t)\chi_F(x_0+x-t)\,dt \\
&= \int_{|t|\le\delta} P_y(t)\chi_F(x_0+x-t)\,dt + \int_{|t|>\delta} P_y(t)\chi_F(x_0+x-t)\,dt \\
&= \int_{|t|\le\delta} P_y(t)\chi_E(x_0+x-t)\,dt + \int_{|t|>\delta} P_y(t)\chi_F(x_0+x-t)\,dt \\
&= \int_{-\infty}^{\infty} P_y(t)\chi_E(x_0+x-t)\,dt - \int_{|t|>\delta} P_y(t)\chi_E(x_0+x-t)\,dt \\
&\qquad + \int_{|t|>\delta} P_y(t)\chi_F(x_0+x-t)\,dt \\
&\ge \int_{-\infty}^{\infty} P_y(t)\chi_E(x_0+x-t)\,dt - \int_{|t|>\delta} P_y(t)\,dt.
\end{aligned}$$

For infinitely many $(x,y)\epsilon\Omega$, we have $\int_{-\infty}^{\infty} P_y(t)\chi_E(x_0+x-t)\,dt > 1-\delta$. Consequently we have $P_y{\star}\chi_F(x_0+x) > (1-\delta) -\delta = 1-2\delta$ for infinitely many $(x,y)\epsilon\Omega$. Since δ was arbitrary, the lim sup must be 1.

To see that the lim inf is 0 for almost every $x_0\epsilon(0,1)$, we do a similar argument and see that $P_y{\star}\chi_F(x_0+x)<2\delta$ for infinitely many $(x,y)\epsilon\Omega$.

We now modify the set E as in the proof of Lemma 1. Let $I_{2k} = [k,k+1)$ and $I_{2k+1} = [-k-1,-k)$, $k=0,1,\cdots$. Place a set E_k of measure $\epsilon/2^{k+1}$ in the kth interval, where E_k has the strong sweeping out property in that interval.

Let $E = \bigcup_{k=0}^{\infty} E_k$. Then $|E| < \epsilon$ and we have (7) and (8) holding for almost every x_0 on the entire real line.

SECTION 2

The maximal ergodic theorem remains true if one considers a subsequence of integers of positive density. (Convergence questions remain, because there is still the problem of convergence on a dense subset.) Thus a natural question was : Can a sequence have density zero, and still satisfy a maximal inequality? This question was addressed by Bellow and Losert in [8] where it was established that sequences which consist of very long blocks of integers, each followed by a very long gap, can be constructed to satisfy a maximal inequality. In this case it is in fact easy to show convergence on a dense set, and hence they show that the averages converge almost everywhere for $f \in L^1(X)$. The proof given there involved showing that a sufficient condition due to Templeman [22] was satisfied. The argument was somewhat combinatorial.

In classical analysis, covering lemmas are often the key ingredient for proving maximal inequalities. It turns out that in fact their theorem can be proven using a covering lemma. This was established by Jones in [16] where a larger maximal function was introduced. For this larger maximal function it was possible to obtain necessary and sufficient conditions for the L^p boundedness of the maximal function. These conditions are very analogous to the conditions used by R. Fefferman and A. Cordoba [14] in their study of general maximal functions in the classical analysis setting.

More precisely, let B be a family of open sets in R^n. The family B has the covering property V_q, $1 \le q \le \infty$, if there exist constants $C < \infty$ and $c > 0$

so that given any subfamily $\{S_i\}_{i\in J}$ of B we can find a sequence $\{\tilde{S}_k\}_{k=1}^{\infty} \subset \{S_i\}_{i\in J}$ satisfying the following conditions:

a) $|\cup_k \tilde{S}_k| \geq c|\cup_{i\in J} S_i|$,

b) $\| \sum \chi_{\tilde{S}_k} \|_q \leq C|\cup_{i\in J} S_i|^{1/q}$.

Let $M(f)(x) = \sup_{x\in S,\ S\in B} \frac{1}{|S|} \int_S |f(y)|\,dy$, then Cordoba and Fefferman prove the following theorem:

Theorem 3. *The maximal operator M is of weak type (p,p) if and only if B has the covering property $V_{q'}$, $1/p + 1/q = 1$, $1<p\leq\infty$.*

Let $\{S_i\}_{i\in N}$ be a collection of finite subsets of integers. Define

$$M^* f(x) = \sup_i \sup_{s\in S_i} \frac{1}{|S_i|} \sum_{k\in(S_i - s)} |f(T^k x)|.$$

Definition: The sequence of sets $\{S_i\}_{i\in N}$ has the property $V_{q'}^{Z}$, $1\leq q\leq\infty$ if there exist constants $C < \infty$ and $c > 0$, such that if U is a finite set of integers, with the property that for each $u \in U$ we have associated a set $S = S(u) \in \{S_i\}_{i\in J}$ and an integer $s = s(u) \in S(u)$, then we can select a subset I of U such that.

a′) $\| \sum_{u\in I} \chi_{\{S(u)+u-s(u)\}} \|_q \leq C| \bigcup_{u\in I} S(u)+u-s(u)|^{1/q}$

b′) $| \bigcup_{u\in I} \{S(u)+u-s(u)| \geq c|U|$.

The following theorem can then be proven:

Theorem 3′. [16] *The maximal function $M^* f$ is weak type (p,p) if and only if the sequence of sets $\{S_i\}_{i\in N}$ has the property $V_{q'}^{Z}$, $1/p + 1/q = 1$, $1<p\leq\infty$.*

With this theorem, it is not difficult to prove the theorem of Bellow and Losert. It is enough to show that the sequences they construct satisfy the condition V_∞, and this follows easily.

SECTION 3

The Hardy–Littlewood maximal function is defined on the real line as

$$M_R f(x) = \sup_{y>0} \frac{1}{y} \int_0^y |f(x+t)|\,dt$$

and on the integers as

$$M_Z \varphi(k) = \sup_{n>0} \frac{1}{n} \sum_{j=0}^{n-1} |\varphi(k+j)|.$$

The ergodic maximal function is defined on a dynamical system (X,Σ,m,T) by

$$f^*(x) = \sup_{n>0} \frac{1}{n} \sum_{k=0}^{n-1} |f(T^k x)|.$$

There is a formal similarity between each of these operators, and indeed the proof given by Hardy and Littlewood in 1930 [15] that $M_R f(x)$ was weak type (1,1) involved the study of $M_Z \varphi$. In 1968 A. P. Calderón [10] showed that there is more than a formal similarity. He showed that it is possible to prove theorems about the ergodic maximal function f^* by first proving an analogous result about the Hardy–Littlewood maximal function, and then transferring the result to the ergodic theory situation. This powerful method deserves to be better known than it appears to be. In fact Calderón showed not only how to use the Hardy–Littlewood maximal function to obtain ergodic theory results, but also showed that the ergodic Hilbert transform which was originally studied by Cotlar [11] can be controlled by using knowledge about the classical Hilbert transform on the real line, and that many other translation invariant operators which satisfy very mild hypothesis can be used to obtain a related ergodic theory results. This has been exploited in [7] to study a certain "square function" which is useful to establish convergence along subsequences.

The basic idea of Calderón's proof can be illustrated by the following important example. Let $\{S_n\}_{n=1}^{\infty}$ be a sequence of finite subsets of integers. We can then define an operator f^* by

$$f_N^*(x) = \sup_{n \leq N} \frac{1}{|S_n|} \sum_{j \in S_n} |f(T^j x)|.$$

Note that this operator is "local" in the sense that the analogous operator on the integers,

$$M_Z^N \varphi(k) = \sup_{n \leq N} \frac{1}{|S_n|} \sum_{j \in S_n} |\varphi(k+j)|$$

does not depend on values too far from the point k where the maximal function is being computed. If we know the operator M_Z^N is bounded from ℓ^p to ℓ^p for some p, with the ℓ^p norm independent of N, then the same is true of the $L^p(X)$ norm of f_N^*. To see this let L denote the maximum integer in $\bigcup_{n \leq N} S_n$.

Then we can argue as follows. For any positive integer M, we have

$$\|f_N^*\|_p^p = \int_X |f_N^*(x)|^p dx = \int_X |f_N^*(T^k x)|^p dx = \frac{1}{M} \sum_{k=1}^{M} \int_X |f_N^*(T^k x)|^p dx$$

$$= \int_X \frac{1}{M} \sum_{k=1}^{M} |f_N^*(T^k x)|^p \, dx. \tag{13}$$

However for each fixed x we can define $\varphi(k) = f(T^k x) \chi_{[0,M+L]}(k)$. For almost every choice of x, this function will be in ℓ^p. By our assumption about the operator M_Z^N on Z, we obtain

$$\sum_{k=1}^{M} |f_N^*(T^k x)|^p = \sum_{k=1}^{M} |M_Z^N \varphi(k)|^p$$
$$\leq C \sum_{k=1}^{M+L} |\varphi(k)|^p$$
$$= C \sum_{k=1}^{M+L} |f(T^k x)|^p.$$

Using this in (13) above, we have

$$\int_X \frac{1}{M} \sum_{k=1}^{M} |f_N^*(T^k x)|^p \, dx \leq \int_X \frac{C}{M} \sum_{k=1}^{M+L} |f(T^k x)|^p \, dx$$
$$= \frac{1}{M} \sum_{k=1}^{M+L} \int_X |f(T^k x)|^p \, dx = \frac{M+L}{M} \int_X |f(x)|^p \, dx,$$

and as $M \to \infty$, the term $\frac{M+L}{M}$ converges to 1, proving the ergodic theory result. The above proof can also be adapted to prove weak type estimates for the operator being studied.

SECTION 4

It is well known, (see for example [2] or [3]) that if we have a sequence of integers $\{n_j\}$ and consider averages of the form $\frac{1}{n} \sum_{j=1}^{n} f(T^{n_j} x)$, for very sparse subsequences, such as $\{2^n\}$, the limit may fail to exist even for some $f \in L^\infty(X)$. Thus it seems reasonable, based on the relationship between ergodic theory and harmonic analysis, that there be a corresponding result in the harmonic analysis setting. A reasonable conjecture is that if we look at a nested sequence of sets $\{E_j\}$, each $E_j \subset R$, with $\{0\}$ as the unique limit point, and if the sets are sufficiently sparse, then the sequence $\frac{1}{|E_j|}\int_{E_j} f(x+t)\,dt$ may fail to have a limit for some $f \in L^1 \cap L^\infty$. In fact it is possible to prove the following:

Theorem 4. *Given a strictly decreasing sequence of points* $\{a_k\}_{k=1}^\infty$ *which converges to zero, there exists a sequence of positive numbers,* $\{\epsilon_k\}_{k=1}^\infty$ *such that* $a_k + \epsilon_k < a_{k-1}$ *and such that if* $E = \bigcup_{k=1}^\infty (a_k, a_k+\epsilon_k)$ *then*

$$\lim_{h\to 0} \frac{1}{|E\cap[0,h)|}\int_{E\cap[0,h)} f(x+t)\,dt$$

fails to exist a.e. for some $f \in L^1 \cap L^\infty$.

We also have the analog of the positive result of Bellow and Losert [8].

Theorem 5. *There exist strictly decreasing sequences of positive numbers* $\{a_k\}_{k=1}^\infty$ *and* $\{d_k\}_{k=1}^\infty$ *with* $a_k + d_k < a_{k-1}$ *and such that if* $E = \bigcup_{k=1}^\infty (a_k, a_k+d_k)$ *then*

$$\lim_{h\to 0} \frac{1}{|E\cap[0,h)|}\int_{E\cap[0,h)} f(x+t)\,dt = f(x) \text{ a.e.}$$

for all $f \in L^1$. *Further,* $\{a_k\}$ *and* $\{d_k\}$ *can be selected so that* 0 *is a point of density of the complement of the set* E.

Proof. Select $\Omega = \{(a_k, d_{k+1})\}_{k=1}^\infty$ such that the linear growth condition of Theorem 1 is satisfied and $d_1 \le 1$. Define $A_{E,h}f(x) = \frac{1}{|E\cap[0,h)|}\int_{E\cap[0,h)} f(x+t)\,dt$ and $S_k = \sum_{j=k}^\infty d_j$.

Case 1. Assume h between a_k+d_k and a_{k-1} for some value of k. In this case, using Corollary 1 of Theorem 1, we have

$$A_{E,h}f(x) = \frac{1}{S_k} \sum_{j=k}^{\infty} d_j \frac{1}{d_j} \int_0^{d_j} f(x+a_j+t)\,dt$$

$$\leq M_{\Omega_1}^r f(x) \frac{1}{S_k} \sum_{j=k}^{\infty} d_j$$

$$= M_{\Omega_1}^r f(x).$$

Case 2. Assume h splits one of the intervals. i.e. $a_k<h<a_k+d_k$. In this case, if $h-a_k$ is less than d_{k+1} integrate over the larger region from a_k to a_k+d_{k+1}. We then have

$$A_{E,h}f(x) \leq \frac{1}{S_{k+1}} \sum_{j=k+1}^{\infty} d_j\, M_{\Omega_1}^r f(x) + \frac{1}{S_{k+1}} \int_0^{d_{k+1}} f(x+a_k+t)\,dt$$

$$\leq M_{\Omega_1}^r f(x) + \frac{d_{k+1}}{S_{k+1}} M_{\Omega_1}^r f(x)$$

$$\leq 2M_{\Omega_1}^r f(x)$$

because $\{(a_k,d_{k+1})\}$ was assumed to satisfy the cone condition. For larger values of h, use the same argument, and note that $(a_k,h-a_k) \in \Omega_1$ for any h such that $h-a_k>d_{k+1}$.

An explicit example of a pair of sequences that work for Theorem 5 can be found in Nagel and Stein [18]; namely let $a_k = 2^k 2^{-k^2}$ and $d_k = 2^{-(k-1)^2}$. In this case the complement of the set $\bigcup_{k=1}^{\infty} (a_k, a_k+d_k)$ has 0 as a point of density.

To obtain the negative result of Theorem 4, we first need a lemma.

Lemma: *Let* $\{a_j\}$ *be a strictly decreasing sequence of real numbers tending to zero. Then given* $\epsilon>0$ *there exists a set* B *with* $|B|<\epsilon$ *such that* $\limsup \chi_B(x+a_j) = 1$ *and* $\liminf \chi_B(x+a_j) = 0$ *a.e. on* R.

Proof: Because the operators we are considering are local, it is enough to construct such a set on the interval $[0,1)$, translate such sets of size $\epsilon/2^k$ to the kth interval, and take the union of these translates for our final set. We get started with the following argument due to Bellow [5]: For each $n>1$ take

$$0 < \alpha < \min\{a_i - a_{i+1} : i=k,k+1,\cdots,k+n-1\}.$$

Let $B_n = [0,\alpha]$. Then the sets $B_n - a_k, B_n - a_{k+1}, B_n - a_{k+2}, \cdots, B_n - a_{k+n}$ are all disjoint. Hence

$$\{\max_{j=k,\,k+1,\cdots,k+n} \chi_{B_n}(x+a_j) \geq 1\} \supset \bigcup_{j=0}^{n} (B_n - a_{k+j})$$

and thus

$$|\{\max_{j=k,\,k+1,\cdots,k+n} \chi_{B_n}(x+a_j) \geq 1\}| \geq |\bigcup_{j=0}^{n} B_n - a_j| = (n+1)|B_n|.$$

This inequality and the fact that the operator $f(x) \to f(x+a_j)$ commutes with the transformation $x \to x+\theta$, θ irrational, means that the conditions of Theorem 3 in [6] are satisfied, and we have the "strong sweeping out property".

Proof of Theorem 4. Let $A_j f(x) = \frac{1}{d_j}\int_0^{d_j} f(x+a_j+t)\,dt$. Let f be χ_B for the set B given by the Lemma. Then define

$$B_j = \{x: |A_j f(x) - f(x+a_j)| > 1/10\}.$$

By the Lebesgue differentiation theorem we can select d_j so small that $|B_j| < 1/2^j$. Note that a.e. x will be in B_j for only a finite number of j. If necessary, decrease d_j still further so that $d_{j+1} < \frac{1}{10} d_j$.

Assume that $x \notin B_k$ and let $h = a_k + d_k$. Define $S_k = \sum_{j=k}^{\infty} d_j$. If $f(x+a_k) = 1$ then have $A_{E,h} f(x) \geq \frac{d_k}{S_k} A_k f(x) \geq \frac{9}{10}\frac{d_k}{S_k}$. Because $d_{j+1} < \frac{1}{10} d_j$ we have $\frac{d_k}{S_k} \geq \frac{9}{10}$ and hence $A_{E,h}f(x) \geq \frac{9}{10} x \frac{9}{10} \geq \frac{8}{10}$.

On the other hand, if $f(x+a_k) = 0$ then

$$A_{E,h}f(x) \leq \frac{S_{k+1}}{S_k} + \frac{d_k}{S_k}(\frac{1}{10}) \leq \frac{1}{9} + \frac{1}{10} \leq \frac{2}{9}.$$

The above lemma implies that for a.e. x there are infinitely many k such that $f(x+a_k) = 1$, and infinitely many k such that $f(x+a_k) = 0$. Thus for a.e. x the lim sup will be at least $\frac{8}{10}$ and the lim inf will be at most $\frac{2}{9}$. Hence no limit can exist.

SECTION 5

It seems clear that a great deal more can be accomplished by exploiting the connection between ergodic theory and harmonic analysis. It would be especially interesting to know if there is some general method of transforming results from one field to the other. Calderón's transfer principal provides a powerful tool in this spirit, but in many cases it is not a simple transformation. For example Theorems 4 and 5 above, were suggested by their ergodic theory analog, but seem to require their own proof. It is not obvious how to use Calderón's transfer principal to obtain these results, despite the fact that the analogous ergodic theory result is known.

As another example, consider the following ergodic theory result of Bellow [3]:

Theorem 6 *Given p, $1<p<\infty$, there exist a nested sequence of subsets of integers, depending on p, such that*

$$\lim_{n\to\infty} \frac{1}{|S_n|} \sum_{k\in S_n} f(T^k x)$$

exists for all $f \in L^p$, *but fails to exist for some* $f \in L^r$ *for each* $r<p$.

This question was suggested from harmonic analysis considerations, in particular a discussion of covering lemmas, however at this time not only is it not known how to "transfer" this result to obtain a proof of the associated harmonic analysis conjecture, but it is not even known if the associated conjecture is true. However, Theorem 6 does suggest the following:

Conjecture: *Given* p, *there exists a set* $E = E(p)$ *such that*

$$\lim_{h\to 0} \frac{1}{|E\cap[0,h)|} \int_{E\cap[0,h)} f(x+t)\,dt$$

exists for all $f \in L^p$, *but fails to exist for some* $f \in L^r$, *for each* $r<p$.

It would be very useful if a more general transfer theorem could be proven which would allow these more general problems to be "transferred".

References

1. Akcoglu, M. and del Junco, A., *Convergence of averages of point transformations*, **Proc. A.M.S. 49**(1975), 265–266.

2. Bellow, A., *On bad universal sequences in ergodic theory (II)*, in "Measure Theory and its Applications", **Lecture Notes in Math, #1033**, Springer, Berlin–New York, 1983, 74–78.

3. Bellow, A., *Perturbation of a sequence*, to appear in **Advances in Math.**

4. Bellow, A., *Sur la structure des suites mauvaises universelles en thorie ergodique*, **C.R. Acad Paris t 294**(1982) 55–58.

5. Bellow, A., *Two problems*, Proc Oberwolfach Conference on Measure Theory (June 1987), **Springer Lecture Notes in Mathematics, #945**.

6. Bellow, A., Jones, R. and Rosenblatt, J., *Convergence for moving averages*, to appear in **Ergodic Theory and Dynamical Systems.**

7. Bellow, A., Jones, R. and Rosenblatt, J., *Almost everywhere convergence of weighted averages*, preprint.

8. Bellow, A. and Losert, V., *On sequences of density zero in ergodic theory*, **Contemporary Mathematics**, **28**(1984) 49–60.

9. Bellow, A. and Losert, V., *The weighted pointwise ergodic theorem and the individual ergodic theorem along subsequences*, **Trans. AMS** **288**(1985) 307–345.

10. Calderón, A. P., *Ergodic theory and translation–invariant operators*, **Proc. Nat. Acad. Sci.** **59**(1968) 349–353.

11. Cotlar, M., *A unified theory of Hilbert transforms and ergodic theorems*, **Rev. Mat. Cuyana** **1**(1955), 105–167.

12. del Junco, A., and Rosenblatt, J., *Counterexamples in ergodic theory and number theory*, **Math. Ann.** **247**(1979) 185–197.

13. Fatou, P., *Series trigonometriques et series de Taylor*, **Acta. Math.** **30** (1906) 335–400.

14. Fefferman, R. and Cordoba, *A geometric proof of the strong maximal theorem*, **Annals of Math.** 102(1975), 95–100.

15. Hardy, G. H. and Littlewood, J. E., *A maximal theorem with function theoretic applications*, Acta Mathematica, 54 (1930), 81–116.

16. Jones, R., *Necessary and sufficient conditions for a maximal ergodic theorem along subsequences*, **Ergodic Theory and Dynamical Systems**, **7**(1987) 203–210.

17 . Littlewood, J. L., *On a theorem of Fatou*, **J. London Math Soc.** **2**(1927), 172–176.

18. Nagel, A. and Stein, E., *On certain maximal functions and approach regions*, **Advances in Mathematics,** **54**(1984), 83–106.

19. Rosenblatt, J., *Ergodic group actions*, **Arch. Math.,** **47**(1986).

20. Schwartz, M., *Polynomially moving ergodic averages*, to appear in **Proc. AMS.**

21. Sueiro, J., *A note on maximal operators of Hardy–Littlewood type*, **Math. Proc. Camb. Phil. Soc.**102(1987), 131–134.)

22. Templeman, A. A., *Ergodic theorems for general dynamical systems*, **Sov. Math. Dokl.** **8**(1967), 1213–1216.

23. Zygmund, A., *On a theorem of Littlewood*, **Summa Brasil, Math.** **2**(1949), 1–7.

Almost Everywhere Convergence of Powers

by

Alexandra Bellow *

Department of Mathematics

Northwestern University

Evanston, IL 60601

Roger Jones

Department of Mathematics

De Paul University

2323 A. Seminary

Chicago, IL 60614

and

Joseph Rosenblatt*

Department of Mathematics

The Ohio State University

Columbus, Ohio 43210

---

* Partially supported by NSF

ISBN 0-12-231050-0

Let (X, β, m) be a separable non-atomic probability space. Let $T : L_1(X) \to L_1(X)$ be a contraction which is also a contraction on $L_\infty(X)$, so that $T : L_p(X) \to L_p(X)$ with $\|T\|_p \leq 1$ for $1 \leq p \leq \infty$. The almost everywhere convergence of the powers $T^n f(x)$ for $f \in L_p(X)$ was studied for symmetric operators by Burkholder and Chow [6], Rota [12], and Stein [13]. See also Akcoglu and Sucheston [14]. The question of what happens when T is not symmetric is the main focus of discussion here. A special case of this is of particular interest. Suppose $\tau : X \to X$ is an ergodic invertible measure-preserving transformation and μ is a probability measure on Z, the integers. On the dynamical system (X, β, m, τ), define $Tf(x) = \mu f(x) = \sum_{k=-\infty}^{\infty} \mu(k) f(\tau^k x)$ for $f \in L_p(X), 1 \leq p \leq \infty$. Various theorems about this special case appear in [1,2,3]. In this article some of these results are summarized with a view to motivating the more general new theorems that are presented here.

Consider first the case where the measure μ is $b = (\delta_0 + \delta_1)/2$. Then $b^n f(x) = (1/2^n) \sum_{k=0}^{n} \binom{n}{k} f(\tau^k x)$. It is easy to see that $\lim_{n\to\infty} \|b^n f - \int f dm\|_p = 0$ for all $f \in L_p(X)$, $1 \leq p < \infty$. The following results are proved in Rosenblatt [11], using del Junco and Rosenblatt [7].

1. Theorem *For all* $\epsilon > 0$, *there exists* $E \in \beta$ *such that* $m(E) < \epsilon$ *and*

$$\limsup_{n\to\infty} b^n 1_E(x) = 1 \quad \text{a.e.}$$

2. Corollary *In the symmetry pseudo-metric on* β, *a dense* G_δ *set* $\mathcal{R}$ *has the property that for* $E \in \mathcal{R}$ *both* $\limsup_{n\to\infty} b^n 1_E(x) = 1$ *a.e. and* $\liminf_{n\to\infty} b^n 1_E(x) = 0$ a.e..

The strong sweeping out of this corollary also shows that generically if $f \in L_p(X), 1 \leq p < \infty, \limsup |b^n f(x)| = \infty$ a.e.. So the convergence of $(b^n f)$ is good in norm, but it is bad as far as a.e. convergence is concerned. For this reason, the following result from [1] may seem quite surprising.

3. Theorem. *For all* $f \in L_1(X)$, $\lim_{m\to\infty} b^{2^{2m}} f(x) = \int f dm$ *a.e..*

This theorem can be thought of as an operator version of the classical fact that a norm convergent sequence in $L_1(X)$ has an a.e. convergent subsequence. It also suggests that there should be a more general result for $L_1 - L_\infty$ contractions T, with suitable conditions on T, that says that for some subsequence (n_m), $T^{n_m}f(x)$ converges a.e..

To see what such subsequence theorems are like in the case that $Tf = \mu f$ as above, we need some definitions and well-known facts.

Definition. A probability measure μ on Z is *adapted* if $supp(\mu)$ generates Z. Also, μ is *strictly aperiodic* if $supp(\mu)$ is not contained in a coset of a proper subgroup of Z.

4. Proposition. *If μ is adapted, then for $1 \leq p < \infty, \{f - \mu f : f \in L_p(X)\}$ is L_p-norm dense in the mean zero functions in $L_p(X)$.*

5. Proposition. 1) *If μ is adapted, then μ is strictly aperiodic if and only if*

$$\lim_{n \to \infty} \|\mu^{n+1} - \mu^n\|_{\ell_1(Z)} = 0.$$

2) *The measure μ is strictly aperiodic, if and only if $|\hat{\mu}(\lambda)| < 1$ for all $\lambda \in \mathbb{C}$, $|\lambda| = 1$, $\lambda \neq 1$.*

6. Corollary. *If μ is strictly aperiodic, then for $1 \leq p < \infty$, if $f \in L_p(X)$,*

$$\lim_{n \to \infty} \|\mu^n f - \int f dm\|_p = 0.$$

Remark. In particular, $\|\mu^n(f-\mu f)\|_\infty \le \|\mu^{n+1}-\mu^n\|_{\ell_1}\|f\|_\infty$ shows that when μ is strictly aperiodic, then $\mu^n F \to \int F dm$ a.e. for all F in $\mathbb{C}+\{f-\mu f : f \in L_\infty\}$. So $\mu^n F(x)$ converges a.e. for all F in a dense subspace of any $L_p(X), 1 \le p < \infty$.

One result of the facts above is that if μ is strictly aperiodic, then a.e. convergence for $(\mu^{n_m} f(x))$ for all $f \in L_p(X)$, for some $p, 1 \le p < \infty$, requires only a maximal estimate. To prove subsequence theorems analogous to Theorem 3, Bellow, Jones, and Rosenblatt [2] use this variation of a theorem of Duoandikoetxea and Rubio de Francia.

7. **Theorem.** *Suppose that (a_m) is a lacunary sequence: $\inf_{m\ge 1} a_{m+1}/a_m > 1$; and suppose (v_m) is a sequence of probability measures on Z such that*

$$|\hat{v}_m(\lambda)-1| \le C a_{m+1}|\lambda - 1|$$

and

$$|\hat{v}_m(\lambda)| \le C/a_m|\lambda - 1|.$$

Then for all $\varphi \in \ell_p(Z), 1 < p \le \infty, \|\sup_m |v_m * \varphi|\|_{\ell_p} \le C\|\varphi\|_{\ell_p}$.

But then Calderón's transfer principle and Theorem 7 can be used to show that if $\hat{\mu}_n(\lambda) \to 0$ uniformly on each $S_\delta = \{\lambda \in \mathbb{C} :$

$|\lambda| = 1, |\lambda - 1| \geq \delta\}, \delta > 0$, then there exists a subsequence (μ_{n_m}) which satisfies a strong maximal estimate on $L_p(X), 1 < p \leq \infty$.

8. Theorem. *If μ is strictly aperiodic, then there is a subsequence (n_m) such that $\lim_{m\to\infty} \mu^{n_m} f(x) = \int f dm$ a.e. for all $f \in L_p(X), 1 < p \leq \infty$.*

If one does a careful computation using the details of the proof of the theorem of Duoandikoetxea and Rubio de Francia, then one can see this.

9. Theorem. *If (n_m) is chosen with $\lim_{m\to\infty} n_{m+1}/n_m^2 log(n_m) = \infty$, then for strictly aperiodic μ with finite support, $\lim_{m\to\infty} \mu^{n_m} f(x) = \int f dm$ a.e. for all $f \in L_p(X), 1 < p \leq \infty$.*

See [2] for the details of Theorem 8 and 9.

The operator viewpoint analogous to the above is clear.

Definition. An L_2-contraction T is said to be *strictly aperiodic* if the spectrum of T, $sp(T)$, satisfies $sp(T) \cap \{\lambda \in \mathbb{C} : |\lambda| = 1\} = \{1\}$.

10. Proposition. *If T is a normal strictly aperiodic operator on $L_2(X)$, then*

$$\lim_{n\to\infty} \|T^n - T^n T^*\|_2 = \lim_{n\to\infty} \left(\sup\{\|T^n(f - T^* f)\|_2 : \|f\|_2 \leq 1\}\right) = 0$$

and

$$\lim_{n\to\infty} \|T^n - T^{n+1}\|_2 = \lim_{n\to\infty} \left(\sup\{\|T^n(f - Tf)\|_2 : \|f\|_2 \le 1\}\right) = 0.$$

Proof. By the spectral theorem, there exists a positive regular Borel probability measure v_f on $sp(T)$ with $\|v_f\|_1 = \|f\|_2^2$ such that

$$\|T^n(f - T^*f)\|_2^2 = \int_{sp(T)} |\lambda^n(1 - \overline{\lambda})|^2 dv_f(\lambda).$$

Let $\delta > 0$ and let $S_\delta = \{\lambda \in sp(T) : |\lambda - 1| \ge \delta\}$. Then, since T is strictly aperiodic, there exists $\epsilon_\delta > 0$ such that $\sup\{|\lambda| : \lambda \in S_\delta\} = 1 - \epsilon_\delta$. So if $4(1 - \epsilon_\delta)^{2n} \le \delta^2$, then

$$\begin{aligned}
&\|T^n(f - T^*f)\|_2^2 \\
&= \int_{S_\delta} |\lambda^n(1 - \overline{\lambda})|^2 dv_f(\lambda) + \int_{sp(T)\setminus S_\delta} |\lambda^n(1 - \overline{\lambda})|^2 dv_f(\lambda) \\
&\le \int_{S_\delta} 4(1 - \epsilon_\delta)^{2n} dv_f(\lambda) + \int_{sp(T)\setminus S_\delta} \delta^2 dv_f(\lambda) \\
&\le \int_{sp(T)} \delta^2 dv_f(\lambda) \le \delta^2 \|f\|_2^2.
\end{aligned}$$

Hence, for all $\delta > 0$, there is $n_\delta \ge 1$ such that if $n \ge n_\delta$, $\|T^n(f - T^*f)\|_2^2 \le \delta^2 \|f\|_2^2$. So $\lim_{n\to\infty} \|T^n - T^nT^*\|_2 = 0$. Similarly, $\lim_{n\to\infty} \|T^n - T^{n+1}\|_2 = 0$. □

Remark. If T is an L_1 - L_∞ contraction which is a normal strictly aperiodic operator on $L_2(X)$, then Proposition 10 and interpolation show $\lim_{n\to\infty} \|T^n - T^{n+1}\|_p = 0$ for $1 < p < \infty$. See [2] for further discussion.

11. Corollary. *If T is a normal strictly aperiodic operator on $L_2(X)$, then for all $f \in L_2(X)$, $\lim_{n\to\infty} T^n f = f_0$ in L_2-norm for a uniquely determined T-invariant function f_0.*

Example. If μ is a probability measure on Z, then the operator $T_\mu f = \mu f$ has $sp(T_\mu) \subset \hat{\mu}(sp(U_\tau))$ where $U_\tau f = f \circ \tau$. So if μ is strictly aperiodic, then T_μ is strictly aperiodic because U_τ is unitary. Clearly T_μ is normal, so Corollary 11 becomes Corollary 6, $p = 2$, in this special case (Note: if one knew the conclusion of Corollary 6 for $p = 2$ for such a μ, it would generally follow that the same held for $1 \le p < \infty$ too).

Another consequence of Proposition 10 is that there is a subsequence (T^{n_m}) such that $\sum_{m=1}^{\infty} \|T^{n_m}(f - T^* f)\|_2^2 < \infty$ for all $f \in L_2(X)$. Hence, $(T^{n_m} f(x))$ converges a.e. for all $f = f_1 + (f_2 - T^* f_2)$ where f_1 is T-invariant and $f_2 \in L_2(X)$. Since such f comprise a dense subspace of $L_2(X)$, a pointwise a.e. convergence theorem holds for (T^{n_m}) on $L_2(X)$ as soon as there is a maximal inequality. Note that since T is normal, $f \in L_2(X)$ is T-invariant if and only if f is T^*-invariant. So the preceding argument works equally well with T replacing T^*. More directly though we have this partial generalization of Theorem 8.

12. Theorem. *If T is a normal strictly aperiodic $L_1 - L_\infty$ contraction, then there is a subsequence (T^{n_m}) such that for all $f \in L_2(X)$, $\lim_{m\to\infty} T^{n_m} f(x)$ exists a.e.. Moreover, if $f^*(x) = \sup_{m\geq 1} |T^{n_m} f(x)|$, then $\|f^*\|_2 \leq C\|f\|_2$ for all $f \in L_2(X)$.*

Proof. By the Dunford-Schwartz ergodic theorem [8], the averages $C_n f(x) = (1/n) \sum_{k=1}^{n} T^k f(x)$ converge a.e. for all $f \in L_1(X)$ and $\|\sup_{n\geq 1} |C_n f|\|_p \leq c_p \|f\|_p$ for $f \in L_p(X), 1 < p \leq \infty$. Consider the differences $D_n f = T^n f - C_n f$. Then by the spectral theorem, $\|D_n f\|_2^2 = \int_{sp(T)} |\lambda^n - (1/n) \sum_{k=1}^{n} \lambda^k|^2 dv_f(\lambda)$ for a positive regular Borel measure v_f with $\|v_f\|_1 = \|f\|_2^2$. For $\delta > 0$, on $S_\delta = \{\lambda \in sp(T) : |\lambda - 1| \geq \delta\}$, $d_n(\lambda) = |\lambda^n - (1/n) \sum_{k=1}^{n} \lambda^k|$ converges uniformly to 0 because T is strictly aperiodic. But $d_n(1) = 0$ too. So as in [2], Lemma 1.10, there is a subsequence (n_m) such that

$$\sum_{m=1}^{\infty} d_{n_m}(\lambda)^2 = \sum_{m=1}^{\infty} |\lambda^{n_m} - (1/n_m) \sum_{k=1}^{n_m} \lambda^k|^2 \leq 5 \text{ for all } \lambda \in sp(T).$$

Thus, $\sum_{m=1}^{\infty} \|D_{n_m} f\|_2^2 = \sum_{m=1}^{\infty} \int_{sp(T)} d_{n_m}(\lambda)^2 dv_f(\lambda) \leq 5v_f(sp(T)) = 5\|f\|_2^2$.

But then for all $f \in L_2(X)$, $\lim_{m\to\infty} D_{n_m} f(x) = 0$ a.e. and so $\lim_{n\to\infty} T^{n_m} f(x)$ exists a.e. Moreover, $f^* \leq \sup_{m\geq 1} |D_{n_m} f| + \sup_{n\geq 1} |C_n f|$ and so $f^* \leq (\sum_{m=1}^{\infty} |D_{n_m} f|^2)^{1/2} + \sup_{n\geq 1} |C_n f|$. Thus, $\|f^*\|_2 \leq c\|f\|_2 + c_2\|f\|_2 \leq C\|f\|_2$ for all $f \in L_2(X)$. □

Remark. a) The proof of Theorem 12 actually shows that if T is a normal strictly aperiodic L_2 contraction such that $C_n f(x)$ converges a.e. for all $f \in L_2(X)$, then, because some subsequence (D_{n_m}) has $\lim_{m\to\infty} D_{n_m} f(x) = 0$ a.e. for all $f \in L_2(X)$, there is a subsequence (T^{n_m}) such that for $f \in L_2(X)$, $\lim_{m\to\infty} T^{n_m} f(x)$ exists a.e..

Remark. b) This theorem can probably be extended to all $L_p(X)$, $1 < p < \infty$.

Example 1. Let $h : X \to \mathbb{C}$ have $|h(x)| \leq 1$. Define $T_0(x) = h(x)f(\tau x)$ for $f \in L_1(X)$. If $|h(x)|$ is constant, then T_0 is a normal $L_1 - L_\infty$ contraction. Consider $T = (1/2)(I + T_0)$.

Because $sp(T_0) \subset \{\lambda : |\lambda| \leq 1\}$, and for $|\lambda| \leq 1, |(1/2)(1+\lambda)| < 1$ unless $\lambda = 1, T$ is strictly aperiodic. Hence, for such h, $T^n f(x) = (1/2^n) \sum_{k=0}^{n} \binom{n}{k} h(x) \cdots h(\tau^{k-1}x) f(\tau^k x)$ satisfies the subsequence theorem above. To choose (n_m) explicitly, it suffices to have $\sup\{ \sum_{m=1}^{\infty} |\lambda^{n_m} - (1/n_m) \sum_{k=1}^{n_m} \lambda^k|^2 : \lambda \in \mathbb{C}, |\lambda| = 1\} < \infty$. An elementary calculation shows this is the case if $(n_m) = (3^{3^m})$. See [2] for similar computations.

Example 2. Let G be a compact abelian group and let $\mu \in M(G), \|\mu\|_1 \leq 1$. Define $T_\mu h = \mu * h$ for $h \in L_1(G)$ where $\mu * h(x) = \int_G h(xy^{-1})\, d\mu(y)$ for all $x \in G$. Then $T_\mu^* = T_{\tilde{\mu}}$ where $d\tilde{\mu}(x) = \overline{d\mu(x^{-1})}$ for $x \in G$. So T_μ is a normal $L_1 - L_\infty$ contraction. It is easy to see that T_μ is strictly aperiodic if and only if $\sup\{|\hat{\mu}(\gamma)| : \gamma \in \hat{G}, \gamma \neq 1\} < 1$ i.e. if and only if μ is a strictly aperiodic measure in G.

If one knows more about the shape of $sp(T)$ near 1, then stronger conclusions than the above are possible.

13. Definition. For a given $\alpha, 0 \leq \alpha < \pi$, define the*Stolz region* $\mathcal{S}_\alpha$ as follows. Let C_α be the closed disc of radius $\sin(\alpha/2)$ with center at the origin. Let T_α be the closed triangle with vertices

$(1,0), (sin^2(\frac{\alpha}{2}), \frac{1}{2}sin\alpha)$, and $(sin^2(\frac{\alpha}{2}), -\frac{1}{2}sin\alpha)$. Let $\mathcal{S}_\alpha = C_\alpha \cup T_\alpha$. See Figure 1.

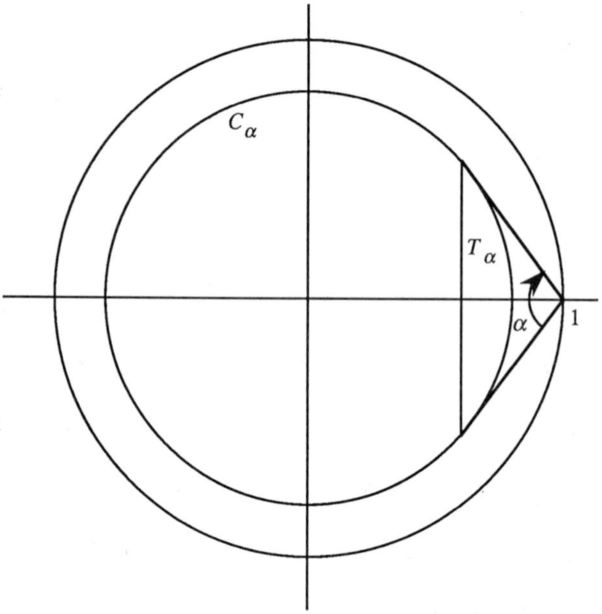

Figure 1

It is easy to see that if T is strictly aperiodic and for sufficiently small $\delta > 0, \{\lambda \in sp(T) : |\lambda - 1| \leq \delta\} \subset T_\alpha$ for some $\alpha, 0 \leq \alpha < \pi$, then $sp(T) \subset \mathcal{S}_{\alpha_0}$ for some (possibly larger) $\alpha_0 < \pi$. Also, if T_0 is an L_2-contraction and $T = T_0^* T_0$, then $sp(T) \subset [0,1] = \mathcal{S}_0$.

14. Theorem. *If T is a normal $L_1 - L_\infty$ contraction such that $sp(T) \subset \mathcal{S}_\alpha$ for some Stolz region $\mathcal{S}_\alpha, 0 \leq \alpha < \pi$, then for all $p, 1 < p \leq \infty$, if $f \in L_p(X)$, then $\lim_{n\to\infty} T^n f(x)$ exists a.e.. Moreover, if $f^*(x) = \sup_{n\geq 1} |T^n f(x)|$, then $\|f^*\|_p \leq c_p \|f\|_p, 1 < p \leq \infty$.*

Proof. First, let us estimate $\|T^n - T^{n+1}\|_2$ more precisely than was done in Proposition 10. By the spectral theorem, $\|T^n - T^{n+1}\|_2^2 \leq \sup\{|\lambda^n(1-\lambda)|^2 : \lambda \in sp(T)\}$. But there is a constant $1 \leq m_\alpha < \infty$ such that $\frac{|\lambda-1|}{1-|\lambda|} \leq m_\alpha$ for $\lambda \in \mathcal{S}_\alpha$. Hence, if $\lambda \in sp(T) \subset \mathcal{S}_\alpha$, $|\lambda^n(1-\lambda)|^2 \leq m_\alpha^2 |\lambda|^{2n}(1-|\lambda|)^2$. For $n \geq 1$, the function $g(x) = x^{2n}(1-x)^2$, $0 \leq x \leq 1$, has its maximum at $x = n/(n+1)$, not exceeding $1/(4(n+1)^2)$. So for $n \geq 1$,

$$\|T^n - T^{n+1}\|_2 \leq m_\alpha/(2(n+1)).$$

Hence, for all $f \in L_2(X), \sum_{n=1}^{\infty} \|T^n(f - Tf)\|_2^2 < \infty$. Thus, for all $f \in L_2(X)$, $T^n(f-Tf)(x) \to 0$ a.e. as $n \to \infty$. Since $T^n f(x)$ obviously converges if $Tf = f$, this shows $(T^n f(x))$ converges a.e.

for all f in the L_2-norm dense subspace of $L_2(X)$ consisting of functions $f = f_1 + (f_2 - Tf_2)$ where $Tf_1 = f_1$ and $f_2 \in L_2(X)$.

Therefore, to prove the theorem only requires giving a strong maximal estimate. This is done in the same manner as Stein [13] except that $sp(T) \subset \mathcal{S}_\alpha$ replaces the symmetry hypothesis. One critical point in the calculation is the computation of

$$\|(\sum_{k=1}^{\infty} k|T^{k+1}f - T^k f|^2)^{1/2}\|_2^2 = \int \sum_{k=1}^{\infty} k|\lambda^{k+1} - \lambda^k|^2 d\nu_f(\lambda).$$

But $\sum_{k=1}^{\infty} k|\lambda^{k+1} - \lambda^k|^2 = |\lambda|^2|\lambda - 1|^2/(1-|\lambda|^2)^2 \leq [|\lambda - 1|/(1-|\lambda|)]^2$ for $|\lambda| < 1$. Because $sp(T) \subset \mathcal{S}_\alpha$, $[|\lambda - 1|/(1-|\lambda|)]^2 \leq m_\alpha^2 < \infty$ independently of $\lambda \in sp(T)\backslash\{1\}$. Thus,

$$\|(\sum_{k=1}^{\infty} k|T^{k+1}f - T^k f|^2)^{1/2}\|_2^2 \leq m_\alpha^2 \|f\|_2^2.$$

Then by partial summation

$$T^{N+1} = (1/N)\sum_{k=1}^{N} T^k + (1/N)\sum_{k=1}^{N} k(T^{k+1} - T^k),$$

and so

$$
\begin{aligned}
\| \sup_{N\geq 1} |T^{N+1} f| \|_2 \leq \| \sup_{N\geq 1} |(1/N) \sum_{k=1}^{N} k(T^{k+1} - T^k f)| \|_2 \\
+ \| \sup_{N\geq 1} (1/N) |\sum_{k=1}^{N} T^k f| \|_2 \\
\leq \| (\sum_{k=1}^{\infty} k |T^{k+1} f - T^k f|^2)^{1/2} \|_2 \\
+ C \|f\|_2 \\
\leq C_\alpha \|f\|_2 .
\end{aligned}
$$

This gives the maximal inequality in $L_2(X)$. A similar calculation gives the maximal inequality

$$
\| \sup_{n\geq r} n^r |\Delta^r T^n f| \|_2 \leq C_{r,\alpha} \|f\|_2
$$

for the r th difference operator $\Delta^r T^n$. The rest of the argument proceeds identically with Stein [12] by using his complex interpolation theorem. □

Remark. Theorem 14 is closely related to the result of Gaposhkin [9]. The proof of Theorem 14 gives another argument for his main theorem: *if* T *is a normal* L_2 *contraction such that* $sp(T) \subset S_\alpha$ *for* *some Stolz region* $S_\alpha, 0 \leq \alpha < \pi$ *, then for* $f \in L_2(X)$*,* $T^N f(x) - \frac{1}{N}\sum_{k=1}^{N} T^k f(x)$ *converges to* 0 *a.e.*. To see this notice that with these hypotheses, the first part of the proof of Theorem 14 shows that if $f = f_1 + (f_2 - Tf_2)$ where $f_1, f_2 \in L_2(X)$, $Tf_1 = f_1$, then $\lim_{N\to\infty} T^{N+1} f(x) = f_1(x)$ a.e.. Hence, for such f, $\lim_{N\to\infty} T^{N+1} f(x) - (1/N)\sum_{k=1}^{N} T^k f(x) = f_1(x) - f_1(x) = 0$ a.e.. But the second part of the proof shows

$$\| \sup_{N\geq 1} |T^{N+1} f - \frac{1}{N}\sum_{k=1}^{N} T^k f| \, \|_2 \leq m_\alpha \|f\|_2 .$$

Thus, $T^{N+1} f(x) - \frac{1}{N}\sum_{k=1}^{N} T^k f(x) \to 0$ a.e. for all $f \in L_2(X)$, and so $T^{N+1} f(x) - \frac{1}{N+1}\sum_{k=1}^{N+1} T^k f(x) = (N/(N+1))(T^{N+1} f(x) - (1/N)\sum_{k=1}^{N} T^k f(x)) \to 0$ a.e. for all $f \in L_2(X)$. This method of proof of Gaposhkin's Theorem is a direct generalization of the method of Stein, which is appropriate because Gaposhkin's theorem generalizes Stein's theorem. Compare this argument with the one in [9].

To see applications for these results, consider this definition.

Definition. A probability measure μ on Z has *bounded angular ratio* if μ is strictly aperiodic and for some $\alpha, 0 \leq \alpha < \pi, \hat{\mu}(\{\lambda \in \mathbb{C} : |\lambda| = 1\}) \subset \mathcal{S}_\alpha$.

That is, μ has bounded angular ratio if $|\hat{\mu}(\lambda)| < 1$ for $|\lambda| = 1, \lambda \neq 1$, and $\sup_{|\lambda|=1,\lambda\neq 1} |\hat{\mu}(\lambda) - 1|/(1 - |\hat{\mu}(\lambda)|) < \infty$.

Now let T_0 be a normal L_1 - L_∞ contraction. For instance, $T_0 f = f \circ \tau$ is a unitary operator of this type. Let $T = \mu T_0 = \sum_{k=1}^{\infty} \mu(k) T_0^k$ for some probability measure μ with $supp(\mu) \subset Z^+$. Then T is a normal $L_1 - L_\infty$ contraction. But μ having bounded angular ratio makes $sp(T) \subset \mathcal{S}_\alpha$ for some α, $0 \leq \alpha < \pi$. Thus Theorem 14 gives this corollary.

15. Corollary. *If T_0 is any normal $L_1 - L_\infty$ contraction and μ is a probability measure with $supp(\mu) \subset Z^+$ such that μ has bounded angular ratio, then $T = \mu T_0$ satisfies $(T^n f(x))$ converges a.e. for all $f \in L_p(X), 1 < p \leq \infty$, and $\|\sup_{n\geq 1} T^n f\|_p \leq c_p \|f\|_p, 1 < p \leq \infty$.*

Remark. a) When $supp(\mu)$ is not restricted to Z^+, the above is true if T_0 is invertible and T_0^{-1} is also an $L_1 - L_\infty$ contraction.

Remark. b) Gaposhkin's theorem can be used above to get a result in $L_2(X)$ when T_0 is not known to be an $L_1 - L_\infty$ contraction. For example, if T_0 is an L_2 contraction, with $T_0 f \geq 0$ when $f \geq 0$, then by the well-known theorem of Akcoglu, $\lim_{N\to\infty} (1/N) \sum_{k=1}^{N} T_0^k f(x)$ exists a.e. for all $f \in L_2(X)$. So for a probability measure μ with supp $(\mu) \subset Z^+$ such that μ has bounded angular ratio, the operator $T = \mu T_0$ satisfies $(T^n f(x))$ converges a.e. for all $f \in L_2(X)$.

Example. Let $h(z) = 1 - (1-z)^{1/2} = \sum_{n=1}^{\infty} c_n z^n$. Then $c_n \geq 0$ and $\sum_{n=1}^{\infty} c_n = 1$. See Brunel and Keane [5] where this series plays an important role. Let T_0 be a normal $L_1 - L_\infty$ contraction. Because $\mu = \sum_{n=1}^{\infty} c_n \delta_n$ has bounded angular ratio, the operator $T = \sum_{h=1}^{\infty} c_n T_0^n$ satisfies the unaveraged convergence Corollary 15 above. Examples like this show how far from any symmetry in T, as assumed in Stein [13], we can deviate and still obtain an a.e. convergence theorem for (T^n).

The behavior of the specific instance of $Tf = \mu f$ for a.e. convergence of the powers can be fairly thoroughly analyzed to determine both necessary and sufficient conditions for a.e. convergence of

$(\mu^n f(x))$ for all $f \in L_p(X), 1 \leq p \leq \infty$. See Bellow, Jones, and Rosenblatt [3] for the details. Theorem 14 above and the remark following it show

16. Theorem. *If* μ *is strictly aperiodic,* $m_2(\mu) = \sum_{k=-\infty}^{\infty} k^2 \mu(k) < \infty$ *and* $E(\mu) = \sum_{k=-\infty}^{\infty} k\mu(k) = 0$, *then for all* $f \in L_p(X), 1 < p \leq \infty$, $\lim_{n \to \infty} \mu^n f(x) = \int f dm$ *a.e..*

Proof. The conditions on μ guarantee that $Tf = \mu f$ is a normal $L_1 - L_\infty$ contraction such that $sp(T) \subset \mathcal{S}_\alpha$ for some α. □

On the other hand, the Central Limit Theorem and the same arguments in [1] used for the version there of Theorem 1 above show this companion to Theorem 16 (see [3] for details).

17. Theorem. *If* μ *has* $m_2(\mu) < \infty$ *and* $E(\mu) \neq 0$, *then there exists* $E \in \beta$ *such that* $(\mu^n 1_E(x))$ *fails to converge on some set of positive measure. Indeed, there is a dense* G_δ *subsets* $\mathcal{R}$ *of* β *such that for* $E \in \mathcal{R}, \limsup_{n\to\infty} \mu^n 1_E(x) = 1$ *a.e. and* $\liminf_{n\to\infty} \mu^n 1_E(x) = 0$ *a.e..*

This result can be carried further using a theorem of Bourgain [4] (see [3] again).

18. Theorem. *If* $\mu \neq \delta_0$ *and* $\lim_{\lambda \to 1} \frac{|\hat{\mu}(\lambda)-1|}{1-|\hat{\mu}(\lambda)|} = \infty$, *then there is* $E \in \beta$ *such that* $(\mu^n 1_E(x))$ *fails to exist on some set of positive measure.*

This theorem, and the relationship of the existence of a maximal inequality for $(\mu^n f)$ if $f \in L_2(X)$ to the recurrence of the random walk on Z determined by the probability measure μ, are both discussed in [3].

In conclusion, the most interesting unresolved problems related to the above theory are ones concerning the behavior of $(T^n f)$ for $f \in L_1(X)$. For instance, here are two questions:

1) If μ is strictly aperiodic, finitely-supported, and $E(\mu) = 0$, does $\lim_{n \to \infty} \mu^n f(x)$ exist a.e. for all $f \in L_1(X)$? What happens if μ is even symmetric?

2) If μ is strictly aperiodic and finitely-supported, does there exist (n_m) such that $\lim_{n \to \infty} \mu^{n_m} f(x) = \int f dm$ a.e. for all $f \in L_1(X)$?

References

1. A. Bellow R. Jones, and J. Rosenblatt, Convergence of moving averages, preprint, 33 pages, to appear in Ergodic Theory and Dynamical Systems.

2. A. Bellow, R. Jones, and J. Rosenblatt, Almost everywhere convergence of weighted averages, preprint, 34 pages.

3. A. Bellow, R. Jones and J. Rosenblatt, Almost everywhere convergence of convolution powers, preprint.

4. J. Bourgain, Almost sure convergence and bounded entropy, Israel J. Math. 62 (1988), to appear.

5. A. Brunel and M. Keane, Sur les operateurs positifs a moyennes bornees, Comptes Rendu Acad. Sci. Paris I 298 (1984) 103-106.

6. D. Burkholder and Y. Chow, Iterates of conditional expectation operators, Proc. AMS 12 (1961) 490-495.

7. A. del Junco and J. Rosenblatt, Counterexamples in ergodic theory and number theory, Math. Annalen 245 (1979) 185-197.

8. N. Dunford and J. Schwartz, Linear Operators, Vol. I, John Wiley and Sons, New York, 1966.

9. V. F. Gaposhkin, Tauberian ergodic theorem for normal contraction operators, Mathematical Notes 35,(1984) 208-212.

10. S. Horowitz, Pointwise convergence of the iterates of a Harris-recurrent Markov operator, Israel J. Math. 33 (1979) 177-180.

11. J. Rosenblatt, Ergodic group actions, Arch. Math. 47 (1986) 263-269.

12. G-C.Rota, An "Alternierende Verfahren" for general positive operators, Bull. AMS 68 (1962) 95-102.

13. E. Stein, On the maximal ergodic theorem, Proc. Nat. Acad. Sci. 47 (1961) 1894-1897.

14. M. A. Akcoglu and L. Sucheston, Pointwise convergence of alternating sequences, Can. J. Math. 40 (1988) 610-632.

Independence Properties of Continuous Flows

V. Bergelson

The Ohio State University, Department of Mathematics

Columbus, Ohio, USA

Abstract

If $\{U_\tau, -\infty < \tau < \infty\}$ is a one-parameter ergodic continuous flow of unitary operators acting on a separable Hilbert space then for all but countably many τ U_τ is ergodic. We show that in measure preserving situation this well known fact can be strengthened: if $\{T_\tau, -\infty < \tau < \infty\}$ is a continuous measure preserving ergodic flow acting on a probability Lebesgue space $(X, \mathcal{B}, \mu)$ then for any $k \in \mathbb{N}$ there exists a set $W \subset \mathbb{R}^k$ of full measure such that for any $f_1, f_2, \ldots, f_k \in L^\infty(X, \mathcal{B}, \mu)$ and for any $(a_1, a_2, \ldots, a_k) \in W$ one has:

$$\lim_{N\to\infty} \left\| \frac{1}{2N+1} \sum_{n=-N}^{N} T_{a_1}^n f_1 T_{a_2}^n f_2 \ldots T_{a_k}^n f_k - \prod_{i=1}^{k} \int f_i d\mu \right\|_{L^2} = 0.$$

As an application we show that if $\{T_\tau\}$ is a continuous (not necessarily ergodic) measure-preserving flow on a probability Lebesgue space then for any $A \in \mathcal{B}$ with $\mu(A) > 0$ and almost every $(a_1, a_2, \ldots, a_k) \in \mathbb{R}^k$ one has:

$$\lim_{N\to\infty} \frac{1}{2N+1} \sum_{n=-N}^{N} \mu(A \cap T_{a_1}^n \cap T_{a_2}^n A \cap \ldots \cap T_{a_k}^n A) \geq (\mu(A))^{k+1},$$

and the equality takes place if and only if $\{T_\tau\}$ is ergodic flow. Finally, non-linear versions of these results are discussed.

ISBN 0-12-231050-0

1. Introduction

A unitary operator U acting on a Hilbert space H is called ergodic if it has no nontrivial invariant vectors: $Uf = f$ implies $f = 0$.

Similarly, if $\{U_\tau = -\infty < \tau < \infty\}$ is a one-parameter flow of unitary operators, then $\{U_\tau\}$ is called ergodic if there are no non-trivial inrariant vectors: $U_\tau f = f$ for all $\tau \in \mathbb{R}$ implies $f = 0$.

It is not hard to construct an ergodic U such that for any nonzero integer n different from ± 1 U^n will not be ergodic. The following (apparently well known) proposition shows that the situation is different in case of an ergodic continuous flow $\{U_\tau\}$ acting on a separable Hilbert space: most of the individual U_τ have to be ergodic. More precisely,

Proposition (cf. [CFS], Lemma 1, §1. Chapter 12). If $\{U_\tau\}$ is a one-parameter ergodic flow of unitary operators on a separable Hilbert space H, then for all but countably many $\tau \in \mathbb{R}$ U_τ is ergodic.

Proof. Let Γ be the set of the eigenfrequencies of the flow $\{U_\tau\}$: $\lambda \in \Gamma$ if and only if there exists $f \in H$ such that $f \neq 0$ and for all $\tau \in \mathbb{R}$ $U_\tau f = e^{i\lambda\tau} f$. Notice that since H is separable, Γ is at most countable.

Let $H_1 \subset H$ be the subspace spanned by all eigenvectors of $\{U_\tau\}$ and let H_2 be the ortho-complement of H_1 in H.

Since Γ is at most countable, all but countably many $\tau_0 \in \mathbb{R}$ satisfy the condition that for all $\lambda \in \Gamma$ $e^{i\lambda\tau_0} \neq 1$ ($0 \notin \Gamma$ since $\{U_\tau\}$ is ergodic). We will show that for any such τ_0 U_{τ_0} is ergodic. Suppose that $f \in H$ is such that $U_{\tau_0} f = f$. Then $f \in H_1$. Let $f = \sum c_n f_n$, where $\{f_n\}$ is an orthonormal base of H_1 made of eigenvectors of $\{U_\tau\}$. Let $\{\mu_n\}$ be the corresponding eigenfrequencies: $U_\tau f_n = e^{i\mu_n\tau} f_n$ for all τ and all n. Notice that since $\{U_\tau\}$ is an ergodic flow, $\mu_n \neq 0$ for all n. Since $U_{\tau_0} f = f$, we have:

$$c_n = e^{i\mu_n\tau_0} c_n \quad \text{for all } n.$$

Since $e^{i\mu_n\tau_0} \neq 1$ for all n, we get $c_n = 0$ for all n, which implies $f = 0$. □

Let now $\{T_\tau, -\infty < \tau < \infty\}$ be a continuous measure preserving flow on a Lebesgue probability space $(X, \mathcal{B}, \mu)$. It follows from the above Proposition that if $\{T_\tau\}$ is an ergodic flow then for all but countably many τ T_τ is an ergodic measure preserving transformation. The following theorem, the proof of which will be given in the next section is a strengthening of this fact.

Theorem 1. Suppose that $\{T_\tau\}$ is a continuous measure preserving flow on a probability Lebesgue space $(X, \mathcal{B}, \mu)$. Fix $k \in \mathbb{N}$. Then for almost any k-tuple $(a_1, a_2, \ldots, a_k) \in \mathbb{R}^k$ the following is true for any $f_1, f_2 \ldots, f_k \in L^\infty(X, \mathcal{B}, \mu)$:

(1) $$\lim_{N\to\infty} \left\| \frac{1}{2N+1} \sum_{n=-N}^{N} T_{a_1}^n f_1 T_{a_2}^n f_2 \ldots T_{a_k}^n f_k - \prod_{i=1}^{k} \int f_i d\mu \right\|_{L^2} = 0$$

Remarks. (i) It will be clear from the proof that the k-tuples $(a_1, a_2, \ldots, a_k)$ for which (1) holds lie outside of at most countable union of hyperplanes in $\mathbb{R}^k$.

(ii) An even stronger result is true when $\{T_\tau\}$ is a weakly mixing flow. In such a case (1) holds for all $(a_1, a_2, \ldots, a_k) \in \mathbb{R}^k$ such that $a_i \neq 0, i = 1, 2, \ldots, k$ and $a_i \neq a_j$, $i \neq j$, $i, j = 1, 2, \ldots, k$.

As a corollary of Theorem 1 one gets the following

Theorem 2. Let $\{T_\tau\}$ be a continuous measure-preserving (not necessarily ergodic) flow on a Lebesgue probability space $(X, \mathcal{B}, \mu)$. Then for any $k \in \mathbb{N}$ and any $A \in \mathcal{B}, \mu(A) > 0$ there exists a set $W \subset \mathbb{R}^k$ having full Lebesgue measure (i.e. the measure of the complement of W in $\mathbb{R}^k$ is zero) such that for any $(a_1, a_2, \ldots, a_k) \in W$ one has:

$$\lim_{N\to\infty} \frac{1}{2N+1} \sum_{n=-N}^{N} \mu(A \cap T_{a_1}^n A \cap T_{a_2}^n \cap \ldots \cap T_{a_k}^n A) \geq (\mu(A))^{k+1}$$

The proofs of Theorems 1 and 2 are given in the next section. Section 3 is devoted to comments and further discussion.

2. Proofs of Theorems 1 and 2

Proof of Theorem 1. We will use some definitions and results from [BB].

Measure preserving transformations $T_1, T_2, \ldots, T_k$ acting on a probability space $(X, \mathcal{B}, \mu)$ are called jointly ergodic if for any $f_1, f_2, \ldots, f_k \in L^\infty(X, \mathcal{B}, \mu)$ one has:

$$\lim_{N\to\infty} \|\frac{1}{N}\sum_{n=1}^{N} T_1^n f_1 T_2^n f_2 \ldots T_k^n f_k - \prod_{i=1}^{k}\int f_i d\mu\|_{L^2} = 0$$

It was proved in [BB] that if $T_iT_j = T_jT_i$ for all $i, j = 1, 2, \ldots k$ then $T_1, T_2, \ldots, T_k$ are jointly ergodic if and only if the cartesian product $T_1 \times T_2 \times \ldots \times T_k$ (acting on the k-th cartesian power of X) and all the transformations $T_iT_j^{-1}$, $1 \le i < j \le k$ (acting on X) are ergodic.

It follows that the set of k-tuples $(a_1, a_2, \ldots, a_k)$ for which Theorem 1 holds coincides with the set of k-tuples such that

(i) $T_{a_1} \times T_{a_2} \times \ldots \times T_{a_k}$

and (ii) all $T_{a_i - a_j}$, $\quad 1 \le i < j \le k$

are ergodic.

Let us examine conditions (i) and (ii) closer. Given an ergodic measure preserving transformation T denote by $\Lambda(T)$ the group of its eigenvalues. It is well known that the cartesian product $T_1 \times T_2 \times \ldots \times T_k$ of the ergodic measure preserving transformations $T_1, T_2 \ldots, T_k$ is ergodic if and only if $\bigcap_{1\le i\le k} \Lambda(T_i) = \{1\}$. In our case $T_i = T_{a_i}$ and it is clear that $T_{a_1} \times T_{a_2} \times \ldots \times T_{a_k}$ will be ergodic if and only if $\gamma_1 a_1 + \gamma_2 a_2 + \ldots + \gamma_k a_k \neq 2\pi n$ for any $\gamma_1, \ldots, \gamma_k \in \Gamma$, $\sum_{i=1}^{k} \gamma_i^2 \neq 0$ and $n \in \mathbb{Z}$, where Γ is the countable group of eigenfrequencies of the flow $\{T_\tau\}$. But this takes place if and only if the k-tuple $(a_1, a_2, \ldots, a_k)$ does not belong to the (countable) union of hyperplanes defined by the equations $\sum_{i=1}^{k} \gamma_i x_i = 2\pi n$, where $\gamma_1, \gamma_2, \ldots, \gamma_k \in \Gamma$, $\sum_{i=1}^{k} \gamma_i^2 \neq 0$, $n \in \mathbb{Z}$.

As for the condition (ii) it is clear that it is satisfied if $e^{i\gamma(a_i - a_j)} \neq 1 \quad \forall \gamma \in \Gamma, \gamma \neq 0$, or $\gamma(a_i - a_j) \neq 2\pi k, \quad k \in \mathbb{Z}, \quad \forall \gamma \in \Gamma, \quad \gamma \neq 0, \quad 1 \le i < j \le k$. Theorem 1 is proved.

Proof of Theorem 2. To treat the case of possibly nonergodic flow we appeal to the theorem on ergodic decomposition (see for example, [R]). According to this theorem, we may assume that X is a disjoint union of sets X_ω each equipped with a σ -algebra $\mathcal{B}_\omega$, and a probability measure μ_ω, such that $\{T_\tau\}$ acts on each $(X_\omega, \mathcal{B}_\omega, \mu_\omega)$ ergodically. Moreover, since $(X, \mathcal{B}, \mu)$ is by assumption a Lebesgue space, each of $(X_\omega, \mathcal{B}_\omega, \mu_\omega)$ is also Lebesgue. The indexing set is another Lebesgue probability space $(\Omega, \mathcal{F}, \nu)$, such that

$$\mu = \int\limits_{\Omega} \mu_\omega d\nu(\omega)$$

(the latter formula means that for any $f \in L^1(X, \mathcal{B}, \mu)$ $\int\limits_X f d\mu = \int\limits_\Omega \int\limits_{X_\omega} f d\mu_\omega d\nu(\omega))$.

Now let $A \in \mathcal{B}, \mu(A) > 0$. We have:

$$\frac{1}{2N+1} \sum_{n=-N}^{N} \mu(A \cap T_{a_1}^n A \cap T_{a_2}^n A \cap \ldots \cap T_{a_k}^n A)$$

$$= \frac{1}{2N+1} \sum_{n=-N}^{N} \int\limits_{\Omega} \mu_\omega(A \cap T_{a_1}^n \cap T_{a_1}^n \cap T_{a_2}^n \cap \ldots \cap T_{a_k}^n A) d\nu(\omega)$$

$$= \int\limits_{\Omega} \left(\frac{1}{2N+1} \sum_{n=-N}^{N} \mu_\omega(A \cap T_{a_1}^n A \cap T_{a_2}^n A \cap \ldots \cap T_{a_k}^n A) \right) d\nu(\omega).$$

Denote

$$f_{N,\bar{a}}(\omega) = \frac{1}{2N+1} \sum_{n=-N}^{N} \mu_\omega(A \cap T_{a_1}^n A \cap T_{a_2}^n A \cap \ldots \cap T_{a_k}^n A),$$

where $\bar{a} = (a_1, a_2, \ldots, a_k)$.

Since the measures μ_ω are ergodic, and, since the strong convergence implies the weak one, it follows from Theorem 1 that for every ω there exists a set $W_\omega \subset \mathbb{R}^k$ such that W_ω has full measure and for every $\bar{a} = (a_1, a_2, \ldots, a_k) \in W_\omega$

$$(2) \qquad f_{N,\bar{a}}(\omega) \to (\mu_\omega(A))^{k+1}$$

It follows that (2) takes place for a set of pairs $(\omega, \bar{a})$ which has full measure in $\Omega \times \mathbb{R}^k$. Hence for almost every $\bar{a} \in \mathbb{R}^k$ there exists a set $S_{\bar{a}} \subset \Omega$ such that $S_{\bar{a}}$ has full measure and for every $\omega \in S_{\overline{a}}$ the formula (2) holds. Let W be the set of all such $\overline{a}$. For any fixed $\overline{a} \in W$ integration of (2) gives:

$$\int_{\Omega} \lim_{N\to\infty} f_{N,\overline{a}}(\omega)d\nu(\omega) = \int_{\Omega} (\mu_\omega(A))^{k+1} d\nu(\omega)$$

$$\geq \left(\int_{\Omega} \mu_\omega(A)d\nu(\omega)\right)^{k+1} = (\mu(A))^{k+1}$$

But

$$\int_{\Omega} \lim_{N\to\infty} f_{N,\overline{a}}(\omega)d\nu(\omega) = \lim_{N\to\infty} \int_{\Omega} f_{N,\overline{a}}(\omega)d\nu(\omega)$$

$$= \lim_{N\to\infty} \frac{1}{2N+1} \sum_{n=-N}^{N} \int_{\Omega} \mu_\omega(A \cap T_{a_1}^n A \cap T_{a_2}^n A \cap \ldots \cap T_{a_k}^n A)d\nu(\omega)$$

$$= \lim_{N\to\infty} \frac{1}{2N+1} \sum_{n=-N}^{N} \mu(A \cap T_{a_1}^n A \cap T_{a_2}^n A \cap \ldots \cap T_{a_k}^n A)$$

Thus $\lim_{N\to\infty} \frac{1}{2N+1} \sum_{n=-N}^{N} \mu(A \cap T_{a_1}^n A \cap T_{a_2}^n A \cap \ldots \cap T_{a_k}^n A) \geq (\mu(A))^{k+1}$. The proof of Theorem 2 is completed.

3. Comments and further refinements

Both Theorem 1 and Theorem 2 can be considered as statements intertwining almost everywhere convergence and multiple recurrence. Before embarking on further discussion let us give a slight reformulation of Theorem 2.

Theorem 3. Let $\{T_\tau, -\infty < \tau < \infty\}$ be a continuous measure preserving flow on a Lebesgue probability space $(X, \mathcal{B}, \mu)$. Let $k \in \mathbb{N}$ and $A \in \mathcal{B}, \mu(A) > 0$. For $\overline{a} = (a_1, a_2, \ldots, a_k) \in \mathbb{R}^k$ denote

$$\mu(A \cap T_{a_1}^n A \cap T_{a_2}^n A \cap \ldots \cap T_{a_k}^n A) = f_n(\overline{a}), \quad n \in \mathbb{Z}.$$

Then for almost every $\overline{a} \in \mathbb{R}^k$ one has:

(3) $$\lim_{N\to\infty} \frac{1}{2N+1} \sum_{n=-N}^{N} f_n(\overline{a}) \geq (\mu(A))^{k+1}$$

The equality takes place if and only if $\{T_\tau\}$ is ergodic. Theorem 3 bears obvious resemblence to the following theorem due to Khintchine.

Theorem 4 ([K]). Let $(X, \mathcal{B}, \mu, T)$ be a measure preserving system with $\mu(X) = 1$. Then for any $A \in \mathcal{B}$ one has:

$$\lim_{N\to\infty} \frac{1}{2N+1} \sum_{n=-N}^{N} \mu(A \cap T^n A) \geq (\mu(A))^2$$

The equality takes place if and only if T is ergodic.

On the other hand, the closest relative of Theorem 3 in the class of $\mathbf{Z}$-actions is the following theorem due to Furstenberg, which is a key element in his proof of Szemeredi's theorem on arithmetic progressions (see [F]).

Theorem 5 (Furstenberg). Let $(X, \mathcal{B}, \mu, T)$ be a weakly mixing system. Then for any $k \in \mathbb{N}$, any pairwise distinct $a_1, a_2, \ldots, a_k \in \mathbf{Z}\backslash\{0\}$ and any $f_1, f_2, \ldots, f_k \in L^\infty(X, \mathcal{B}, \mu)$ one has

$$\text{(4)} \qquad \lim_{N\to\infty} \left\|\frac{1}{2N+1} \sum_{n=-N}^{N} T^{a_1 n} f_1 T^{a_2 n} f_2 \ldots T^{a_k n} f_k - \prod_{i=1}^{k} \int f_i d\mu\right\|_{L^2} = 0.$$

There exists however one important difference between Theorem 3 and Theorem 5. Theorem 2 shows that high degree of independence is inherent in any ergodic dynamical system with continuous time. Since ergodicity is in a sense the mildest possible independence assumption which one can make about a dynamical system, Theorem 2 leads to a rather general Khintchine type theorem (Theorems 2 or 3) which is valid for any dynamical system with continuous time.

On the other hand the high degree of independence which is described by Theorem 5 is possible only for weakly mixing $\mathbf{Z}$-actions. As a matter of fact, a dynamical system $(X, \mathcal{B}, \mu, T)$ is weakly mixing if and only if (4) holds for any $f_1, f_2, \ldots, f_k \in L^\infty(X, \mathcal{B}, \mu)$ (where $k \geq 2$). Notice also that replacing in the formulation of Theorem 4 the condition: "for any pairwise distinct $a_1, a_2, \ldots, a_k \in \mathbf{Z}\backslash\{0\}$" by a condition: "for almost all k-tuples $(a_1, a_2, \ldots, a_k) \in \mathbf{Z}^k$" (with any reasonable notion of "almost all" – say in sense of density) will not change the situation. Indeed, one can easily construct an ergodic system $(X, \mathcal{B}, \mu, T)$ such that for any $a \in \mathbf{Z},\ a \neq \pm 1$

the system $(X, \mathcal{B}, \mu, T^a)$ is not ergodic. Thus for any such $(X, \mathcal{B}, \mu, T)$ the formula (4) will hold only if $k = 1$ and $a_1 = \pm 1$.

One of the reasons for so nice independence properties of continuous flows is that continuous actions contain a lot of subactions – discrete and continuous, whereas $\mathbb{Z}$-actions have very few (only countably many) subactions.

Independence properties of continuous flows are further illustrated by the following continuous versions of Theorems 1 and 2.

Theorem 6. Let $\{T_\tau, -\infty < \tau < \infty\}$ be a continuous ergodic measure preserving flow on a probability Lebesgue space $(X, \mathcal{B}, \mu)$. Then for any $k \in \mathbb{N}$ there exists a set $W \subset \mathbb{R}^k$ which is the complement of at most countable family of hyperplanes such that for any $(a_1, a_2, \ldots, a_k) \in W$ and any $f_1, f_2, \ldots, f_k \in L^\infty(X, \mathcal{B}, \mu)$ one has

$$\lim_{t\to\infty} \|\frac{1}{2t}\int_{-t}^{t} T^{a_1\tau} f_1 T^{a_2\tau} f_2 \ldots T^{a_k\tau} f_k d\tau - \prod_{i=1}^{k} \int f_i d\mu\|_{L^2} = 0$$

Theorem 7. Let $\{T_\tau, -\infty < \tau < \infty\}$ be a continuous measure preserving flow on a probability Lebesgue space $(X, \mathcal{B}, \mu)$. Let $k \in \mathbb{N}$ and $A \in \mathcal{B}, \mu(A) > 0$. For $\overline{a} = (a_1, a_2, \ldots, a_k) \in \mathbb{R}^k$ denote

$$\mu(A \cap T^{a_1\tau} A \cap T^{a_2\tau} A \cap \ldots \cap T^{a_k\tau} A) = f_\tau(\overline{a}), \tau \in \mathbb{R}.$$

Then for almost every $\overline{a} \in \mathbb{R}^k$

$$\lim_{t\to\infty} \frac{1}{2t}\int_{-t}^{t} f_\tau(\overline{a}) d\tau \geq (\mu(A))^{k+1}$$

The equality takes place if and only if $\{T_\tau\}$ is ergodic.

The proofs of Theorems 6 and 7 are similar to those of 1 and 2. The conditions for joint ergodicity of commuting $\mathbb{R}$-actions which are needed for the proof of Theorem 6 are analogous to those for $\mathbb{Z}$-actions (cf. [BRo], where joint ergodicity criterion is given for commuting actions of amenable groups).

The analogy between the multiple recurrence properties of continuous ergodic $\mathbb{R}$-actions and weakly mixing $\mathbb{Z}$-actions extends, with certain restrictions, further to non-linear theorems.

This becomes clear from the comparison of the following two PETs (Polynomial Ergodic Theorems). We will say that the polynomials $p(t)$ and $q(t)$ are essentially distinct if $p(t) - q(t) \not\equiv$ const.

Theorem 8 ([B]). Suppose that $(X, \mathcal{B}, \mu, T)$ is a weakly mixing system and let $p_1(t), p_2(t), \dots, p_k(t)$ be pairwise essentially distinct polynomials with rational coefficients taking on integer values on the integers. Then for any $f_1, f_2, \dots, f_k \in L^\infty(X, \mathcal{B}, \mu)$ one has

$$\lim_{N\to\infty} \|\frac{1}{2N+1} \sum_{n=-N}^{N} T^{p_1(n)}f_1 T^{p_2(n)}f_2 \dots T^{p_k(n)}f_k - \prod_{i=1}^{k} \int f_i d\mu\|_{L^2} = 0.$$

Theorem 9. Let $\{T_\tau, -\infty < \tau < \infty\}$ be a continuous ergodic measure preserving flow on a probability Lebesgue space $(X, \mathcal{B}, \mu)$. Let $p_1(t), p_2(t), \dots, p_k(t), t \in \mathbb{R}$ be pairwise essentially distinct and linearly independent polynomials. Then for any $f_1, f_2, \dots, f_k \in L^\infty(X, \mathcal{B}, \mu)$ one has

$$\lim_{t\to\infty} \|\frac{1}{2t} \int_{-t}^{t} T^{p_1(\tau)}f_1 T^{p_2(\tau)}f_2 \dots T^{p_k(\tau)}f_k d\tau - \prod_{i=1}^{k} \int f_i d\mu\|_{L^2} = 0.$$

Again, one can derive from Theorem 9 a Khintchine type theorem, analogous to Theorem 7. The proof of Theorem 9 will appear elsewhere.

The phenomena discussed in this paper for $\mathbb{R}$- and $\mathbb{Z}$-actions are quite general and deserve perhaps more careful study. Whereas continuous actions of connected locally compact non-compact groups seem to exhibit more or less the same kind of independence, the situation with actions of discrete groups is more complicated. For example Dan Rudolph and the author were able to show that weakly mixing actions of groups like $\mathbb{Z}_p^\infty$ (direct sum of countably many copies of cyclic group of prime order) always have subactions which are Bernoulli ([BRu]). There is no counterpart to this theorem within the class of $\mathbb{Z}$-actions.

References

[B] V. Bergelson. Weakly mixing PET. Erg. Theory and Dyn. Systems 7, 1987, 337-349.

[BB] D. Berend & V. Bergelson. Jointly ergodic measure preserving transformations. Israel Journal of Math. 49, 1984, 307-314.

[BRo] V. Bergelson & J. Rosenblatt. Joint ergodicity for group actions. Erg. Theory and Dyn. Systems 8, 1988, 351-364.

[BRu] V. Bergelson & D. Rudolph. Weakly mixing actions of F^{∞} have infinite subgroup actions which are Bernoulli. Dynamical Stytems. Proceedings, University of Maryland 1986-87. Springer Lecture Notes #1342, Springer, 1988, 7-22.

[CFS] I. Cornfeld, S. Fomin, Y. Sinai. Ergodic Theory, Springer, 1982.

[F] H. Furstenberg. Ergodic behavior of diagonal measures and a theorem of Szemeredi on arithmetic progressions. J. d'Analyse Math. 31, 1977, 204-256.

[K] A. Khintchine. Eine Vershärfung des Poincaréschen "Wiederkehrsatzes." Comp. Math. 1, 1934, 177-179.

[R] V. Rokhlin. Selected topics from the metric theory of dynamical systems. Amer. Math. Soc. Transl. Series 2, 49, 1966, 171-209.

MOVING AVERAGES

N. H. BINGHAM

Mathematics Department, Royal Holloway & Bedford New College,

Egham Hill, Egham, Surrey TW20 0EX, England

§ 1. Laws of large numbers.

Throughout, write $X, X_1, X_2, \ldots$ for independent and identically distributed (iid) random variables, S_n for $\Sigma_1^n X_k$.

We begin with the classical Kolmogorov strong law of large numbers:

$$E(|X|) < \infty \ \& \ EX = \mu \Leftrightarrow \frac{1}{n} \Sigma_1^n X_k \to \mu \quad \text{a.s.}$$

The question underlying the present work is how one may refine this theorem, given more information on the law of X.

We recall two classical answers to this question. The first is the strong law of Marcinkiewicz & Zygmund (1937): for $0 < p < 2$, there exists c with

$$(S_n - nc)/n^{1/p} \to 0 \quad \text{a.s.}$$

iff $X \in L^p$ (and then we can take c as 0 if $0 < p < 1$, EX if $1 \leq p < 2$). Observe that $p < 2$ is forced here by the central limit theorem. This drawback is avoided by the second classical answer, the strong law of Baum & Katz (1965): take $\alpha > \frac{1}{2}$, $\alpha p \geq 1$, $EX = 0$ if $\alpha \leq 1$ (for instance, $\alpha = 1$, $p \geq 1$, $EX = 0$). The following are equivalent:

ISBN 0-12-231050-0

(i) $X \in L^p$,

(ii) $\Sigma\, n^{\alpha p-2}\, P(|S_n| > \varepsilon n^{\alpha}) < \infty \quad \forall \varepsilon > 0$,

(iii) $\Sigma\, n^{\alpha p-2}\, P(\max_{k\le n} |S_k| > \varepsilon n^{\alpha}) < \infty \quad \forall \varepsilon > 0$.

Here (iii) may be rephrased, for $\alpha p > 1$, as

(iii') $S_n/n^{\alpha} \to 0 \quad (\alpha p-1)$-quickly

in the language of Strassen and Lai; see e.g. Bingham (1986a) for background and discussion.

§2. Cesàro methods.

In both the Marcinkiewicz-Zygmund and Baum-Katz strong laws, μ does not appear conveniently on the right-hand side as it does in the Kolmogorov strong law. One may thus ask for a strong law for L^p $(p \geq 1)$ sharing the structure of the Kolmogorov law.

To this end, observe that Kolmogorov's strong law links membership of L^1 with almost-sure Cesàro convergence of sample means. Recall (Hardy (1949)) the classical Cesàro summability methods C_α, $\alpha > 0$ ($\alpha > -1$ is possible, but we shall not need this):

$$s_n \to s \qquad (C_\alpha)$$

means

$$\frac{1}{A_n^{\alpha}} \Sigma_0^n A_{n-k}^{\alpha-1} \to s,$$

where

$$A_n^{\alpha} := (\alpha+1)..(\alpha+n)/n! \sim n^{\alpha}/\Gamma(1+\alpha) \qquad (n \to \infty).$$

If $0 < \alpha < \beta$, $C_\alpha \subset C_\beta$: $s_n \to s$ (C_α) implies $s_n \to s$ (C_β) (Hardy (1949), Th. 43). So convergence under C_α is stronger than the Kolmogorov convergence under $C = C_1$ if $\alpha < 1$, weaker if $\alpha > 1$.

The key rôle of the Kolmogorov strong law is reflected in a discontinuity of the behaviour of C_α across $\alpha = 1$:

THEOREM 1. (i) For $0 < \alpha \leq 1$,

$$X_n \to \mu \quad \text{a.s.} \quad (C_\alpha) \Leftrightarrow E(|X|^{1/\alpha}) < \infty \ \& \ EX = \mu,$$

(ii) For $\alpha \geq 1$,

$$X_n \to \mu \quad \text{a.s.} \quad (C_\alpha) \Leftrightarrow E(|X|) < \infty \quad \& \ EX = \mu.$$

<u>Proof</u>. For (i), the case $\alpha = 1$ is Kolmogorov's result.

For $\frac{1}{2} < \alpha < 1$ (the L^p case with $1 < p < 2$, for $p := 1/\alpha$), the result is Theorem 3 of Lorentz (1955).

For $0 < \alpha < \frac{1}{2}$ (the L^p case with $p > 2$), the result follows from Theorem 1 of Chow & Lai (1973): in the zero-mean case, $X \in L^{1/\alpha}$ iff

$$n^{-\alpha} \Sigma_0^n c_{n-k} X_k \to 0 \quad \text{a.s.}$$

for some (all) (c_n) with $\Sigma c_n^2 < \infty (c_n := A_n^{\alpha-1}$: $(c_n) \in \ell^2$ for $\alpha < \frac{1}{2})$.

For $\alpha = \frac{1}{2}$, the result is due to Déniel & Derriennic (1988+), who pointed out the link with the Lorentz and Chow-Lai theorems.

For (ii), we use a result of Lai (1974a): for iid random variables, the Cesàro methods C_α, $\alpha \geq 1$ (and the Abel method A) are all equivalent.

<u>§3. Riesz methods</u>.

In Theorem 1 one may ask whether a.s. C_α-convergence may be replaced by some formally stronger statement (a.s. convergence under a more stringent summability method). One may also ask whether moment conditions more general than $E(|X|^p) < \infty$ may be handled. It turns out that both questions have a positive answer; to formulate it we need some background from summability theory.

Recall the Riesz (typical) mean $R(\lambda_n,1)$ of order 1, based on a sequence $\lambda_n \uparrow \infty$ (Hardy (1949), §§4.16, 5.16): writing $s_n := \Sigma_0^n a_k$,

$$s_n \to s \qquad R(\lambda_n,1)$$

means

$$\frac{1}{x}\int_0^x \{\sum_{n:\lambda_n \le y} a_n\}\, dy \to s \qquad (x \to \infty).$$

We consider first the case $\lambda_n = \exp(n^{1-1/p})$, $p > 1$. This was considered by Bingham & Tenenbaum (1986):

$$s_n \to s \qquad R(\exp(n^{1-1/p}),1) \qquad (p > 1)$$

holds iff the following 'moving averages' converge:

$$\frac{1}{\varepsilon n^{1/p}} \sum_{n \le k < n+\varepsilon n^{1/p}} s_k \to s \qquad (n \to \infty) \qquad \forall \varepsilon > 0.$$

In an obvious notation, we may write this as

$$s_n \to s \qquad M(n^{1/p}).$$

One can extend this to include $p = 1$: the Riesz method $R(n,1)$ is the Cesàro method $C = C_1$ (Hardy (1949), Th. 58), and the 'moving-average method' $M(n^{1/p})$ is equivalent to the Riesz method

$$R_p := R(\exp(\int_1^n x^{-1/p} dx),1) \qquad (p \ge 1).$$

We recall (Bingham & Tenenbaum (1986)) the following alternative to Theorem 1(i):

THEOREM 2. For $p \ge 1$, $E(|X|^p) < \infty$ & $EX = \mu$ iff

$$X_n \to \mu \quad \text{a.s.} \qquad (R_p \text{ or } M(n^{1/p})).$$

The case $p = 2$ is particularly important. Here the method R_2 may be replaced by the Borel method B, the Euler method E_p (Chow (1973)), or certain Valiron methods (Bingham & Tenenbaum (1986)).

We now have two candidates for the 'natural' summability method for the L^p case ($p \geq 1$), the Cesàro method $C_{1/p}$ and the Riesz method R_p. For $p = 1$, the methods coincide. For $p > 1$, it turns out that Riesz convergence is strictly stronger than Cesàro convergence (and thus that Theorem 2 is the stronger result):

THEOREM 3. For $p > 1$, $R_p \subsetneqq C_{1/p}$.

Proof. This follows from results of Jurkat, Kratz & Peyerimhoff (1975) ('JKP'), who give a thorough study of the links between different Riesz means $R(\lambda,k)$ for $0 \leq k \leq 1$ (extended to $k > 1$ in Kratz (1978)). One writes $\Lambda(x) := \lambda(x)/\lambda'(x)$. Use Theorem C of JKP (p. 255) with

$$R_p = R(\exp(\textstyle\int_1^n x^{-1/p}dx),1) = R(\lambda_1,k_1)\colon\ k_1 = 1,\ \Lambda_1(x) = x^{1/p},$$

$$C_{1/p} = R(x,1/p) \qquad\qquad = R(\lambda_3,k_3)\colon\ k_3 = 1/p,\ \Lambda_3(x) = x$$

(and $\lambda_2 = \lambda_1$, $k_2 = k_1$). As $k_i \in [0,1]$, $k_3 < k_1$, $\Lambda_1^{k_1}(x) \leq \Lambda_3^{k_3}(x)$ (in fact, equality holds), Theorem C gives the inclusion $R_p \subset C_{1/p}$.

It remains to disprove the reverse inclusion. To do this, we use the minimality (best-possible nature) of the bounds in Theorem 5, the main result of JKP. This time, take

$$k_3 = 1,\ \lambda_3(x) = \exp(x^{1-1/p}),\ \Lambda_3(x) = x^{1/p},$$

$$k_1 = 1/p,\ \lambda_1(x) = x,\qquad \Lambda_1(x) = x$$

(and $\lambda_2 = \lambda_1$, $k_2 = k_1$). From Theorem LC, (42), one obtains that a bound c on the $R(\lambda_3,k_3)$-mean $(A_{\lambda_3}^{k_3}(x)/\lambda_3^{k_3}(x))$ for large x implies a bound $c.x^{(1-1/p).1/p}$ on the $R(\lambda_1,k_1)$-mean, and (Theorem 5) that this is best-possible. The factor $x^{(1-1/p).1/p}$ being unbounded for $p > 1$, the inclusion $R(\lambda_1,k_1) \subset R(\lambda_3,k_3)$, i.e. $C_{1/p} \subset R_p$, is false, as required.

§4. Self-neglecting functions.

As we have seen, Riesz means R_p (or moving averages $M(n^{1/p})$) give a stronger a.s. convergence result than the Cesàro methods $C_{1/p}$, in the L^p case. Another advantage of the Riesz/moving average formulation is the flexibility provided by our freedom to choose the sequence $\lambda_n \uparrow \infty$ in $R(\lambda_n,1)$; this will enable us to handle moment conditions much more general than the $E(|X|^p) < \infty$ of the L^p case.

Some restriction on (λ_n) - or on the length $\phi(x)$ of the averaging interval - is needed. It turns out that the appropriate condition is supplied by the concept of Beurling slow variation, also known as the self-neglecting property. Recall (Bingham, Goldie & Teugels (1987), §2.11) that a function $\phi:\mathbb{R} \to \mathbb{R}_+$ is <u>self-neglecting</u> ($\phi \in SN$) if it is continuous, $o(x)$ at ∞, and

$$\phi(x+t\phi(x))/\phi(x) \to 1 \qquad (x \to \infty) \qquad \forall t \in \mathbb{R}.$$

The following link between Riesz means and moving averages extends that of §3 from ϕ x) $= x^{1/p}$ to the general case.

THEOREM 4 (Bingham & Goldie (1988)). If $\phi \in SN$, $\phi \uparrow \infty$ and

$$\gamma(x) := \exp(\textstyle\int_1^x dt/\phi(t)),$$

$$s_n \to s \qquad R(\gamma,1)$$

is equivalent to the moving-average convergence

$$s_n \to s \qquad M(\phi),$$

that is,

$$\frac{1}{\varepsilon\phi(x)} \sum_{x \le k < x+\varepsilon\phi(x)} s_k \to s \qquad (x \to \infty) \qquad \forall \varepsilon > 0.$$

Since $\phi \uparrow \infty$ and is continuous, it has a continuous inverse function $\psi := \phi^{\leftarrow} \uparrow \infty$. To extend Theorem 2, we will need a restriction on ϕ, ψ. Recall first (Bingham et al. (1987)) that ϕ may be represented as $\phi(x) = c(x) \int_0^x \varepsilon(y)dy$ with $c(x) \to c \in (0,\infty)$, $\varepsilon(x) \to 0$ at ∞. For our purposes, $c(.)$ may be replaced by its limit c, which may be absorbed in $\varepsilon(.)$: $\phi(x) = \int_0^x \varepsilon(y)dy$, where $\varepsilon(x) > 0$ as $\phi \uparrow$. Then $\psi = \phi^{\leftarrow}$ has derivative $1/\varepsilon(\psi(.)) \to \infty$.

Recall also that $f(.) > 0$ has <u>bounded increase</u>, $f \in BI$, if

$$f(\lambda x)/f(x) \le C\,\lambda^{\alpha} \qquad (1 \le \lambda \le \Lambda,\ x \ge x_0)$$

for suitable $\Lambda > 1$, C, x_0, α. The restriction needed is bounded increase (roughly, polynomial growth) on ψ':

THEOREM 5 (Bingham & Goldie (1988)). If ϕ is as above, $\psi := \phi^{\leftarrow}$ and $\psi' \in BI$, the following are equivalent:

(i) $\quad E\psi(|X|) < \infty$ & $EX = \mu$,

(ii) $\quad X_n \to \mu \quad$ a.s. $\quad R(\gamma)$ or $M(\phi)$,

(iii) $\quad \sum n^{-1}(\psi(n+1)-\psi(n))\, P(\max_{k \le n} |S_k - k\mu| > \varepsilon n) < \infty \quad \forall \varepsilon > 0.$

The crux of the proof is a maximal inequality due to Asmussen & Kurtz (1980), Kurtz (1972).

To see the role of the condition $\psi' \in BI$, note that the result fails badly without it. The classical case of this is the law of large numbers of Erdős & Rényi (1970). Here $\phi(x) = c \log x$, $\psi(x) = e^{x/c}$, $\psi'(x) = c^{-1} e^{x/c} \notin BI$. If X has analytic characteristic function, its cumulant-generating function

$$k(t) := \log E \exp\{tX\}$$

exists in a neighbourhood of the origin. It is convex; form its Fenchel dual

$$k^*(\alpha) := \sup_t \{t\alpha - k(t)\}$$

(dually,

$$k(t) = \sup_\alpha \{t\alpha - k^*(\alpha)\}).$$

If $c = c(\alpha)$ and $\alpha = \alpha(c)$ are linked by

$$1/c = k^*(\alpha),$$

the Erdős-Rényi law states that

$$\max_{0 \le i \le n - c \log n} \frac{1}{c \log n} \sum_{i \le k < i + c \log n} X_k \to \alpha \quad \text{a.s.}$$

The point here is that the limit $\alpha = \alpha(c)$ on the right determines $k^*(.)$, $k(.)$ and so the law F of X: the entire distribution of X can be recovered from the limit, by varying c. This is in contrast to Theorem 5(ii), where the limit determines only the mean, $\mu = EX$. Behaviour of the latter type has the character of an almost-sure invariance principle (ASIP): the limit is unchanged if the law F is altered preserving the mean μ. By contrast, the former is an almost-sure non-invariance principle (ASNIP). The ASIP/ASNIP interface, together with refinements of the Erdős-Rényi law and its relatives, have recently been studied in detail by Steinebach (1984), Deheuvels (1985), Deheuvels, Devroye & Lynch (1986), Deheuvels & Devroye (1987), Deheuvels & Steinebach (1987), (1988).

One should compare the following result of Chow & Lai (1973) with their Theorem 1(i) ($\alpha < \frac{1}{2}$): if X_n are iid with mean 0 and entire characteristic function,

$$X_n = o(\log n) \qquad \text{a.s.}$$

iff

$$\frac{1}{\log n} \Sigma_1^n c_{n-k} X_k \to 0 \qquad \text{a.s.}$$

for some (all) (c_n) with $\Sigma c_n^2 < \infty$.

§5. Results of iterated-logarithm type.

The maximum occurring in the Erdős-Rényi law indicates that it is of the type of the law of the iterated logarithm (LIL) rather than the law of large numbers (LLN). The LIL analogue for moving averages complementing Theorem 5 is due to de Acosta & Kuelbs (1983). Their Theorem 1(A) specialised to the real-valued case, plus Remarks (I) and (X), gives the equivalence of

$$EH(|X|) < \infty, \ EX = 0 \ \& \ \mathrm{var}\, X = \sigma^2 < \infty$$

and

$$\limsup \frac{1}{(2\phi(x)\log(x/\phi(x)))^{\frac{1}{2}}} \sum_{x \le n < x+\phi(x)} X_n = \sigma \qquad \text{a.s.,}$$

$$\liminf \quad \ldots\ldots \qquad = -\sigma \qquad \text{a.s.,}$$

subject to suitable restrictions on ϕ (e.g. $\phi(x) = \int_0^x \varepsilon(y)dy$, $\varepsilon(x) \downarrow 0$, $\varepsilon(x) = o(1/\log x)$, $\psi := \phi^{\leftarrow} \in BI$) and with

$$H^{\leftarrow}(x) = (\phi(x)\log x)^{\frac{1}{2}}.$$

In particular, taking $\phi(x) = x^{\alpha}$ $(0 < \alpha < 1)$, one recovers a result of Lai (1974b):

$$E(|X|^{2/\alpha}/\log^{1/\alpha}|X|) < \infty, \ EX = 0 \ \& \ \text{var } X = \sigma^2$$

iff

$$\limsup \ [\liminf] \ \frac{1}{(2(1-\alpha)n^{\alpha}\log n)^{\frac{1}{2}}} \sum_{n\leq k<n+n^{\alpha}} X_k = \sigma \ [-\sigma] \ \text{a.s.}$$

('law of the single logarithm', or LSL). For comparison, the same integrability condition in the context of Theorem 5 yields

$$E(|X|^{2/\alpha}/\log^{1/\alpha}|X|) < \infty \ \& \ EX = 0$$

iff

$$\frac{1}{(n^{\alpha}\log n)^{\frac{1}{2}}} \sum_{n\leq k<n+(n^{\alpha}\log n)^{\frac{1}{2}}} X_k \to 0 \quad \text{a.s.}$$

In Theorems 1-4 of de Acosta & Kuelbs (1983), the length a_n of the averaging interval is, as above, larger than logarithmic and the a.s. cluster set is $[-\sigma,\sigma]$. Their Theorem 5 includes the Erdős-Rényi case: $a_n = [c \log n]$, and the a.s. cluster set determines (through c) the law of X. Their Theorems 7-9 are the first to handle the case $a_n = o(\log n)$.

For further background, and references, see e.g. Bingham (1986b), §14.

§6. Complements.

1. The ergodic case. It is interesting to compare the results here, of 'iff' character and with sharp integrability conditions, with those obtaining for moving averages in the ergodic case,

$$A_k f(x) := \ell_k^{-1} \sum_{n_k\leq n<n_k+\ell_k} f(T^n x).$$

For the case $n_k = k$ and $\ell_k = \sqrt{k}$ of the moving-average method

$M(\sqrt{n})$ (or the Riesz mean $R(e^{\sqrt{n}},1)$), no moment condition suffices for a.s. convergence: this may fail even for f the indicator of a measurable set, as was shown by Akcoglu & del Junco (1975). A similar failure occurs for Euler weights (Rosenblatt (1986); for background on the close links between the Euler and $R(e^{\sqrt{n}},1)$-methods, see e.g. Bingham (1984)). A thorough study of this phenomenon is given by Bellow, Jones & Rosenblatt (1989) in this volume.

For the failure of the Marcinkiewicz-Zygmund law in this context, see Jain, Jogdeo & Stout (1975), 135-137, and of the Baum-Katz law, Baum & Katz (1965), 117-118.

2. The dependent case. Despite the failure of the iid results for the general ergodic case, good results can still be obtained if independence is weakened to an appropriate mixing condition; see for instance Peligrad (1985), (1989). In particular, using results of Peligrad (1989) and Bingham (1986a), Theorem 2 can be extended to (X_n) ϕ-mixing with any rate.

3. The Banach-valued case. For background on the case where the X_n are random vectors taking values in a Banach space, see de Acosta & Kuelbs (1983) for the 'LIL' results of §5, and for LLN, e.g. Bingham (1986a) and the references cited there.

Postscript. The work presented here may be regarded as bringing to fruition the ideas in a sentence of great prescience, written (à propos of his $\frac{1}{2} < \alpha < 1$ case of Theorem 1(i)) by G. G. Lorentz (1955): "There are similar theorems with Cesàro means replaced by Riesz means $R(\lambda_n,\alpha)$ and classes L^p replaced by Orlicz classes L^{ω}."

Acknowledgements. I thank Yves Derriennic and Werner Kratz for their comments.

References.

1. A. de Acosta & J. Kuelbs (1983): Limit theorems for moving averages of random vectors. Z. Wahrschein. 64, 67-123.

2. M. Akcoglu & A. del Junco (1975): Convergence of averages of point transformations. Proc. Amer. Math. Soc. 49, 265-266.

3. S. Asmussen & T. G. Kurtz (1980): Necessary and sufficient conditions for complete convergence in the law of large numbers. Ann. Probab. 8, 176-182.

4. L. E. Baum & M. Katz (1965): Convergence rates in the law of large numbers. Trans. Amer. Math. Soc. 120, 108-123.

5. A. Bellow, R. Jones & J. Rosenblatt (1989): Convergence for moving averages. This volume.

6. N. H. Bingham (1984): On Euler and Borel summability. J. London Math. Soc. (2) 29, 141-146.

7. N. H. Bingham (1986a): Extensions of the strong law. Adv. Appl. Probab. Supplement, 27-36 (G. E. H. Reuter Festschrift, ed. D. G. Kendall).

8. N. H. Bingham (1986b): Variants on the law of the iterated logarithm. Bull. London Math. Soc. 18, 433-467.

9. N. H. Bingham & C. M. Goldie (1988): Riesz means and self-neglecting functions. Math. Z.

10. N. H. Bingham, C. M. Goldie & J. L. Teugels (1987): Regular variation. Cambridge Univ. Press (Encycl. Math. Appl. Vol. 27).

11. N. H. Bingham & G. Tenenbaum (1986): Riesz and Valiron

means and fractional moments. Math. Proc. Cambridge Phil. Soc. 99, 143-149.

12. Y.-S. Chow (1973): Delayed sums and Borel summability of independent identically distributed random variables. Bull. Inst. Math. Acad. Sinica 1, 207-220.

13. Y.-S. Chow & T.-L. Lai (1973): Limiting behaviour of weighted sums of independent random variables. Ann. Probab. 1, 810-824.

14. P. Deheuvels (1985): On the Erdős-Rényi theorem for random fields and sequences and its relationships to the theory of runs and spacings. Z. Wahrschein. 70, 91-115.

15. P. Deheuvels & L. Devroye (1987): Limit theorems of Erdős-Rényi-Shepp type. Ann. Probab. 15, 1363-1386.

16. P. Deheuvels, L. Devroye & J. Lynch (1986): Exact convergence rates in the limit theorems of Erdős-Rényi and Shepp. Ann. Probab. 14, 209-223.

17. P. Deheuvels & J. Steinebach (1987): Exact convergence rates in strong approximation laws for large increments of partial sums. Probab. Th. Rel. Fields 76, 369-393.

18. P. Deheuvels & J. Steinebach (1988): Limit laws for the modulus of continuity of the partial-sum process and for the Shepp statistic. Stoch. Proc. Appl. 29, 223-245.

19. Y. Déniel & Y. Derriennic (1988+): Sur la convergence presque sure, au sens de Cesàro d'ordre α, $0 < \alpha < 1$, des variables aléatoires indépendantes et identiquement distribuées. Probab. Th. Rel. Fields.

20. P. Erdős & A. Rényi (1970): On a new law of large numbers. J. Analyse Math. 23, 103-111.

21. G. H. Hardy (1949): Divergent series. Oxford Univ. Press.

22. N. C. Jain, K. Jogdeo & W. F. Stout (1975): Upper and lower functions for martingales and mixing processes. Ann. Probab. 3, 119-145.

23. W. B. Jurkat, W. Kratz & A. Peyerimhoff (1975): The Tauberian theorems which interrelate different Riesz means. J. Approximation Theory 13, 235-266.

24. W. Kratz (1978): The Tauberian theorems which interrelate different Riesz means, II. J. Indian Math. Soc. 42, 45-66.

25. T. G. Kurtz (1972): Inequalities for the law of large numbers. Ann. Math. Statist. 43, 1874-1883.

26. T.-L. Lai (1974a): Summability methods for independent, identically distributed random variables. Proc. Amer. Math. Soc. 45, 253-261.

27. T.-L. Lai (1974b): Limit theorems for delayed sums. Ann. Probab. 2, 432-440.

28. G. G. Lorentz (1955): Borel and Banach properties of methods of summation. Duke Math. J. 22, 129-141.

29. J. Marcinkiewicz & A. Zygmund (1937): Sur les fonctions indépendantes. Fundam. Math. 29, 60-90.

30. M. Peligrad (1985): Convergence rates of the strong law for stationary mixing sequences. Z. Wahrschein. 70, 307-314.

31. M. Peligrad (1989): In this volume.

32. J. Rosenblatt (1986): Ergodic group actions. Arch. Math. 47, 263-269.

33. J. Steinebach (1984): Between invariance principles and Erdős-Rényi laws. Limit theorems in probability and statistics II (ed. P. Révész), 981-1005, North Holland, Amsterdam.

ALMOST SURE CONVERGENCE IN ERGODIC THEORY

J. Bourgain(*)

0. INTRODUCTION

My purpose in this exposé is to give a brief summary of some recent research on the almost sure convergence of certain ergodic averages. The exposé is mainly a report on the author's personal work on the subject, by no means a general survey. Most of the results mentionned below were discussed in lectures at the Columbus meeting, June 1988. Some of them had a further development or were improved. References are [B-L], [$Bo_{1,2,3}$], [Fu].

Let Λ be an infinite subset of the positive integers $\mathbb{Z}_+$ and denote $\Lambda_N = \Lambda \cap [0, N]$ (which we assume non-void). Our aim is to study the behaviour of the averages

$$A_N f \equiv \frac{1}{|\Lambda_N|} \sum_{n \in \Lambda_N} T^n f \tag{0.1}$$

where (Ω, β, μ, T) is a dynamical system (DS for short) and $f \in L^\infty(\mu)$, i.e. a bounded measurable function on Ω.

It follows from spectral theory of unitary operators that (1.1) will converge in the mean for $N \to \infty$, provided the sequence of polynomials

$$P_N(z) \equiv \frac{1}{|\Lambda_N|} \sum_{n \in \Lambda_N} z^n \tag{0.2}$$

converges pointwise on the unit circle $\mathbb{C}_1 = \left\{ z \in \mathbb{C} \mid |z| = 1 \right\}$.

Our interest goes here to a more refined form of convergence, namely the almost sure (a.s) convergence. It is indeed a general fact that the pointwise behaviour of (averages of) a sequence of discrete observables may only have significance, provided at least a.s. convergence holds. This is for instance the case if we let $\Lambda = \mathbb{Z}_+$. The a.s. convergence of $A_N f$ for any $f \in L^1(\mu)$ is then ensured by Birkhoff's ergodic theorem. This well known result has a variety of applications.

(*) IHES - Bures sur Yvette.

ISBN 0-12-231050-0

The a.s. convergence of averages of the form (1.1) is in general not implied by purely spectral theory considerations. In $[Bo_{1,2}]$, I developed a method, based on Fourier Analysis techniques, to subtitute spectral theory in the context of a.s. convergence. The most striking application of this approach is an extension of Birkhoff's theorem to the case of "arithmetic sets", for instance sets Λ of the form

$$\Lambda = \left\{ \varphi(n) \mid n = 1, 2, \dots \right\}$$

where $\varphi(x)$ is a polynomial with integer coefficients. The particular case $\varphi(n) = n^2$ answered questions raised by Bellow, Erdős, Furstenberg and Herman.

Precise formulations of these results will be given later. I want to observe now however that the condition on f appears to be $f \in L^r(\mu)$ for some $r > 1$ rather than $f \in L^1(\mu)$ (here and in the sequel, the measure μ is assumed finite).

There are further applications of this Fourier-Analysis method to a.s. convergence for averages of compositions of commuting transformations and to multiple recurrence.
They will also be reported on here.

Summary

1. Polynomial sequences
2. Commuting transformations
3. Return-time sequences
4. Multiple recurrence
5. Remarks and problems

1. POLYNOMIAL SEQUENCES

To conclude a.s. convergence of (0.1) requires more information on the sequence (0.2) than just the convergence on $\mathbb{C}_1$. Let $d \geq 2$, $\varphi(x) = a_1 x + a_2 x^2 + \dots + a_d x^d$, $a_j \in \mathbb{Z}$ and $a_d > 0$. Redefining

$$p_N(z) = \frac{1}{N} \sum_{1}^{N} z^{\varphi(n)} \tag{1.1}$$

the fine behaviour of these polynomials (= exponential sums) is one of the concerns of analytic number theory. Using this information, we succeeded in proving the following

THEOREM 1.2 : *If φ is as above, then*

$$\frac{1}{N} \sum_{1}^{N} T^{\varphi(n)} f \tag{1.3}$$

converges a.s. for any measure preserving transformation T *on a finite measure space* (Ω, μ) *and* $f \in L^r(\mu)$, $r > 1$.

The proof of (1.2) for $r = 2$ appears in $[Bo_1]$ while the extension to $r > 1$ is the object of $[Bo_2]$.

Remarks

1. It is presently not known whether (1.2) holds for $r = 1$, even in the case of the squares $\varphi(n) = n^2$
2. Using the same method, the analogue of (1.2) is obtained for positive (not necessarily invertible) isometries. In particular, there is the following extension of the Riesz-Raikov theorem.

THEOREM 1.4 : *Let* φ *be as above and* $f \in L^r(\mathbb{T})$, $r > 1$, *where* $\mathbb{T} = \mathbb{R}/\mathbb{Z}$ *stands for the circle. Then* $\frac{1}{N}\sum_1^N f\left(2^{\varphi(n)}x\right) \to \int_0^1 f$ *for almost all* $x \in \mathbb{T}$.

3. The usual averages $\frac{1}{N}\sum_1^N T^n f$ converge to $\int f$ if T is ergodic. This is not the case any more for the averages (1.3). If T has no non-trivial rational point-spectrum, in particular for weakly mixing T, the same conclusion holds however. Let us notice that in proving (1.2), there is no identification of the limit.

2. COMMUTING TRANSFORMATIONS

Let (Ω, μ) be a probability space and $T_1, T_2, \ldots, T_\ell$ commuting measure preserving transformations. Let $\varphi_1, \ldots, \varphi_\ell$ be polynomials as in Th. 1.2.

THEOREM 2.1 : *The averages* $\frac{1}{N}\sum_1^N T_1^{\varphi_1(n)} \ldots T_\ell^{\varphi_\ell(n)} f$ *converges a.s., for* $f \in L^r(\mu)$, $r > 1$

The proof of (2.1) for $r = 2$ may be found in $[Bo_1]$. Extending the result to $r > 1$ may be done using the techniques of $[Bo_2]$.

Remarks

1. It suffices to prove Th. 2.1 letting $\varphi_1(n) = n, \varphi_2(n) = n^2, \ldots, \varphi_\ell(n) = n^\ell$. The "real Analysis" component in the proof relates then to Radon transforms along the moment curve $t \to (t, t^2, \ldots, t^\ell)$.

2. Th. 2.1 permits a refinement of Birkhoff's theorem when considering certain skew-products. The mapping of $\mathbb{T}^2 = \mathbb{T} \times \mathbb{T}$

$$T(x, y) = (x + \alpha, y + x) \qquad \alpha \in \mathbb{R}\backslash\mathbb{Q} \tag{2.2}$$

for instance, is a skew extension of $\mathbb{T}$ by itself. Birkhoff's theorem asserts that for $f \in L^\infty(\mathbb{T})$

$$\frac{1}{N}\sum_1^N f\left(y + nx + \frac{n(n-1)}{2}\alpha\right) \to \int f \tag{2.3}$$

a.s with respect to the product variable (x, y). Applying now (2.1) with $\ell = 2$, $T_1 = R_x, T_2 = R_\alpha, \varphi_1(n) = n, \varphi_2(n) = n = \frac{n(n-1)}{2}$, yields a.s. convergence of (2.3) in y, for each individual value of x.

3. RETURN-TIME SEQUENCES

Let T be an ergodic transformation and $A \in B$, $\mu(A) > 0$. For $x \in \Omega$, the return-time sequence $\Lambda_x = \{n \in \mathbb{Z}_+ \mid T^n x \in A\}$ has positive density, by Birkhoff's theorem. The following result completes an investigation started in [B-L].

THEOREM 3.1 : *With above notation, Λ_x is a summing sequence for almost all $x \in \Omega$.*

Agree to call $\Lambda \subset \mathbb{Z}_+$ a summing sequence provided (0.1) converges a.s. for any DS (Ω', μ', T') and $f \in L^1(\mu')$. In [B-L], the particular case was studied when the transformation T, generating the sequence Λ_x, has Lebesgue spectrum. Th. (3.1) may be seen as a refinement of Poincaré's recurrence principle and improves on the Wiener-Wintner ergodic theorem. In [Bo_2], Appendix, a simple proof of (3.1) may be found, simplifing an earlier argument.

4. MULTIPLE RECURRENCE

The multiple recurrence theorem appearing in Furstenberg's book [Fu] states that averages

$$\frac{1}{N}\sum_1^N T^n f_1 . T^{2n} f_2 \ldots T^{\ell n} f_\ell \tag{4.1}$$

converge in the mean. Here ℓ is an arbitrary positive integer and $f_1, \dots f_\ell \in L^\infty(\mu)$. A generalization due to Furstenberg and Katznelson [F-K] permits to replace $T, T^2, \dots, T^\ell$ by any system of commuting transformations. The techniques involved are very much L^2 and it seems difficult to reach a.s. results along these lines, at least when more than 2 factors in (4.1) appear. The author obtained recently following result, settling a question from [Fu]

THEOREM 4.2 : *Averages of the form* $\frac{1}{N}\sum_1^N T_1^n f_1 . T_2^n f_2$, *where* T_1, T_2 *are powers of a same transformation* T, *converge* a.s.

Observe indeed that the mean convergence of (4.1) with $\ell = 2$ (or, more generally, with 2 factors) is considerably more elementary than in the general case (it does not require the theory of extensions). From a Number Theory point of view, K. Roth's argument for the existence of triples in arithmetic progression in sets of positive density has so far not be extended beyond ternary structures. From a DS point of view, if $\ell = 2$, (4.1) converges to o if either f_1 or f_2 is orthogonal on the eigenfunctions of T. This is not the case any more for $\ell > 2$.

Presently, the proof of (4.2) is rather complicated. Some ingredients are similar to the techniques used in proving (1.2). In particular, there is a shift reduction and a Fourier Analysis approach.

Remark :

There does not seem to be an "elementary" proof of the L^2-convergence of averages $\frac{1}{N}\sum_1^N T_1^n f_1 . T_2^n f_2$ for a general pair of commuting transformation T_1, T_2. It looks hopeless to try to turn the existent argument into an a.s. convergence result.

5. REMARKS AND PROBLEMS

1. An obvious question araising from section 1 in whether Th (1.2) has an L^1-version. Weak-type 1.1 results are difficult to reach by means of Fourier Analysis and interpolation methods. In various contexts in Real Analysis, their validity is undecided. There is a clear lack of methods to deal with L^1-functions.

2. In $[Bo_4]$, a criterium is given enabling to disprove a.s. convergence and the validity of certain maximal inequalities. In the context of (0.1), a.s. convergence of $A_N f$, f bounded measurable, implies uniform estimates on the metrical entropy numbers in L^2 of the sets

$$\{A_N f \mid N = 1, 2, \dots\}$$

when f runs in the unit ball of $L^2(\mu)$.

The method of $[Bo_4]$ unifies certain counterexamples, including for instance, Marstrand's (negative) solution to the Khintschine conjecture and Rudin's disproof of a.s. convergence of Riemann-sums.

The following two question were posed by R. Salem and are still unsolved (to my knowledge).

<u>Q1</u> : *Let* $\theta > 1$ *and* f *a bounded 1-periodic function on* $\mathbb{R}$. *Does* $\frac{1}{N}\sum_{1}^{N} f(\theta^n x)$ *converge a.s. ?*

If θ is an integer, an affirmative answer follows from the Riesz-Raikov theorem mentioned above.

<u>Q2</u> : *Let* f *be as above and consider the operators (Riemann sums)*

$$R_n f(x) = \frac{1}{n}\left[f(x) + f\left(x + \frac{1}{n}\right) + f\left(x + \frac{2}{n}\right) + \dots + f\left(x + \frac{n-1}{n}\right)\right]$$

Put

$$A_N f = \frac{1}{N}\sum_{1}^{N} R_n f$$

Does $A_N f$ *converge a.s. ?*

W. Rudin proved that the sequence $R_N f$ itself need not. The author verified that if N_k is choosen sufficiently rapidly increasing ($N_k = 2^{2^k}$ would do), the sequence $A_{N_k} f$ does converge a.s. for $f \in L^\infty(\mathbb{T})$. The argument is based on distributional results for prime factors in integer factorization . (Dickman's theorem).

In searching for a counter example, both Q_1 and Q_2 turn out to escape the entropy method from $[Bo_4]$.

For Salem's work related to Q_1, Q_2, the reader should consult [Sa] (p. 414 for Q_1 and p.97 for Q_2)

REFERENCES

[B-L] A. BELLOW, V. LOSERT : The weighted pointwise ergodic theorem and the individual ergodic theorem along subsequences, TAMS 1985, Vol 288, 307-345

[Bo_1] J. BOURGAIN : On the maximal ergodic theorem for certain subsets of the integers, Israel J. Math., Vol 61, N1, 39-72 (1988)

[Bo_2] J. BOURGAIN : Pointwise ergodic theorems on arithmetic sets, with an Appendix on Return-time Sequences (jointly with H. Furstenberg, Y. Katznelson, D. Ornstein) preprint IHES (1989)

[Bo3] J. BOURGAIN : Double recurrence and almost sure convergence, preprint 1988

[Fu] H. FURSTENBERG : Recurrence in ergodic theory and combinatorial number theory, Princeton UP (1981)

[Sa] R. SALEM : Collected works, Hermann 1967.

[Bo_4] J. BOURGAIN : Almost sure convergence and bounded entropy, Israel J. Math., Vol 63, N1, 79-97 (1988)

[F-K] H. FURSTENBERG, Y. KATZNELSON : An ergodic Szemerédi theorem for commuting transformations. J. d'Analyse Math. 34 (1978), 275-291

A pointwise ergodic theorem for positive, Cesaro-bounded operators on L_p $(1 < p < \infty)$

by

Antoine Brunel
Professor at the Université Paris VI

In 1975 M.A. Akcoglu [1] solved a conjecture left open during a long period, proving the following result: T being a linear positive contraction on L_p $(1 < p < \infty)$, and letting $M_n = \frac{1}{n}\sum_{j=0}^{n-1} T^j$, we have

1. $\lim_n M_n f$ exists a.e. and in L_p for every $f \in L_p$.
2. $\| \sup_n |M_n f| \, \|_p \leq (\frac{p}{p-1}) \|f\|_p$.

The maximal inequality (2) was obtained by Mrs A. Bellow [2] when T is a positive isometry and M.A. Akcoglu used a clever method of dilation to get (2) from the A. Bellow's result.

We could generalize Akcoglu's result recently, proving

Theorem 1 *The dominated ergodic theorem holds for a linear positive operator on L_p $(1 < p < \infty)$ if, and only if, T is Cesaro-bounded.*

Actually, this theorem does not generalize Akcoglu's maximal inequality because we have not given any estimate of the constant k such that $\| \sup_n |M_n f| \, \|_p \leq k\|f\|_p$.

Recall that an operator T is Cesaro-bounded if $\sup_n \|M_n(T)\| < +\infty$.

The complete proof of theorem 1 will appear in "Ergodic theory and dynamical systems". Here I want to give the main features of that proof.

Let $A(T)$ be the operator $\sum_{j=0}^{\infty} \alpha_j^1 T^j$, where α_j^1's are the coefficients of the Taylor's expansion of $\frac{1}{x}(1 - \sqrt{1-x}) = \sum_{j=0}^{\infty} \alpha_j^1 x^j$. $A = A(T)$ is defined when T is a contraction, but here we have

(1) A is defined and A is power-bounded iff T is Cesaro-bounded.

(2) $\forall m, n \in \mathbb{N}$ $(A^m - I)(A^n - I) \leq I$.

ISBN 0-12-231050-0

(3) There is a universal constant $\chi > 0$ such that $\forall n\ M_n(T) \leq \chi\ M_{q(n)}(A)$ if $q(n) = [\sqrt{n}] + 1$.

Inequalities (2) and (3) and more generally $P \geq 0$ mean $Pf \geq 0$ if f in L_p is not negative.

The basic properties of A and the computation of the α_j^n, defined by $A^n = \sum_{j=0}^{\infty} \alpha_j^n T^j$, are given in the book "Ergodic theorems" [7] and in a note [3] jointly made by A. Brunel and R. Emilion.

T^*(resp A^*) denote the adjoint operator of T(resp A) acting on the dual space L_q ($\frac{1}{p} + \frac{1}{q} = 1$). Observe that $A^* = A(T^*)$.

Some reductions permit to assume a canonical situation where the measure space is the unit interval, eliminating the atomic part of the inital space. We may also assume there exists a measurable subset of positive measure without absorbing subsets. ($F, \mu(F) > 0$ is *A-absorbing* if $1_{F^c} A(1_F) = 0$). If not, every set is absorbing and T would be an operator of the form

$$Tf(x) = f(x)\theta(x)$$

where θ is a positive, measurable function, bounded by 1 if T is Cesaro bounded, which proves trivially the theorem 1. Furthermore recall that R. Emilion [6] proved the mean-ergodic theorem for a positive Cesaro-bounded T through a Tauberian theorem by reduction to the power-bounded case.

The following proposition is the key of the proof,

Proposition 2 *Let T be a positive, Cesaro-bounded operator on the L_p-space of the unit interval ($1 < p < +\infty$). Assume existence of an interval $[0, r[$ without absorbing subsets. Then at least one of the following conditions holds*

(a) $\sup_n A^{*n} 1 \in L_1$.

(b) $A1 \leq 1$.

To establish this we are going to apply an optimization method which needs some preliminary definitions: let $z = (U, \phi)$ an element in $L_p^+ \times L_p$ such that

$$\begin{cases} U = \sum_{k=0}^{m} u_k, & u_k \in L_p^+ \\ \phi = \sum_0^m (A^k - I) u_k \end{cases}$$

The set of these elements z is a convex cone Γ and the closure of Γ in the Banach space $L_p^+ \times L_p$ is also a convex cone. In some part of the proof we need also some constraints and a z of the form $z = (t, U, \phi) \in \mathbb{R}_+ \times L_p^+ \times L_p$ endowed with the product topology. The constraints are

$$\{U \leq t,\ U \leq v \int Uh,\ (U, \phi) \in \Gamma\},$$

where $h(x) = \frac{2}{(1-x)^\alpha}$, $v(x) = \frac{c}{h(x)}$, $\alpha > 1$ and $1 < c < 2$. The constants α and c must satisfy some conditions.

This new set of z is also a cone Γ', and Z is its closure.

Let us describe only one step of the optimizations processes when $p = 2$.

The functional to optimize depends on one parameter $k \in \mathbb{N}^*$, $R(z) = \|1 - U\|^2 + \|\phi\|^2 + t^2 + k\|G_+\|^2$, with $G = \frac{t^2}{t-U} - \phi$ and $z = (t, U, \phi) \in Z$. Let $\tilde{U} = \frac{1}{t}U$, $\tilde{\phi} = \frac{1}{t}\phi$, $\gamma = \frac{1}{t}G = \frac{1}{1-\tilde{U}} - \tilde{\phi}$.

We want to minimize $R(z)$ when z runs in Z. First $\inf_z R(z) < 1$, because $U_0 = t_0 v$, $\phi_0 = 0$ gives a $z_0 \in \Gamma'$, if t_0 is small enough;

$$R(z_0) = \|1 - t_0 v\|^2 + t_0^2 + kt_0^2\|\frac{1}{1-v}\|^2 < 1,$$

if $0 < t_0 \int v < \frac{1}{2}$. We may assume that the set E without absorbing subsets is an interval $[0, r[$ (eventually making once more an automorphism of measure space on the unit interval) and we choose c and α such that $2(1 - \frac{1}{c}) < r$ and $\frac{\alpha-1}{\alpha+1} > 2 - c$. That induces the relation $\{\tilde{U} \geq \frac{1}{2}\} \subset [0, r[$ and the set $\{\tilde{U} \geq \frac{1}{2}\}$ is without absorbing subsets.

Then we show existence and unicity of an optimum $\bar{z} = ((\bar{U}, \bar{t}, \bar{\phi})$; $R(\bar{z}) = inf\{R(z) | z \in Z\} = \beta < 1$. For any optimizing sequence $z_n = (t_n, U_n, \phi_n)$, if $R(z_n) < \beta + \frac{1}{n}$, it is easy to show that (z_n) is a Cauchy sequence, because using the convexity of the function $z \mapsto \|G^+(z)\|^2$, we have

$$\|\frac{1}{2}(z_n - z_{n+k})\|^2 + \beta \leq \|\frac{1}{2}(z_n - z_{n+k})\|^2 + R(\frac{1}{2}(z_n + z_{n+k})) \leq \beta + \frac{1}{n}.$$

The optimality of $\bar{z}$ implies in particular the conditions:

(i) $\frac{\bar{t}}{k} + 2\int \frac{\bar{\gamma}}{1-\tilde{U}} \geq \int \frac{\bar{\gamma}}{(1-\tilde{U})^2}$.

(ii) $\|\bar{U}\|^2 + t^2 + \|\bar{\phi}\|^2 + k\|\bar{G}_+\|^2 = \int \bar{U}$.

(iii) $\sup_j (A^{*j} - I)[-\phi + k\bar{t}\bar{\gamma}] \leq \frac{\bar{t}\bar{\gamma}}{(1-\tilde{U})^2} + \tilde{U} - 1$.

The last inequality is proved as follows

Let J be a Borel set, $Y = 1_B$ then $U' = \bar{U} + sY$, $\phi' = \phi + s(A^j - I)Y$ where $s > 0$ and $sY < \bar{t} \wedge v \int \bar{U}h - \bar{U}$ give a new $z' \in Z$ and $R(z') \geq R(z)$ whatever be s and Y. That gives the desired result. We know that $\bar{U} < \bar{t}$ a.e. but $\bar{U}$ could be equal to $v \int \bar{U}h$ on some not negligable set. That is the reason to have two former steps of minimizations with other parameters. The last step is $k \to +\infty$ and discussion of the different cases which may occur. Without going into details we will get,

either a $g \in L_2$, $g > 0$, given by (iii) such that $\sup_n A^{*n} g \in L^1$,

or $\sup_n A^j 1 \leq W$, with $W < \infty$ a.e. and $W \leq 2$ on $[r, 1]$. That comes from $\bar{\gamma}_k = \frac{1}{1-\tilde{u}_k} - \tilde{\phi}_k \to 0$ with $\frac{1}{k}$ and $\frac{1}{1-\tilde{U}_k} \leq 2$ on $[\dot{r}, 1]$ then, if $\psi = \underline{\lim}_k \tilde{\phi}_k$ $\sup_j A^j \phi_k \leq \phi_k + 1$ gives for each j, $A^j \psi \leq \psi + 1$ and $1_{\{\psi < \infty\}} A(1_{\{\psi = \infty\}}) = 0$, but $\{\psi = \infty\} \subset [0, r]$ and could not be absorbing. So ψ is finite and we have also $A^j 1 \leq \psi + 1 = W < \infty$. Now taking a sequence of constants c, $c = 1 + \frac{1}{n}$, we may see that if (a) does not occur $r_n \downarrow 0$ and we get $A1 \leq 2$. Another change of terms like replacing $\frac{t^2}{t-U}$ by $\frac{t^2}{t-mU}$ and looking at the set $\{U \geq \frac{1}{m^2}\}$, will give, when (a) never occurs, the desired inequality $A1 \leq 1$.

At this stage of the proof, it remains to replace $g > 0$ by the constant function 1. That is done using another step of optimization with another functional, along a path very close to the precedent one.

That finishes the key proposition.

Back to the main theorem let $f_0 \in L_2$, $f_0 > 0$ (we will write $f_0 \in L_2^{++}$). If $Af_0 \not\leq f_0$, let $\tilde{T}$ and $\tilde{\mu}$, new operator and new probability, be defined by

$$\tilde{T}(h) = \frac{1}{f_0} T(f_0 h), \quad \tilde{\mu} = f_0^2 \mu$$

assuming $\|f_o\| = 1$. Then if $\hat{f}_0 = \sup_n A^{*^n} f_0$, the maximal function, we have

$$\int \hat{f}_0 f_0 d\mu = \int \sup_n \tilde{A}^{*^n} 1 d\tilde{\mu} < \infty \text{ because } \tilde{A}1 \not\leq 1$$

and the key proposition. If $g \in L_2^{++}$ $\int \widehat{(f_0 + g)}(f_0 + g) < +\infty$, then $\int \hat{f}_0 g d\mu < +\infty$ giving $\hat{f}_0 \in L_2$. Banach Steinhaus theorem proves existence of a constant K such that

$$\| \sup_n |A^n f| \| \leq K \|f\|$$

which implies $\| \sup_n \|M_n f| \| \leq \chi K \|f\|$, according to the dominating property (3). Now the proof ends as in the classical case when T is power bounded, and using the Emilion's Tauberian theorem in the general case, by the Banach principle.

If $Af_0 \leq f_0$, take another $f_1 \geq f_0$ in L_2 which is not A-subinvariant to get $\int \hat{f}_0 f_0 d\mu \leq \int \hat{f}_1 f_1 d\mu < +\infty$. If every $f_1 \geq f_0$ in L_2 is A-subinvariant, it is easy to see that any function in L_2^+ would be A-subinvariant and then would be A^*-subinvariant. In such a case the conclusions of the main theorem are obvious.

Remark: the case of a positive Cesaro-bounded operator which is not power-bounded is non empty. Such an example was obtained by Y. Derriennic and M. Lin [5].

Bibliography

[1] Akcoglu, Mustafa A. "A pointwise ergodic theorem in L_p-spaces". CAN. JN. 27, 1075-1082.

[2] (Ionescu Tulcea) Bellow, Alexandra, "Ergodic properties of isometries in L_p-spaces", Bull AMS 70(1964), 366-371.

[3] Brunel, Antoine and Emilion, Richard, "Sur les opérateurs positifs ā moyennes bornées", C.R.A.S. Paris t 298,I, no. 6, 1984.

[4] Burkhölder, Donald,L. "maximal inequalities as necessary conditions for a.e. convergence". Z.W., 3, 75-88.

[5] Derriennic, Yves and Lin, Michael, "On invariant measures and ergodic Theorems for positive operators". Jour. Funct. Anal. 13, 1973, 252-267.

[6] Emilion, Richard, "Mean bounded operators and mean ergodic theorems". J. Functional Analysis, 61(1985), 1-14.

[7] Krengel, Ulrich, "Ergodic theorems". de gruyter Studies in Mathematics, 6.

ON THE NUMBER OF ESCAPES OF A MARTINGALE AND ITS GEOMETRICAL SIGNIFICANCE

Donald L. Burkholder

University of Illinois, Department of Mathematics

Urbana, Illinois, USA

INTRODUCTION

Our main goal here is to establish probability bounds on the number of escapes of a martingale, bounds that immediately imply the convergence. These bounds, which throw new light on the behavior of B-valued martingales even in the case $B = \mathbb{R}$, are of interest in their own right and reflect the geometry of the Banach space B.

We shall begin by rephrasing Cauchy's criterion for the convergence of a sequence in B.

Let B be a real or complex Banach space with norm $|\cdot|$ and consider *the number of* ε*-escapes* of the sequence $x = (x_n)_{n\geq 0}$ where $\varepsilon > 0$ and $x_n \in B$. We denote this number by $C_\varepsilon(x)$ and define it as follows. Let

$$\nu_0(x) = \inf\{\, n \geq 0: \ |x_n| \geq \varepsilon \,\}.$$

If $\nu_0(x) = \infty$, that is, if the set on the right is empty, let $C_\varepsilon(x) = 0$ and $\nu_j(x) = \infty$ for $j \geq 1$. If $\nu_0(x) < \infty$, let

ISBN 0-12-231050-0

$$\nu_1(x) = \inf\{ n > \nu_0(x): |x_n - x_{\nu_0(x)}| \geq \varepsilon \}.$$

Let $C_\varepsilon(x) = 1$ and $\nu_j(x) = \infty$ for $j \geq 2$ if $\nu_0(x) < \nu_1(x) = \infty$. On the other hand, if $\nu_1(x) < \infty$, let

$$\nu_2(x) = \inf\{ n > \nu_1(x): |x_n - x_{\nu_1(x)}| \geq \varepsilon \}$$

and continue as above. Note that $C_\varepsilon(x) \leq j$ if and only if $\nu_j(x) = \infty$. The Cauchy criterion for convergence yields at once that x converges if and only if $C_\varepsilon(x)$ is finite for all $\varepsilon > 0$.

As we shall see, the counting function $C_\varepsilon(\cdot)$ can be used to give new information about the convergence of B-valued martingales. In the case $B = \mathbb{R}$, there are several counting methods for proving martingale convergence, for example, Doob's upcrossing method [9] and Dubins's method of rises [10]. The counting function $C_\varepsilon(\cdot)$, related to one introduced by Davis [8] in the real case, is dimension free.

Let $(\Omega, \mathcal{F}, P)$ be a probability space and $(\mathcal{F}_n)_{n\geq 0}$ a nondecreasing sequence of sub-σ-fields of $\mathcal{F}$. Suppose that $f = (f_n)_{n\geq 0}$ and $g = (g_n)_{n\geq 0}$ are B-valued martingales with respect to the filtration $(\mathcal{F}_n)_{n\geq 0}$. Let d be the difference sequence of f and e the difference sequence of g: $f_n = \sum_{k=0}^{n} d_k$ and $g_n = \sum_{k=0}^{n} e_k$ for all $n \geq 0$. The martingale g is *differentially subordinate* to f if

$$|e_k(\omega)| \leq |d_k(\omega)| \tag{1}$$

for all $\omega \in \Omega$ and $k \geq 0$.

In the following theorem, $B = H$ where H is a real or complex Hilbert space, the $\mathcal{F}$-measurable function $C_\varepsilon(g)$ is defined by $C_\varepsilon(g)(\omega) = C_\varepsilon(g(\omega))$, and $\|f\|_1 = \sup_{n\geq 0}\|f_n\|_1$.

THEOREM 1. *Let* f *and* g *be* H-*valued martingales with respect to the same filtration. If* g *is differentially subordinate to* f *then, for all* $j \geq 1$,

$$P(C_\varepsilon(g) \geq j) \leq 2\|f\|_1/\varepsilon j^{\frac{1}{2}}. \tag{2}$$

Both the constant 2 *and the exponent* 1/2 *are best possible.*

Consequently, the finiteness of $\|f\|_1$ implies that $C_\varepsilon(g)$ is finite almost everywhere, hence that g converges almost everywhere. Note that g can be equal to f.

Davis [8] shows that the value of his counting function applied to a real-valued martingale f is at least j with probability no greater than $44\|f\|_1/\varepsilon j^{\frac{1}{2}}$. The proof does not yield information about g.

The proof of Theorem 1 will be given in the next section.

If B is a Banach space and (2) holds for all B-valued martingales f and g as above, then B is a Hilbert space. This can be seen as follows. Let g^* denote the maximal function of g: $g^*(\omega) = \sup_{n\geq 0}|g_n(\omega)|$. Then

$$P(g^* > \varepsilon) \leq P(C_\varepsilon(g) \geq 1) \tag{3}$$

so (2), together with a continuity argument, gives

$$P(g^* \geq \varepsilon) \leq 2\|f\|_1/\varepsilon. \tag{4}$$

But by Theorem 2.2 of [4] , the inequality (4) holds only if B is a Hilbert space. So (2) characterizes Hilbert space.

There is a less precise version of (2) that holds for spaces isomorphic to some Hilbert space.

THEOREM 2. *Suppose that* B *is a Banach space isomorphic to some Hilbert space. There is a real number* $\beta = \beta(B)$ *such that if* f *and* g *are* B*-valued martingales with respect to the same filtration,* g *is differentially subordinate to* f , *and* $j \geq 1$, *then*

$$P(C_\varepsilon(g) \geq j) \leq \beta \|f\|_1 / \varepsilon j^{\frac{1}{2}}. \tag{5}$$

If, in addition, $\|f\|_1$ *is finite, then* g *converges almost everywhere.*

The convergence follows from the inequality as before; for a different proof, see [3] . Convergence in the real-valued case is one of the results of [1] .

Theorem 2 is as far as one can go in this direction with differentially subordinate B-valued martingales. Even the last assertion of the theorem does not hold if B is not isomorphic to a Hilbert space ([3] , Section 5). However, if g is assumed to be the transform of a B-valued martingale f by a scalar-valued predictable sequence v uniformly bounded in absolute value by 1 , then g has good behavior for a much larger class of Banach spaces, the UMD class [3]. Here we shall show that an inequality of nearly the same form as (5) is satisfied.

Recall that such a transform g of f has the form $e_n = v_n d_n$ where v_n is $\mathcal{F}_{n-1}$-measurable (assume $\mathcal{F}_{-1} \subset \mathcal{F}_0$) so that the condition $|v_n(\omega)| \leq 1$ implies that g is a martingale with respect to $\mathcal{F}$ and (1) is satisfied. Thus, g is differentially subordinate to f but more is true: $e_n(\omega)$ is in the linear manifold determined by $d_n(\omega)$.

THEOREM 3. *Let* B *be UMD . There are strictly positive real numbers* $\alpha = \alpha(B)$ *and* $\beta = \beta(B)$ *such that if* f *is a* B*-valued martingale and* g *is the transform of* f *by a scalar-valued predictable sequence* v *uniformly bounded in absolute value by* 1 *, then, for all* $j \geq 1$,

$$P(C_\varepsilon(g) \geq j) \leq \beta \|f\|_1 / \varepsilon j^\alpha. \tag{6}$$

If, in addition, $\|f\|_1$ *is finite, then* g *converges almost everywhere.*

For a sharper version of (6) , see (18) or (24) depending on whether v is real-valued or complex-valued.

The convergence of g in the real case was first proved in [1] ; the UMD case is in [3] . If B is not UMD then the inequality (6) cannot hold, for otherwise, convergence would hold, contradicting Theorem 1.1 of [3] . Therefore, the class of UMD spaces is characterized by (6) .

Nevertheless, by further restricting g , we can say something more. In fact, we let g = f and consider the larger class of superreflexive Banach spaces.

THEOREM 4. *Let* B *be superreflexive. There are strictly positive real numbers* $\alpha = \alpha(B)$ *and* $\beta = \beta(B)$ *such that if* f *is a* B*-valued martingale and* j *is a positive integer, then*

$$P(C_\varepsilon(f) \geq j) \leq \beta \|f\|_1 / \varepsilon j^\alpha. \tag{7}$$

For more information about α and β, see the proof of this theorem near the end of the paper.

If B is not superreflexive, then this inequality does not hold. Suppose otherwise. Let $0 < \varepsilon < 1$ and choose j so large that $\beta/\varepsilon j^{\alpha} < 1$. By the work of James [11] and Pisier [14], there is a B-valued martingale f with $\|f\|_1 \le \|f\|_\infty \le 1$ such that $|d_k(\omega)| \ge \varepsilon$ for all $\omega \in \Omega$ and $k < j$. Therefore, $P(C_\varepsilon(f) \ge j) = 1$, contradicting (7).

So each of the inequalities (2), (5), (6), and (7) characterizes a class of Banach spaces. Inequality (2) characterizes the class of Hilbert spaces, (5) the class of spaces isomorphic to a Hilbert space, (6) the UMD class, and (7) the class of superreflexive spaces.

PROOF OF THEOREM 1

The proof rests on two lemmas of interest in their own right.

LEMMA 1. *Let* f *and* g *be* H*-valued martingales with respect to the same filtration. If* g *is differentially subordinate to* f, *then, for* $\lambda > 0$ *and* $n \ge 1$,

$$P(|g_n| \ge \lambda) \le 2\|f_n\|_1/\lambda. \tag{8}$$

The proof is contained in [7] where more is proved. Here (8) is sufficient for our purposes. The idea of the proof is to find a function $L\colon H \times H \longrightarrow \mathbb{R}$ with the following properties:

$$L(x, y) \le 2|x| \quad \text{if } |x| + |y| \ge 1, \tag{9}$$

$$L(x, y) \le 2|x| + 1, \tag{10}$$

$$L(x, y) \geq 1 \quad \text{if } |y| \leq |x|, \tag{11}$$

$$EL(f_n, g_n) \geq EL(f_{n-1}, g_{n-1}). \tag{12}$$

In fact, L can be the function on $H \times H$ defined by

$$L(x, y) = 2|x| \quad \text{if } |x| + |y| \geq 1,$$
$$= 1 + |x|^2 - |y|^2 \quad \text{if } |x| + |y| < 1.$$

Then, by (9) and (10),

$$\begin{aligned} P(|f_n| + |g_n| \geq 1) &\leq P(2|f_n| \geq L(f_n, g_n)) \\ &= P(2|f_n| - L(f_n, g_n) + 1 \geq 1) \\ &\leq E[2|f_n| - L(f_n, g_n) + 1]. \end{aligned}$$

By (11) and (12),

$$EL(f_n, g_n) \geq EL(f_0, g_0) = EL(d_0, e_0) \geq 1.$$

Therefore, $P(|g_n| \geq 1) \leq 2\|f_n\|_1$ and (8) follows.

The above lemma yields an inequality for

$$S(g, \tau) = \left[|g_{\tau_0}|^2 + \sum_{j=0}^{\infty} |g_{\tau_j} - g_{\tau_{j-1}}|^2\right]^{\frac{1}{2}} \tag{13}$$

where $\tau = (\tau_j)_{j \geq 0}$ is a nondecreasing sequence of stopping times with values in the set of nonnegative integers.

LEMMA 2. *Let* f *and* g *be* H-*valued martingales with respect to the same filtration and* $\tau = (\tau_j)_{j \geq 0}$ *a nondecreasing sequence of finite stopping times. If* g *is differentially subordinate to* f, *then, for all* $\lambda > 0$,

$$P(\, S(g, \tau) \geq \lambda \,) \leq 2\|f\|_1/\lambda. \tag{14}$$

PROOF. We can assume that the τ_j are uniformly bounded by a positive integer N. Let $K = \ell^2_H$, the Hilbert space of sequences $x = (x_0, x_1, \cdots)$ with $x_j \in H$ and

$$|x|_K = \left(\sum_{j=0}^{\infty} |x_j|^2\right)^{\frac{1}{2}} < \infty.$$

Define K-valued martingales F and G, with respective difference sequences D and E, by

$$\begin{aligned} D_k(\omega) &= (d_k(\omega), 0, 0, \cdots), \\ E_k(\omega) &= (e_k(\omega), 0, 0, \cdots) \quad \text{if } k \leq \tau_0(\omega), \\ E_k(\omega) &= (0, \cdots, 0, e_k(\omega), 0, \cdots) \quad \text{if } \tau_{j-1}(\omega) < k \leq \tau_j(\omega) \text{ and } j \geq 1, \end{aligned}$$

where $e_k(\omega)$ is in the j-th position. Note that G is differentially subordinate to F since

$$|E_k(\omega)|_K = |e_k(\omega)| \leq |d_k(\omega)| = |D_k(\omega)|_K.$$

Also,

$$F_N = (f_N, 0, 0, \cdots)$$
$$G_N = (g_{\tau_0}, g_{\tau_1} - g_{\tau_0}, \cdots)$$

so $|F_N|_K = |f_N|$ and $|G_N|_K = S(g, \tau)$. Therefore, by Lemma 1,

$$P(S(g, \tau) \geq \lambda) = P(|G_N|_K \geq \lambda) \leq 2\|F_N\|_1/\lambda \leq 2\|f\|_1/\lambda.$$

This completes the proof of Lemma 2.

We can now prove Theorem 1. To prove the inequality (2), we can assume that the martingale f is of finite length: there is a positive integer N such that $f_n = f_N$ for all $n \geq N$. Using $\nu_0, \nu_1, \cdots$ as defined in the introduction, we let $\tau_j = \nu_j(g) \wedge N$ where $\nu_j(g)(\omega) = \nu_j(g(\omega))$. Then $\tau = (\tau_j)_{j \geq 0}$ is a nondecreasing sequence of stopping times bounded by N and, by (13),

$$\varepsilon^2 C_\varepsilon(g) \leq S^2(g, \tau). \tag{15}$$

Therefore, by Lemma 2,

$$P(C_\varepsilon(g) \geq j) \leq P(S^2(g, \tau) \geq \varepsilon^2 j) \leq 2\|f_N\|_1/\varepsilon j^{\frac{1}{2}} \leq 2\|f\|_1/\varepsilon j^{\frac{1}{2}}.$$

So the proof of (2) rests on (8). The order can be reversed: (8) follows from (4) which follows from (2). One consequence is that 2 is the best possible constant in (2); see [2]. To see that the exponent 1/2 is best possible, consider simple random walk stopped at time j.

Using some of the inequalities of [6] in place of (8), we get L^p and exponential inequalities for $C_\varepsilon(g)$.

PROOF OF THEOREM 2

Suppose that B is isomorphic to H where H is a Hilbert space with norm $\|\cdot\|$. So there is a linear map T from B onto H and strictly positive real numbers γ and δ such that

$$\gamma\|Tx\| \le |x| \le \delta\|Tx\| \tag{16}$$

for all $x \in B$. Suppose that f and g are B-valued martingales with respect to $(\mathcal{F}_n)$ where g is differentially subordinate to f and consider the H-valued martingales F and G defined by

$$F_n(\omega) = \delta T f_n(\omega),$$
$$G_n(\omega) = \gamma T g_n(\omega).$$

To see that F_n and G_n are $\mathcal{F}_n$-measurable and integrable, approximate f_n and g_n by simple $\mathcal{F}_n$-measurable functions and then use (16). The martingale condition can also be checked by using approximation: T commutes with expectation. Furthermore, G is differentially subordinate to F: $\|E_n(\omega)\| = \|\gamma T e_n(\omega)\| \le |e_n(\omega)| \le |d_n(\omega)| \le \|\delta T d_n(\omega)\| = \|D_n(\omega)\|$. By a similar argument, $C_\varepsilon(g) \le C_{\gamma\varepsilon/2\delta}(G)$ and $\|F_n\| \le \delta\gamma^{-1}|f_n|$ so, by Theorem 1,

$$P(C_\varepsilon(g) \ge j) \le P(C_{\gamma\varepsilon/2\delta}(G) \ge j) \le 2\,\|F\|_1/(\gamma\varepsilon/2\delta)j^{\frac{1}{2}} \le 4\delta^2\|f\|_1/\gamma^2\varepsilon j^{\frac{1}{2}}.$$

This implies Theorem 2.

Let $\beta_p(B)$ be the least $\beta \leq \infty$ such that if d is a B-valued martingale difference sequence, then

$$\|\sum_{k=0}^{n} \varepsilon_k d_k\|_p \leq \beta \|\sum_{k=0}^{n} d_k\|_p \tag{17}$$

for all choices of $\varepsilon_k \in \{1, -1\}$ and all $n \geq 0$. Here $1 < p < \infty$. If, in addition, $2 \leq q < \infty$, let $\gamma_{p,q}(B)$ be the least $\gamma \leq \infty$ such that

$$\|S(f, q)\|_p \leq \gamma \|f\|_p$$

for all B-valued martingales f where $S(f, q) = (\sum_{k=0}^{\infty} |d_k|^q)^{\frac{1}{q}}$.

THEOREM 5. *If* f *is a* B-*valued martingale and* g *is the transform of* f *by a real predictable sequence* v *uniformly bounded in absolute value by* 1 *, then, for all* $j \geq 1$ *,*

$$P(C_\varepsilon(g) \geq j) \leq 2\beta_2(B)\gamma_{2,q}(B)\|f\|_1/\varepsilon j^{\frac{1}{q}}. \tag{18}$$

If $B = H$ and $q = 2$, the constant on the right is best possible: $\beta_2(H) = \gamma_{2,2}(H) = 1$ and the example showing that 2 is the best constant in Theorem 1 shows that 2 is also best here. So (18), as well as (24) below, contains for transforms the full strength of (2) in the special case $B = H$.

PROOF OF THEOREM 5

To prove (18), we can assume that $f_0 = v_0 \equiv 0$ and, if $k \geq 1$, then

$$v_k = \varphi_k(f_0, \cdots, f_{k-1}) \tag{19}$$

where $\varphi_k\colon B \times \cdots \times B \to [-1, 1]$ is continuous. To see this, let $\delta > 0$ and fix $z \in B$ with $|z| = 1$. Assume without loss of generality that $(r_k)_{k \geq 0}$ is an independent sequence on $(\Omega, \mathcal{F}, P)$ satisfying

$$P(r_k = 1) = P(r_k = -1) = 1/2$$

and that $(r_k)_{k \geq 0}$ is independent of $\vee_n \mathcal{F}_n$. Define a new martingale F and its transform G as follows: $F_n = \sum_{k=0}^{n} D_k$ and $G_n = \sum_{k=0}^{n} V_k D_k$ where $D_0 = V_0 \equiv 0$ and, for $k \geq 1$, $V_{3k-2} = V_{3k-1} \equiv 0$, $V_{3k} = v_{k-1}$, and

$$\begin{aligned} D_{3k-2} &= 2^{-k}\delta\, r_{3k-2}\, v_{k-1}^{+}\, z, \\ D_{3k-1} &= 2^{-k}\delta\, r_{3k-1}\, v_{k-1}^{-}\, z, \\ D_{3k} &= r_0\, d_{k-1}. \end{aligned}$$

Here $v_{k-1} = v_{k-1}^{+} - v_{k-1}^{-}$ and $|v_k| = v_{k-1}^{+} + v_{k-1}^{-} \leq 1$. It is easy to check that D is a martingale difference sequence relative to the filtration that it generates. Note that $F_0 \equiv 0$ and, if $n \geq 1$, then

$$|F_{3n}| = \Big|\sum_{k=1}^{3n} D_k\Big| \leq \Big|\sum_{k=1}^{n} d_{k-1}\Big| + \Big|\sum_{k=1}^{n} 2^{-k}\delta(v_{k-1}^{+} + v_{k-1}^{-})\Big| \leq |f_{n-1}| + \delta.$$

Thus, $\|F\|_1 \leq \|f\|_1 + \delta$. Also,

$$G = (G_0, G_1, G_2, G_3, \cdots)$$
$$= (0, 0, 0, r_0g_0, r_0g_0, r_0g_0, r_0g_1, r_0g_1, r_0g_1, \cdots)$$

so that $C_\varepsilon(G) = C_\varepsilon(g)$. To check that F and V satisfy (19) , let $\varphi(t)$ be the number in $[-1, 1]$ nearest to t for $t \in \mathbb{R}$. If $k \geq 1$, then

$$V_{3k} = v_{k-1} = \varphi(v_{k-1}) = \varphi(v_{k-1}^+ - v_{k-1}^-) = \varphi(|2^k \delta^{-1} D_{3k-2}| - |2^k \delta^{-1} D_{3k-1}|).$$

So, clearly, F and V satisfy (19) . If the inequality (18) holds for F and G , then

$$P(C_\varepsilon(g) \geq j) = P(C_\varepsilon(G) \geq j)$$
$$\leq 2\beta_2(B)\ \gamma_{2,q}(B)\ \|F\|_1/\varepsilon j^{\frac{1}{q}}$$
$$\leq 2\beta_2(B)\ \gamma_{2,q}(B)\ (\|f\|_1 + \delta)/\varepsilon j^{\frac{1}{q}}.$$

Consequently, we can assume, and do assume in the remainder of the proof, that f and v satisfy the condition stated in the first sentence of this paragraph.

The next step is to prove an inequality analogous to Lemma 2. Let $\tau = (\tau_j)_{j\geq 0}$ be a nondecreasing sequence of finite stopping times and set

$$S(g, q, \tau) = \left[|g_{\tau_0}|^q + \sum_{j=1}^{\infty} |g_{\tau_j} - g_{\tau_{j-1}}|^q\right]^{\frac{1}{q}}.$$

Then, for all $\lambda > 0$,

$$P(S(g, q, \tau) \geq \lambda) \leq 2\beta_2(B)\ \gamma_{2,q}(B)\ \|f\|_1/\lambda. \tag{20}$$

Theorem 5 follows from this inequality just as Theorem 1 follows from (14) . Note that the inequality (15) is replaced by

$$\varepsilon^q C_\varepsilon(g) \leq S^q(g, q, \tau).$$

For convenience, let $\beta_p = \beta_p(B)$ and $\gamma_{p,q} = \gamma_{p,q}(B)$. Then, for $1 < p < \infty$ and $2 \leq q < \infty$,

$$\|S(g, q, \tau)\|_p \leq \beta_p\ \gamma_{p,q}\ \|f\|_p. \tag{21}$$

By Remark 2.1 of [5] and the assumption that v is a *real* predictable sequence uniformly bounded by 1 in absolute value,

$$\|g\|_p \leq \beta_p \|f\|_p. \tag{22}$$

To prove (21) , we can assume there is a positive integer N such that the nondecreasing sequence τ of finite stopping times is uniformly bounded by N ; otherwise, replace τ_n by $\tau_n \wedge N$. Then $h = (g_{\tau_n})_{n\geq 0}$ is a B-valued martingale and by the definition of $\gamma_{p,q}$ we have

$$\|S(g, q, \tau)\|_p = \|S(h, q)\|_p \leq \gamma_{p,q}\|h\|_p \leq \gamma_{p,q}\|g\|_p \leq \beta_p\ \gamma_{p,q}\|f\|_p.$$

This gives (21) , which we shall need later.

To prove (20) , we shall use some ideas from [2] and [13] . By a theorem of

McConnell [13] , there is a probability space $(\Omega', \mathscr{F}', P')$ on which is defined a B-valued martingale $M = (M_t)_{t\geq 0}$ with continuous paths and stopping times $0 \equiv \sigma_0 \leq \sigma_1 \leq \cdots$ such that f has the same distribution as $(M_{\sigma_n})_{n\geq 0}$. Let $\delta > 0$ and define additional stopping times $0 \leq \mu_0 \leq \mu_1 \leq \cdots$ by

$$\mu_0 = \inf\{\, t > 0\colon |M_t| > \delta/2 \,\},$$
$$\mu_1 = \inf\{\, t > \mu_0\colon |M_t - M_{\mu_0}| > \delta/2 \,\},$$

and so forth. Let $0 \equiv \rho_0 \leq \rho_1 \leq \cdots$ be the stopping times obtained by interlacing $(\sigma_n)_{n\geq 0}$ and $(\mu_n \wedge \sigma_\infty)_{n\geq 0}$ where $\sigma_\infty = \lim_{j\to\infty} \sigma_j$. That is, ρ_n is the n-th order statistic associated with the random variables $\sigma_0, \sigma_1, \cdots, \mu_0 \wedge \sigma_\infty, \mu_1 \wedge \sigma_\infty, \cdots$. Let $F_n = M_{\rho_n}$. Then $F = (F_n)_{n\geq 0}$ is a martingale with a difference sequence D satisfying $D^* \leq \delta$, where D^* denotes the maximal function of D , and $\|F\|_1 \leq \|f\|_1$. Note that f has the same distribution as $(F_{\eta_n}) = (M_{\sigma_n})$ where the η_n are stopping times associated with a filtration with respect to which F is a martingale. By (19) ,

$$g_n = \sum_{k=1}^{n} \varphi_k(f_0, \cdots, f_{k-1})d_k.$$

Therefore, g has the same distribution as

$$\Big(\sum_{k=0}^{\eta_n} V_k D_k \Big)_{n\geq 0}$$

where V is predictable and is uniformly bounded by 1 in absolute value. Let $G_n = \sum_{k=0}^{n} V_k D_k$. Accordingly, there is a nondecreasing sequence $T = (T_n)_{n\geq 0}$ of finite stopping times such that (g_{τ_n}) has the same distribution as (G_{T_n}). In particular, $S(g, q, \tau)$ has the same distribution as $S(G, q, T)$. The advantage of F and G is that $D^* \leq \delta$.

Now let $\lambda > 0$ and $b > 0$. Define the stopping time R by

$$R = \inf\{ n \geq 0: |F_n| > b\lambda \}$$

and let F^R and G^R be the martingales $(F_{R\wedge n})_{n\geq 0}$ and $(G_{R\wedge n})_{n\geq 0}$ respectively. Note that G^R is the transform of F^R by V. We have that

$$\begin{aligned} P'(S(G, q, T) \geq \lambda, F^* \leq b\lambda) &= P'(S(G^R, q, T) \geq \lambda, R = \infty) \\ &\leq \lambda^{-p} \|S(G^R, q, T)\|_p^p, \end{aligned}$$

which, by (21), is less than or equal to $\lambda^{-p} \beta_p^p \gamma_{p,q}^p \|F^R\|_p^p$. Using the definition of R and the fact that $D^* \leq \delta$, we see that

$$\|F^R\|_p^p \leq (b\lambda + \delta)^{p-1} \|F^R\|_1 \leq (b\lambda + \delta)^{p-1} \|F\|_1.$$

Therefore,

$$\begin{aligned} P(S(g, q, \tau) \geq \lambda) &= P'(S(G, q, T) \geq \lambda) \\ &\leq P'(S(G, q, T) \geq \lambda, F^* \leq b\lambda) + P(F^* > b\lambda) \\ &\leq \lambda^{-1} \left[\beta_p^p \gamma_{p,q}^p (b + \delta\lambda^{-1})^{p-1} + b^{-1} \right] \|F\|_1. \end{aligned}$$

Substituting $\|f\|_1$ for $\|F\|_1$ and letting $\delta \to 0$, we obtain

$$P(S(g, q, \tau) \geq \lambda) \leq \lambda^{-1}\left[b^{p-1}\beta_p^p\,\gamma_{p,q}^p + b^{-1}\right]\|f\|_1.$$

If $p = 2$ and $b = \beta_2^{-1}\,\gamma_{2,q}^{-1}$, then this inequality is the desired inequality (20).

This completes the proof of Theorem 5.

PROOF OF THEOREM 3

Here the condition on B is the UMD condition: $\beta_p(B) < \infty$ for some (equivalently, for all) $p \in (1, \infty)$. This condition implies that B is superreflexive (Maurey [12]). By a theorem of Pisier [14], superreflexivity implies that $\gamma_{p,q}(B)$ is finite for some $q \in [2, \infty)$ and all $p \in (1, \infty)$.

Therefore, Theorem 5 gives Theorem 3 with $\alpha = 1/q$ and $\beta = 2\beta_2(B)\,\gamma_{2,q}(B)$ in the special case that the predictable sequence v is real. The complex case follows with the use of the inequality

$$C_{4\varepsilon}(x + y) \leq C_\varepsilon(x) + C_\varepsilon(y) \tag{23}$$

in the obvious way. Here $x = (x_n)_{n\geq0}$ and $y = (y_n)_{n\geq0}$ where $x_n, y_n \in B$. The number 4 cannot be replaced by a smaller number as the following example shows. Let $B = R$, $\varepsilon = 1$, and $0 < \delta < 1/2$. Then

$$x = (0, 2 - \delta, 3 - 2\delta, 1, 3 - 2\delta, 1, \cdots),$$
$$y = (0, 0, 1 - \delta, -1 + \delta, 1 - \delta, -1 + \delta, \cdots)$$

satisfy $C_1(x) = 1$, $C_1(y) = 0$, but $C_{4(1-\delta)}(x + y) = \infty$.

We shall not prove (23) here since there is alternative approach to Theorem 3 that yields a better constant, in fact, the best constant if $B = H$.

Let $\beta_p(B, \mathbb{R}) = \beta_p(B)$, the UMD constant defined in (17) , and recall the inequality (22) in which v is real. Let $\beta_p(B, \mathbb{C})$ be the best constant in the complex case: B is a Banach space over the complex field $\mathbb{C}$ and $\beta_p(B, \mathbb{C})$ is the least $\beta \leq \infty$ such that if f is a B-valued martingale and g is its transform by a complex predictable sequence v uniformly bounded in modulus by 1 , then

$$\|g\|_p \leq \beta\|f\|_p.$$

If $B = H$, then $\beta_p(B, \mathbb{C}) = \beta_p(B, \mathbb{R})$ as is shown in [6] . In any case, $\beta_p(B, \mathbb{C}) \leq 2\beta_p(B, \mathbb{R})$ as is obvious.

The analogue of (18) holds:

$$P(C_\varepsilon(g) \geq j) \leq 2\beta_2(B, \mathbb{C})\ \gamma_{2,q}(B)\ \|f\|_1/\varepsilon j^{\frac{1}{q}}. \tag{24}$$

The proof of this has the same pattern as the proof of (18) . Only slight modifications are needed. For example, one can assume that (19) holds where φ_k is a continuous function from $B \times \cdots \times B$ to the closed unit disk of the complex plane. Here, for $k \geq 1$, $V_{5k-4} = \cdots = V_{5k-1} \equiv 0$, $V_{5k} = v_{k-1}$, and

$$\begin{aligned}
D_{5k-4} &= 2^{-k}\,\delta\, r_{5k-4}\,(\operatorname{Re} v_{k-1})^+\, z,\\
D_{5k-3} &= 2^{-k}\,\delta\, r_{5k-3}\,(\operatorname{Re} v_{k-1})^-\, z,\\
D_{5k-2} &= 2^{-k}\,\delta\, r_{5k-2}\,(\operatorname{Im} v_{k-1})^+\, z,\\
D_{5k-1} &= 2^{-k}\,\delta\, r_{5k-1}\,(\operatorname{Re} v_{k-1})^-\, z,\\
D_{5k} &= r_0\, d_{k-1}.
\end{aligned}$$

Let φ be the map on $\mathbb{C}$ that takes each point of $\mathbb{C}$ to the nearest point in the closed unit disk. Then $V_{5k} = v_{k-1} = \varphi(v_{k-1})$ can be written as

$$\varphi((|2^k\delta^{-1}D_{5k-4}| - |2^k\delta^{-1}D_{5k-3}|) + i(|2^k\delta^{-1}D_{5k-2}| - |2^k\delta^{-1}D_{5k-1}|)),$$

so $V_k = \varphi_k(F_0, \cdots, F_{k-1})$ where φ_k is a continuous function from $B \times \cdots \times B$ to the closed unit disk. The remainder of the proof of (24) is the same as in the proof of (18). This completes the proof of Theorem 3.

PROOF OF THEOREM 4

The proof of Theorem 5 simplifies in the case $v \equiv (1, 1, \cdots)$ and gives

$$P(C_\varepsilon(f) \geq j) \leq 2\, \gamma_{2,q}(B)\, \|f\|_1/\varepsilon j^{\frac{1}{q}}. \tag{25}$$

Again, by [14], the constant $\gamma_{2,q}(B)$ is finite for some $q \in [2, \infty)$. Thus, (25) gives Theorem 4 with $\alpha = 1/q$ and $\beta = 2\gamma_{2,q}(B)$ for this choice of q.

REFERENCES

1. D. L. Burkholder, Martingale transforms, *Ann. Math. Statist.* 37 (1966), 1494 – 1504.

2. D. L. Burkholder, A sharp inequality for martingale transforms, *Ann. Probab.* 7 (1979), 858 – 863.

3. D. L. Burkholder, A geometrical characterization of Banach spaces in which martingale difference sequences are unconditional, *Ann. Probab.* 9 (1981), 997 – 1011.

4. D. L. Burkholder, Martingale transforms and the geometry of Banach spaces, Proceedings of the Third International Conference on Probability in Banach Spaces, Tufts University, 1980, *Lecture Notes in Mathematics* 860 (1981), 35 – 50.

5. D. L. Burkholder, Boundary value problems and sharp inequalities for martingale transforms, *Ann. Probab.* 12 (1984), 647 – 702.

6. D. L. Burkholder, Sharp inequalities for martingales and stochastic integrals, Colloque Paul Lévy, *Astérisque* 157 – 158 (1988), 75 – 94.

7. D. L. Burkholder, Differential subordination of harmonic functions and martingales, Proceedings of the Seminar on Harmonic Analysis and Partial Differential Equations (El Escorial, Spain, 1987), *Lecture Notes in Mathematics*, to appear.

8. Burgess Davis, A comparison test for martingale inequalities, *Ann. Math. Statist.* 40 (1969), 505 – 508.

9. J. L. Doob, *Stochastic Processes*, Wiley, New York, 1953.

10. Lester E. Dubins, Rises and upcrossings of nonnegative martingales, *Illinois J. Math.* 6 (1962), 226 – 241.

11. Robert C. James, Some self-dual properties of normed linear spaces, *Ann of Math. Studies* 69 (1972), 159 – 175.

12. B. Maurey, Système de Haar, *Séminaire Maurey-Schwartz*, 1974 – 1975, Ecole Polytechnique, Paris, 1975.

13. Terry R. McConnell, A Skorohod-like representation in infinite dimensions, Probability in Banach Spaces V, *Lecture Notes in Mathematics*, 1153 (1985), 359 – 368.

14. Gilles Pisier, Martingales with values in uniformly convex spaces, *Israel J. Math.* 20 (1975), 326 – 350.

LOSS OF RECURRENCE IN REINFORCED RANDOM WALK

Burgess Davis

Purdue University, Department of Statistics

West Lafayette, Indiana, USA

In this note it is shown that a necessary and sufficient condition on the initial weighting a_i, $-\infty < i < \infty$, of a reinforced random walk $X_0, X_1, \ldots$, to guarantee that $P(\lim_{n\to\infty} |X_n| = \infty) = 0$, is that both $\sum_{i=1}^{\infty} a_i^{-2} = \infty$ and $\sum_{i=1}^{\infty} a_{-i}^{-2} = \infty$. Together with an old result of T. E. Harris, this characterizes those initial weightings which, if unreinforced, correspond to a recurrent process, but which, if suitably reinforced, yield a process converging to infinity with positive probability, and in particular shows that there are such initial weightings.

It is easily seen that any Markov process, with state space the integers and stationary transition probabilities $p_{i,j}$ which satisfy $p_{i,i+1} + p_{i,i-1} = 1$, $p_{i,i+1} > 0$, and $p_{i,i-1} > 0$, is associated with a set w_i, $-\infty < i < \infty$, of positive numbers, unique up to multiplication by constants, by the relations

$$p_{i,i+1} = 1 - p_{i,i-1} = \frac{w_i}{w_{i-1} + w_i}. \tag{1}$$

We will refer to w_i as the weight of the interval $(i, i+1)$. We have $p_{i,i+1}/p_{i,i-1} = w_i/w_{i-1}$, that is, the transition probabilities from i to $i+1$ and $i-1$ are proportional

ISBN 0-12-231050-0

to the weights of the connecting intervals. For example, to get ordinary fair coin tossing random walk, the weights of all intervals must be equal.

Recently Coppersmith and Diaconis introduced a model in which the weight of each interval $(i, i+1)$ is initially 1 and is increased by 1 each time the process jumps across it, so that its weight at time n is one plus the number of indices $k \leq n$ such that (X_k, X_{k+1}) is either $(i, i+1)$ or $(i+1, i)$, where X_n is the state of the process at time n. Given $\{X_0 = i_0, \ldots, X_n = i_n\}$, X_{n+1} is either $i_n + 1$ or $i_n - 1$ with probabilities proportional to the weights at time n of $(i_n, i_n + 1)$ and $(i_n - 1, i_n)$ respectively. See [2] or [4] for a more detailed description of this and related processes, as well as a description of the methods of exchangeability theory and random walk in a random environment which may be used to study them.

In [2] the author considers the more general setting in which any positive initial weights are permitted and any nonnegative reinforcement which does not depend on the future is allowed. In this situation the methods mentioned above can not be used, and martingales are the principal tool. Formally, let I stand for the integers. We define a reinforced random walk (RRW) on the integers to be a sequence $X_0, X_1, \ldots = \boldsymbol{X}$ of integer valued random variables and a collection $\boldsymbol{w} = \{w(n, j),\ n = 0, 1, 2, \ldots,\ j \in I\}$ of positive random variables such that

i) $w(n, i) \geq w(n-1, i)$, $0 < n < \infty$, $i \in I$, with equality unless (X_{n-1}, X_n) is either $(i, i+1)$ or $(i+1, i)$, and

ii) $$P(X_{n+1} = i+1 | X_n = i, \mathcal{F}_n) = 1 - P(X_{n+1} = i-1 | X_n = i, \mathcal{F}_n) = w(n, i)/[w(n, i-1) + w(n, i)],\ n \geq 0,\ i \in I,$$

where $\mathcal{F}_n = \sigma\{X_j,\ 0 \le j \le n,\ w(k,i),\ 0 \le k \le n,\ i \in I\}$. Usually we will just use $\boldsymbol{X}$ to designate a RRW, and do not explicitly mention the weights. We call $\{w(0,j)\colon\ j \in I\}$ the initial weights of $\boldsymbol{X}$, which from now on are assumed to be constants; if they are not, $\boldsymbol{X}$ may be analyzed by conditioning on them. We call $\boldsymbol{X}$ *initially fair* if all the initial weights are equal and *initially recurrent* if the Markov process with transition probabilities $p_{i,j}$ satisfying $p_{i,i+1} = 1 - p_{i,i-1} = w(0,i)/(w(0,i-1) + w(0,i))$, $i \in I$, is recurrent. We set $R(\boldsymbol{X}) = \{j \in I\colon\ X_n = j \text{ for some } n\}$, and call $\{R(\boldsymbol{X}) = I\}$ the set where $\boldsymbol{X}$ is recurrent and $\{R(\boldsymbol{X})$ has finite cardinality$\}$ the set where $\boldsymbol{X}$ has finite range. In [2] initially fair processes are studied, and the following theorem is proved.

Theorem 1. *If $\boldsymbol{X}$ is initially fair then*

$$P(\text{range of } \boldsymbol{X} \text{ is finite}) + P(\boldsymbol{X} \text{ is recurrent}) = 1.$$

Of course, fair random walk is recurrent, so one way to interpret this result is to say that this property is preserved under reinforcement unless the reinforcement is so strong as to trap the process in a finite number of states. This paper addresses the question of whether this remains true for any initially recurrent RRW. Explicit criteria are available for initial recurrence. An old result of T. E. Harris (see [2], or [3] p. 109) is equivalent to the statement that a Markov process like that discussed in the first paragraph of this paper is recurrent if and only if both the sums $\sum_{i=1}^{\infty} w_i^{-1}$ and $\sum_{i=1}^{\infty} w_{-i}^{-1}$ are infinite, where the w_i are given by (1). Thus a RRW $\boldsymbol{X}$ is initially recurrent if and only if both $\sum_{i=1}^{\infty} w(0,i)^{-1}$ and $\sum_{i=1}^{\infty} w(0,-i)^{-1}$ are infinite.

The purpose of this paper is to prove the following theorem.

Theorem 2. *Let $\{a_i, \ -\infty < i < \infty\}$ be a set of positive numbers. A necessary and sufficient condition in order that for each* RRW $\boldsymbol{X}$ *with* $w(0,i) = a_i, \ -\infty < i < \infty$, *we have*

$$P(\boldsymbol{X} \text{ is recurrent}) + P(\boldsymbol{X} \text{ has finite range}) = 1,$$

is that both $\sum_{i=1}^{\infty} a_i^{-2}$ *and* $\sum_{i=1}^{\infty} a_{-i}^{-2}$ *are infinite.*

Proof. Our proof of sufficiency resembles the proof of Theorem 1 given in [2], although that proof did not use the martingale square function while this one, necessarily as far as we can tell, does. See also the proof of Theorem 3.3 of [2]. We assume that $EX_0 < \infty$. If this is not true we can just condition on the value of X_0.

Let $a_i, \ -\infty < i < \infty$, satisfy $\sum_{i=1}^{\infty} a_i^{-2} = \sum_{i=1}^{\infty} a_{-i}^{-2} = \infty$, and let $\boldsymbol{X}$ have initial weighting $w(0,i) = a_i, \ -\infty < i < \infty$. We will show

$$P(\overline{\lim}_{n\to\infty} X_n = +\infty, \underline{\lim}_{n\to\infty} X_n > -\infty) = 0, \tag{2}$$

which immediately, by reflection, implies $P(\underline{\lim}_{n\to\infty} X_n = -\infty, \overline{\lim}_{n\to\infty} X_n < \infty) = 0$, which together with (2) establishes sufficiency. To prove (2) it suffices to show that, given $n \in \{0, 1, 2 \ldots\}$ and $k \in I$,

$$P(X_n > k, \ \sup_i X_i = \infty, \ \inf_{i \geq 0} X_{n+i} > k) = 0, \tag{3}$$

and to prove (3) it suffices to prove it in the special case $k = 0$, since our condition on a_i, $i \in I$, is equivalent to $\sum_{i=1}^{\infty} w(0,i)^{-2} = \infty$ and $\sum_{i=1}^{\infty} w(0,-i)^{-2} = \infty$, which is clearly invariant under change of origin.

Now let n be fixed and put $A = \{X_n > 0\}$, $T = \inf\{k \geq n\colon \ X_k \leq 0\}$, $B = A \cap \{T = \infty\} \cap \{\sup_i X_i = \infty\}$. To prove (3) in the case $k = 0$ we must show

$P(B) = 0$. Note that $\sum\limits_{i=0}^{\infty} w(n,i)^{-2} = \infty$ a.s. since $w(n,k) = w(0,k)$ for all but at most n integers k.

For $m \geq 0$ put

$$G(m,j) = \sum_{i=0}^{j-1} w(n+m,i)^{-1}, \quad j \geq 1,$$
$$= 0, \quad j \leq 0,$$

and define

$$M_\lambda = G((n+\lambda) \wedge T, X_{(n+\lambda)\wedge T}), \quad \lambda = 0,1,2,\ldots$$

where $\wedge$ denotes minimum.

For $\lambda \geq 0$ put

$$H_\lambda = M_\lambda + \sum_{i=n+1}^{n+\lambda} [w(i-1, X_{i-1})^{-1} - w(i, X_{i-1})^{-1}] I(X_i > X_{i-1}, i < T), \tag{4}$$

where the sum is taken to be zero if $\lambda = 0$. Then H_λ, $\lambda = 0,1,2,\ldots$ is a nonnegative martingale, and furthermore, if $j \geq X_n$, $H_{i+1} - H_i = w(n,j)^{-1}$ on

$$D_{i,j} = \{X_{n+i+1} = j+1, X_{n+i} = j, \ 1 \leq X_\gamma \leq j, \ n \leq \gamma \leq n+i\}.$$

For a proof of this see the end of the proof of Lemma 3.1 of [2], although the reader will probably be able to construct the proof, which is just a calculation. Note since we assume $j \geq X_n$, $\bigcup_{i=0}^{\infty} D_{i,j} \supset B$. Thus

$$\begin{aligned}
S(H) &= H_0^2 + \sum_{i=1}^{\infty}(H_{i+1} - H_i)^2 \\
&\geq \sum_{i=0}^{\infty} \sum_{j=X_n}^{\infty} w(n,j)^{-2} I(D_{i,j}) \\
&= \sum_{j=X_n}^{\infty} \sum_{i=0}^{\infty} w(n,j)^{-2} I(D_{i,j}) \\
&\geq \sum_{j=X_n}^{\infty} w(n,j)^{-2} I(B) \\
&= \infty I(B),
\end{aligned}$$

where I denotes indicator function. Now since H is a nonnegative it is L^1 bounded, and thus a result of D. G. Austin ([1]) gives $P(S(H) = \infty) = 0$, implying $P(B) = 0$, concluding the proof of sufficiency.

To prove necessity it suffices, with no loss of generality, to let $b_0, b_1, b_2 \ldots$ be a sequence of positive numbers satisfying $\sum_{i=1}^{\infty} b_i^{-2} < \infty$ and to construct a RRW, such that for $k \geq 0$ the initial weighting of $(k, k+1)$ is b_k, which converges to $+\infty$ with positive probability. We will use $\boldsymbol{Y} = Y_0, Y_1, \ldots$ to stand for this example, and the associated interval weights will be designated $\gamma(n,k)$, defined by

$$\gamma(0,k) = b_k, \quad k \geq 0,$$

$$\gamma(n,k) - \gamma(n-1,k) = b_k I(Y_{n-1} = k+1, Y_n = k), \quad k \geq 0,\ n \geq 1,$$

$$\gamma(n,k) = 1 \quad \text{if } k < 0.$$

That is, if $k \geq 0$ the weight of $(k, k+1)$ is increased by b_k each time the process downcrosses this interval. We also specify $Y_0 = 1$. In the discussion below, the weights of $(k-1, k)$, $k \leq 0$, are irrelevant.

To show this example has the desired properties we first prove the following lemma. Much stronger results are known.

Lemma 1. *Let $f_0, f_1, \ldots$ be a nonnegative martingale satisfying $P(f_0 = 1) = 1$ and $\lim_{n\to\infty} f_n = 0$. Then $P(\sum_{i=1}^{\infty}(f_i - f_{i-1})^2 > \lambda) > 0,\ \lambda > 0$.*

Proof. Suppose to the contrary that there is a $\lambda_0 > 0$ such that $P(\sum_{i=1}^{\infty}(f_i - f_{i-1})^2 > \lambda_0) = 0$. Then $E(f_n - f_i)^2 = E\sum_{i=1}^{n}(f_i - f_{i-1})^2 \le \lambda_0^2$, so that $f_0, f_1, \ldots$ is an L^2 bounded martingale, implying $1 = Ef_0 = \lim_{n\to\infty} Ef_n = E\lim_{n\to 0} f_n = 0$, a contradiction.

Now put $\tau = \inf\{j\colon\ Y_j = 0\}$, define

$$
\begin{aligned}
G(m,j) &= \sum_{i=0}^{j-1}\gamma(m,i)^{-1}, \quad j > 0 \\
&= 0, \quad j = 0,
\end{aligned}
$$

and put $\Gamma_n = G(n \wedge \tau, Y_{n\wedge\tau})$, $n \ge 0$. Then Γ_i, $i \ge 0$, is a nonnegative martingale. The proof of this follows in a manner similar to the proof that the process $H_0, H_1, \ldots$ is a nonnegative martingale together with the fact that on $\{Y_n > Y_{n-1}\}$, $\gamma(n, Y_{n-1}) = \gamma(n-1, Y_{n-1})$, so that the expression corresponding to the sum in (4) is zero. See [2], especially the comment after the proof of Theorem 3.1, for a fuller discussion of this.

Now

$$
\begin{aligned}
S(\Gamma)^2 &= \Gamma_0^2 + \sum_{i=1}^{\infty}(\Gamma_i - \Gamma_{i-1})^2 \qquad (5)\\
&= 1 + \sum_{i=1}^{\infty}(\Gamma_i - \Gamma_{i-1})^2 \\
&\le 1 + \sum_{k=0}^{\infty}\sum_{i=1}^{\infty}(\Gamma_i - \Gamma_{i-1})^2 I\{(Y_{i-1}, Y_i) = (k, k+1) \text{ or } (k+1, k)\}
\end{aligned}
$$

$$= 1 + \sum_{k=0}^{\infty} Z_k,$$

where Z_k is just defined to be the second sum in the inequality above, and the inequality holds since $\Gamma_n = 0$ if $Y_k = 0$ for some $k < n$, so that Z_j for $j < 0$, if we defined it analogously, would be 0.

Now

$$Z_k \leq \sum_{i=0}^{\infty} \gamma(i,k)^{-2} I\{(Y_i, Y_{i+1}) = (k, k+1) \text{ or } (k+1, k)\}$$
$$= \gamma(\tau_1, k)^{-2} + \gamma(\tau_2, k)^{-2} + \ldots,$$

where Y crosses $(k, k+1)$ for the ith time between times τ_i and $\tau_i + 1$. For $k \geq 1$ we have, recalling that $Y_0 = 1$, that $\gamma(\tau_1, k) = b_k$, $\gamma(\tau_2, k) = b_k$, $\gamma(\tau_3, k) = 2b_k$ (for in fact $\gamma(\tau_2 + 1, k) = 2b_k$, since the first downcrossing of $(k, k+1)$ occurs between times τ_2 and $\tau_2 + 1$), and in general $\tau_{2n-1} = \tau_{2n} = nb_k$. Thus

$$Z_k \leq 2 \sum_{n=1}^{\infty} (nb_k)^{-2} < 4b_k^{-2}, \quad k \geq 1,$$

and, similarly, $Z_0 < 4b_0^{-2}$, so that by (5) we have $S(\Gamma)^2 < 4 \sum_{k=0}^{\infty} b_k^{-2}$. Lemma 1 now implies that

$$P(\lim_{n\to\infty} \Gamma_n = 0) < 1,$$

so that

$$P(Y_n > 0 \text{ for all } n) > 0. \tag{6}$$

We will now prove

$$P(0 < \overline{\lim}_{n\to\infty} Y_n < \infty) = 0, \tag{7}$$

which, together with (6), proves $P(\lim Y_n = \infty) > 0$. To prove (7) it suffices to show that, if $\lambda > 0$ is an integer and n is an integer, and $A(n, \lambda) = A = \{Y_{n+i} = \lambda$ for infinitely many i, $Y_{n+i} \leq \lambda$, $i \geq 0\}$, then $P(A) = 0$. To establish this we assume that $\lambda \geq Y_n$, and note that there is a (random) integer ψ such that $\gamma(n, \lambda - 1) = \psi b_{\lambda-1}$ and a (random) integer θ such that $\gamma(n, \lambda) = \theta b_\lambda$. Let $T_1 = \inf\{j \geq n\colon Y_j = \lambda\}$, and $T_i = \inf\{j \geq T_{i-1}\colon Y_j = \lambda\}$, $i > 1$, and put $A_k = \{T_k < \infty\} \cap \{\max_{n \leq j \leq T_k} Y_j = \lambda\}$ so that $A_1 \supseteq A_2 \supseteq \ldots$ and $\bigcap_{i=1}^{\infty} A_i = A$. Now $\gamma(T_k, \lambda) = \theta b_\lambda$ on A_k, while

$$\gamma(T_k, \lambda - 1) = \psi b_{\lambda-1} + (k-1) b_{\lambda-1} \text{ on } A_k.$$

Thus

$$P(Y_{T_k+1} = \lambda + 1 | A_k, \psi, \theta) = \frac{\theta b_\lambda}{\psi b_{\lambda-1} + (k-1) b_{\lambda-1}}, \quad k \geq 1,$$

and, since $Y_{T_k+1} = \lambda - 1$ on A_{k+1}, this implies

$$P(A_{k+1}^c | A_k, \psi, \theta) \geq \frac{\theta b_\lambda}{\psi b_{\lambda-1} + (k-1) b_{\lambda-1}},$$

where E^c stands for the complement of E, so that

$$P(\bigcap_{k=1}^{\infty} A_k | \psi, \theta) \leq \prod_{k=1}^{\infty} \left(1 - \frac{\theta b_\lambda}{\psi b_{\lambda-1} + (k-1) b_{\lambda-1}}\right) = 0.$$

Thus $P(A) = 0$, completing the proof of (7) and thus the proof that Y_n approaches ∞ with positive probability.

We remark that it is not difficult to show that in fact $P(\lim_{n \to \infty} Y_n = \infty) = 1$.

References

[1] Austin, D. G. (1966). A sample function property of martingales. *Ann. Math. Statist.* **37**, 1396–1397.

[2] Davis, B. Reinforced random walk. Purdue Department of Statistics Technical Report #88-22.

[3] Karlin, S., and Taylor. A first course in stochastic processes, second edition. Academic Press, New York, 1975.

[4] Pemantle, R. Phase transition in reinforced random walk and RWRE. To appear, *Ann. Prob.*

ON POINTWISE CONVERGENCE IN RANDOM WALKS

Yves Derriennic
Université de Bretagne Occidentale
Brest, France

Michael Lin
Ben-Gurion University of the Negev
Beer Sheva, Israel

Abstract

Let G be a locally compact σ-compact group, and μ a probability on G. We show, for G non-compact, that if the Markov operator defined on $L_1(G)$ by $Tf = \mu * f$ is completely mixing, then $\|\mu^n * f\|_\infty \to 0$ for every $f \in C_0(G)$.

1 Pointwise convergence

Let G be a locally compact σ-compact group with right Haar measure m. A probability μ on G defines a right random walk on G, and a Markov operator on $L_1(G)$ by $Tf = \mu * f$ where $\mu * f(s) = \int f(st)d\mu(t)$.

We call μ *ergodic* if T is ergodic, i.e, for $f \in L_1(G)$ with $\int f dm = 0$ we have $\|\frac{1}{N}\sum_{n=1}^{N} \mu^n * f\|_1 \to 0$. A necessary condition for ergodicity is that μ be *adapted*: the closed subgroup generated by the support of G is all of G. This condition is sufficient if G is Abelian, but not in general [G]. Ergodic probabilities exist only on amenable groups. J. Rosenblatt [R$_1$] proved that on any amenable group there exist ergodic probabilities.

ISBN 0-12-231050-0

The following result was obtained independently by Derriennic [D] and by Mukherjea [M]:

Theorem 1.1 *If G is not compact and μ is adapted, then for every $f \in C_0(G)$ we have $\mu^n * f(x) \to 0$ everywhere.*

The simple example of the shift on Z shows uniform convergence cannot be expected, even if μ is ergodic.

Definition 1.2 μ is *completely mixing* if T is, i.e., for any $f \in L_1(G)$ with $\int f dm = 0$ we have $\|\mu^n * f\|_1 \to 0$.

Our main result is the following.

Theorem 1.3 *If G is not compact and μ is completely mixing, then for every $f \in C_0(G)$ we have $\|\mu^n * f\|_\infty \to 0$.*

Proof: The translation operators $T(t)f(s) = f(st)$ yield a continuous representation of G by contractions in $C_0(G)$, and $\mu * f = \int T(t) f d\mu$ is an average of the representation. The required result follows now from the general theorem which we prove in the next section.

It is well known that if μ is completely mixing, then μ is *strictly aperiodic*, i.e., G is the smallest *closed* normal subgroup a class of which contains the support of μ. Thus, μ completely mixing is ergodic and strictly aperiodic.

Problem A If μ is ergodic and strictly aperiodic, is μ completely mixing?

S. Glasner [G] showed that if μ is also *spread-out* (for some n, μ^n is not singular), then the answer is positive. He also showed that μ spread-out, *adapted* and strictly aperiodic need not be completely mixing. Thus, the following result, obtained by the authors in [DL], does not follow from theorem 1.3.

Theorem 1.4 *If G is not compact and μ is spread-out, adapted and strictly aperiodic, then for every $f \in C_0(G)$ we have $\|\mu^n * f\|_\infty \to 0$.*

Since theorem 1.3 applies to non spread-out probabilities, it is not implied by theorem 1.4.

Problem B Is theorem 1.4 true without assuming μ to be spread-out?

Some partial solutions to this problem, with the additional assumption of *irreducibility* of the Markov operator, were given by Hoffmann and Mukherjea [MH]. Other special cases are discussed in [DL].

Theorem 1.5 *If G is not compact and μ is spread-out and adapted, then for every $f \in L_1(G)$ we have $\mu^n * f(x) \to 0$ a.e.*

The theorem follows from the fact that for μ adapted, the random walk is either transient, and then the result is trivial, or recurrent, and then the result follows from S. Horowitz's result [Ho] for Harris recurrent Markov operators (since μ is spread-out).

Problem C is the conclusion of theorem 1.5 true for μ recurrent and ergodic?

Remark If G is *compact* and μ is completely mixing and spread-out, then $\mu^n * f(x) \to \int f dm$ a.e. for any $f \in L_1(m)$, by [Ho]. On the other hand, if μ is not spread-out, a.e. convergence may fail even for G Abelian and $f \in L_\infty(m)$: let G be the unit circle, $\alpha \in G$ an irrational rotation, $\mu = \frac{1}{2}(\delta_1 + \delta_\alpha)$. By [R2] there is a measurable set A with $0 < m(A) < 1$ and $\mu^n * 1_A$ does not converge in a very strong sense: $\limsup \mu^n * 1_A = 1$ a.e. (and not $m(A)$, which is the only possible limit).

2 Convergence of the iterates of representation averages

Given a locally compact σ-compact group G, an *operator representation* is a function T from G into the space of bounded linear operators of a Banach space B such that $T(ts) = T(t)T(s)$ for $t, s \in G$. Since $T(e)$ is a projection and $T(e)B$ is invariant under all $T(t)$, we may assume $T(e) = I$. The representation is *continuous* if $t \to T(t)x$ is norm-continuous for each $x \in B$.

For a regular proability measure μ on G we define the μ-average of a bounded continuous representation $T(\cdot)$ by $P_\mu x = \int T(t)x d\mu(t)$. If $\|T(t)\| \leq 1$ for every t (and then in fact $T(t)$ is an isometry for each t), then $\|P_\mu\| \leq 1$. We also have $P_{\mu * \nu} = P_\mu P_\nu$. When μ is understood, we'll write P for P_μ.

Main Theorem: *Let μ be a completely mixing probability on G. If $T(\cdot)$ is a continuous representation of G by contractions on a Banach space B, then $\|P^n(I-P)x\| \to 0$ for every $x \in B$.*

Proof: The hypothesis on μ is equivalent to the following: If $\{g_n\}$ in $L_\infty(G)$ satisfies $\|g_n\|_\infty \le 1$, $\check{\mu}^n * g_n = g_0$, then g_0 =const. a.e. [L] (the dual operator, in $L_\infty(G)$, of the convolution operator by μ in $L_1(G)$, is the convolution by $\check{\mu}$). Hence, if g_n are continuous and satisfy the above, then g_0=const.

The assertion of the theorem is equivalent to saying that $\|P^n x\| \to 0$ whenever $\|\frac{1}{N}\sum_{n=1}^N P^n x\| \to 0$, which is equivalent, by [L], to: If $\{x_n^*\}$ in B^* satisfy $\|x_n^*\| \le 1$, $P^{*n}x_n^* = x_0^*$, then $P^* x_0^* = x_0^*$.

Let $\{x_n^*\}$ satisfy the above. Fix $x \in B$, and define $g_n(t) = \langle T(t^{-1})x, x_n^*\rangle$. Then $g_n(t) \in C(G)$, and $\|g_n\|_\infty \le \|x\|$. Now

$$\begin{aligned}\check{\mu}^n * g_n(t) &= \int g_n(ts^{-1})d\mu^n(s) = \int \langle T((ts^{-1})^{-1})x, x_n^*\rangle d\mu^n(s)\\ &= \int \langle T(st^{-1})x, x_n^*\rangle d\mu^n(s) = \int \langle T(s)T(t^{-1})x, x_n^*\rangle d\mu^n(s)\\ &= \langle P^n T(t^{-1})x, x_n^*\rangle = \langle T(t^{-1})x, x_0^*\rangle = g_0(t).\end{aligned}$$

Hence, by the hypothesis on μ, $g_0(t) \equiv$ const. Since $T(e) = I$ can be assumed, we to obtain

$$\langle x, T^*(t)x_0^*\rangle = \langle T(t)x, x_0^*\rangle = g_0(t^{-1}) = g_0(e) = \langle x, x_0^*\rangle$$

for every $t \in G$. Since this is for any $x \in B$, $T^*(t)x_0^* = x_0^*$ for every t, and $P^* x_0^* = x_0^*$.

Remark If $T(\cdot)$ is a continuous bounded representation (i.e., $\|T(t)\| \le M$ for $t \in G$), we have (assuming $T(e) = I$) that $|||T(t)||| = \sup\{\|T(t)x\| : t \in G\}$ is an equivalent norm with $|||T(t)||| = 1$; hence the main theorem applies in this case too.

References

[D] Y. Derriennic, Lois "zéro ou deux" pour les processus de Markov. Applications aux marches aléatoires. *Ann. Inst. Poincaré (B)*, 12(1976), 111-129.

[DL] Y. Derriennic and M.Lin, Convergence of iterates of averages of certain operator representations and of convolution powers. *J. Functional Anal.*, to appear.

[G] S. Glasner, On Choquet-Deny measures. *Ann. Inst. Poincaré (B)*, 12(1976), 1-10.

[Ho] S. Horowitz, Pointwise convergence of the iterates of a Harris-recurrent Markov operator *Israel J. Math.* 33(1979), 177-180.

[HM] K. Hofmann and A. Mukherjea, Concentration functions and a class of non-compact groups. *Math. Ann.* 256(1981), 535-538.

[L] M.Lin, Mixing for Markov operators. *Z. Wahrscheinlichkeitsth. verw. Geb.* 19(1971), 231-242.

[M] A.Mukherjea, Limit theorems for probability measures on non compact groups. *Z. Wahrscheinlichkeitsth. verw. Geb* 33(1976), 273-284.

[R1] J. Rosenblatt, Ergodic and mixing random walks on locally compact groups. *Math. Ann.* 257(1981), 31-42.

[R2] J. Rosenblatt, Ergodic group actions, *Arch. Math.* 47(1986), 263-269.

BANACH SPACES WITH THE ANALYTIC RADON-NIKODÝM PROPERTY AND COMPACT ABELIAN GROUPS

G. A. EDGAR
The Ohio State University, Department of Mathematics
Columbus, Ohio, U.S.A.

A complex Banach space E has the analytic Radon-Nikodým property iff any bounded analytic function, defined on the open unit disk in the complex plane, with values in the space E, has radial limits almost everywhere on the bounding circle. In this paper, the definition is generalized. We replace the circle group $\mathbb{T}$ in the conventional definition by a compact metrizable abelian group G. (Metrizability can probably be eliminated by those who care about such things.) To replace the interior of the unit disk, we replace the Poisson kernel by a good approximate identity on the group G. We replace the nonnegative integers (used to distinguish an analytic function from a

This research was supported in part by National Science Foundation grant DMS 87-01120.

ISBN 0-12-231050-0

harmonic function) by a subset of the dual group of G. Some of the equivalent formulations of the analytic Radon-Nikodým property are seen to carry over to the generalization. These formulations enable us to give a characterization of Riesz subsets of a coutable discrete abelian group in terms of operators.

THE ANALYTIC RADON-NIKODÝM PROPERTY

The analytic Radon-Nikodým property was defined by Bukhvalov and Danilevich in [**2**]. Let E be a complex Banach space. We say that E has the **analytic Radon-Nikodým property** iff every bounded analytic function $\varphi\colon D \to E$ has radial limits almost everywhere. Here, D is the open unit disk in the complex plane $\mathbb{C}$. A function $\varphi\colon D \to E$ is analytic iff there is a convergent expansion

$$\varphi(z) = \sum_{k=0}^{\infty} a_k z^k,$$

where $a_k \in E$.

Several equivalent formulations have been proved (see [**2**], [**4**], [**5**], [**8**], [**10**]). Some of them will be stated here, so that they can be compared to the generalization to be discussed below.

There is a formulation in terms of measures. This is the reason for the name "analytic Radon-Nikodým property". Let $\mathbb{T}$ be the unit circle in the complex plane, and let λ be normalized arc length measure on $\mathbb{T}$. Let E be a Banach space. If $f\colon \mathbb{T} \to E$ is a vector-valued function on $\mathbb{T}$, then its Fourier coefficients are

$$\widehat{f}(k) = \int f(z) z^{-k}\, d\lambda(z) = \int_0^{2\pi} f(e^{i\theta}) e^{-ik\theta}\, \frac{d\theta}{2\pi}.$$

Similarly, if μ is a vector measure on $\mathbb{T}$, its Fourier coefficients are

$$\widehat{\mu}(k) = \int z^{-k}\, d\mu(z).$$

THEOREM ([**2**]). *A complex Banach space E has the analytic Radon-Nikodým property if and only if any vector measure μ with values in E, bounded variation, and $\widehat{\mu}(k) = 0$ for all $k < 0$, has a Radon-Nikodým derivative.*

There is a formulation in terms of martingales. Let $\Omega = \mathbb{T}^{\mathbb{N}}$ with product measure. For $\omega \in \Omega$ we will write $\omega = (\omega_1, \omega_2, \cdots)$. An **analytic martingale** with values in a complex Banach space E is a sequence (f_n) of measurable functions of the form

$$f_n(\omega) = \sum_{j=1}^{n} d_j(\omega_1, \omega_2, \cdots, \omega_{j-1})\, \omega_j,$$

where $d_j\colon \mathbb{T}^{j-1} \to E$ is measurable.

THEOREM [**6**]. *A complex Banach space E has the analytic Radon-Nikodým property if and only if every L^1-bounded E-valued analytic martingale converges a.e.*

There is a formulation in terms of operators. We write L^1 for $L^1(\mathbb{T}, \lambda)$, and H_0^1 for the subspace

$$\left\{ f \in L^1 : \widehat{f}(n) = 0 \text{ for all } n \leq 0 \right\}.$$

Write Q for the natural quotient map $Q\colon L^1 \to L^1/H_0^1$. Let E be a Banach space. An operator $T\colon L^1 \to E$ is **representable** iff there exists $f \in L^\infty(\mathbb{T}, \lambda; E)$ with

$$T(g) = \int g\, f\, d\lambda \qquad \text{for all } g \in L^1.$$

(See [**3**, §III.1].)

THEOREM ([**4**], [**10**]). *A complex Banach space E has the analytic Radon-Nikodým property if and only if $TQ\colon L^1 \to E$ is representable for each operator $T\colon L^1/H_0^1 \to E$.*

A slight variant [**5**] states: E has the analytic Radon-Nikodým property if and only if every operator $T\colon L^1 \to E$ that factors through L^1/H_0^1 is representable.

We begin by fixing the notation to be used. Let G be a compact metrizable abelian group. We will write it multiplicatively. Let $\Gamma = \widehat{G}$ be the dual group, the set of continuous homomorphisms $\gamma \colon G \to \mathbb{C}$. Then Γ is a countable (discrete) group. We will write λ for the normalized Haar measure on G; but an integral $\int_G f(x)\, d\lambda(x)$ will often be written simply $\int_G f(x)\, dx$. Let $\mathcal{B}(G)$ be the σ-algebra of Borel subsets of G.

Next let E be a Banach space. The set of all E-valued measures on $\mathcal{B}(G)$ with bounded variation will be denoted $V^1(G;E)$. That is: a vector measure μ belongs to $V^1(G;E)$ iff there is a constant C such that for any partition $A_1, A_2, \cdots, A_n$ of G,

$$\sum_{j=1}^{n} \|\mu(A_j)\| \leq C.$$

(See [**3**, p. 2].) The least such constant is the **variation** of μ, written $\|\mu\|_1$. The space $L^1(G;E)$ of E-valued Bochner integrable functions is naturally isometric to a subspace of $V^1(G;E)$; this is obtained by identifying a function with the Radon-Nikodým density of a measure.

The set of all E-valued measures with bounded average range (with respect to Haar measure) will be denoted $V^\infty(G;E)$. That is: a vector measure μ belongs to $V^\infty(G;E)$ iff there is a constant C such that for any Borel set A,

$$\|\mu(A)\| \le C\lambda(A).$$

The least such constant is the norm $\|\mu\|_\infty$. The space $L^\infty(G;E)$ of bounded E-valued Bochner measurable functions is naturally isometric to a subspace of $V^\infty(G;E)$. Of course, if G has the Radon-Nikodým property, then $V^\infty(G;E) = L^\infty(G;E)$. The space V^∞ was introduced by Bochner and Taylor [**1**], who proved the isometric identification

$$L^1(G;E)^* = V^\infty(G;E^*).$$

If $\mu \in V^1(G;E)$ and $\gamma \in \Gamma$, then the **Fourier coefficient** is

$$\widehat{\mu}(\gamma) = \int_G \overline{\gamma}\, d\mu.$$

If $\Lambda \subseteq \Gamma$ is a set of characters, let

$$L^1_\Lambda(G;E) = \left\{ f \in L_1(G;E) : \widehat{f}(\gamma) = 0 \text{ for all } \gamma \notin \Lambda \right\}.$$

Similar definitions will be used for V^1_Λ, V^∞_Λ, and so on.

The classical examples that we have in mind are:

$$G = \mathbb{T}, \text{ the circle,}$$

$$\Gamma = \mathbb{Z}, \text{ the integers,}$$

$$\Lambda = \mathbb{N} \cup \{0\}, \text{ the non-negative integers,}$$

$$\text{or } \Lambda = \mathbb{N}, \text{ the positive integers.}$$

A subset Λ of Γ is called a **Riesz set** iff $V^1_\Lambda(G;\mathbb{C}) = L^1_\Lambda(G;\mathbb{C})$. For example, the F & M Riesz theorem states that $\mathbb{N}$ is a Riesz set in $\mathbb{Z} = \widehat{\mathbb{T}}$. Note that $V^\infty_\Lambda(G;\mathbb{C}) = L^\infty_\Lambda(G;\mathbb{C})$ for any Λ by the usual Radon-Nikodým theorem.

Definition. Let G be a locally compact metrizable abelian group, let $\Gamma = \widehat{G}$, let $\Lambda \subseteq \Gamma$, and let E be a Banach space. Then E has the **Λ-Radon-Nikodým property** iff $V^\infty_\Lambda(G;E) = L^\infty_\Lambda(G;E)$.

Some special cases are easy to identify. If $G = \mathbb{T}$, then $\widehat{\mathbb{T}} = \mathbb{Z}$; the $\{0\}$-Radon-Nikodým property holds for all Banach spaces; the $\mathbb{N}$-Radon-Nikodým property is equivalent to the analytic Radon-Nikodým property; the $\mathbb{Z}$-Radon-Nikodým property is equivalent to the Radon-Nikodým property.

For $\Lambda \subseteq \Gamma$, let $\Lambda' = \{\, \gamma \in \Gamma : \overline{\gamma} \notin \Lambda \,\}$. Thus $\int \gamma_1 \gamma_2 \, d\lambda = 0$ if $\gamma_1 \in \Lambda$ and $\gamma_2 \in \Lambda'$. In the example $G = \mathbb{T}$, if $\Lambda = \mathbb{N} \cup \{0\}$, then $\Lambda' = \mathbb{N}$.

The Poisson kernel has an important place in the theory of the analytic Radon-Nikodým property. It is used to relate boundary values of a vector valued function to the values inside the disk of the corresponding analytic function. For the generalization, we will consider a "good" approximate identity. (A compact metric abelian group always has an approximate identity of this kind.) A good approximate identity is a sequence $(i_n)_{n=1}^{\infty}$ of measurable functions $i_n : G \to \mathbb{R}$ with

$$i_n \geq 0;$$

$$\int_G i_n(x)\, dx = 1;$$

$$\sum_{\gamma \in \Gamma} \widehat{i_n}(\gamma) < \infty;$$

$$\int_U i_n(x)\, dx \to 1 \quad \text{for all neighborhoods } U \text{ of } 1 \text{ in } G.$$

It follows that $i_n * f \to f$ in $L^1(G; E)$-norm for all $f \in L^1(G; E)$. (A proof is included below.) Also, if $(a_\gamma)_{\gamma \in \Gamma} \subseteq E$ is a bounded family of vectors, then

$$\sum_{\gamma \in \Gamma} \widehat{i_n}(\gamma)\, a_\gamma\, \gamma$$

converges uniformly on G, hence is in $L^\infty(G; E)$.

EQUIVALENT FORMULATIONS

Our main theorem shows how some of the properties discussed above for the analytic Radon-Nikodým property can be generalized to this situation.

THEOREM. *Let G be a compact metrizable abelian group, let $\Gamma = \widehat{G}$, let $\Lambda \subseteq \Gamma$, let (i_n) be a good approximate identity on G, and let E be a Banach space. Then the following are equivalent:*

(a) *E has the Λ-Radon-Nikodým property.*

(b) *If $(a_\gamma)_{\gamma\in\Lambda}$ is a bounded family of vectors in E, and the sequence*

$$f_n = \sum_{\gamma\in\Lambda} \widehat{i_n}(\gamma)\, a_\gamma\, \gamma$$

is bounded in $L^\infty_\Lambda(G;E)$, then there exists $f \in L^\infty_\Lambda(G;E)$ with $\widehat{f}(\gamma) = a_\gamma$ for all $\gamma \in \Lambda$.

(c) *If $(a_\gamma)_{\gamma\in\Lambda}$ is a bounded family of vectors in E, and the sequence (f_n) defined above is bounded in $L^\infty_\Lambda(G;E)$, then (f_n) converges in $L^1(G;E)$ norm.*

(d) *If $T\colon L^1(G)/L^1_{\Lambda'}(G) \to E$ is a bounded operator, and*

$$Q\colon L^1(G) \to L^1(G)/L^1_{\Lambda'}(G)$$

is the natural quotient, then TQ is a representable operator.

Notice that (a) and (d) do not depend on (i_n), so if (b), (c) hold for *some* choice of (i_n), then they hold for all other choices, too. In the special case $G = \mathbb{T}$, we could choose (i_n) as the Poisson kernel (at radii $1 - 2^{-n}$) to discuss boundary values of analytic functions, or we could choose the Fejer kernel to discuss Fourier series convergence.

PROOF: (b) $\Longrightarrow$ (c) is a standard argument. Let

$$\mathcal{R} = \left\{\, f \in L^1(G; E) : \|i_n * f - f\|_1 \to 0 \,\right\}.$$

Then $\mathcal{R}$ contains all $x\,\gamma$ for $x \in E$ and $\gamma \in \Gamma$. These functions have dense span in $L^1(G; E)$. The operator norms of the convolutions by the functions i_n are all ≤ 1; therefore $\mathcal{R}$ is closed. So $\mathcal{R} = L^1(G; E)$. By (b), every sequence (f_n) as given in (c) has the form $f_n = i_n * f$ for some $f \in L^\infty_\Lambda(G; E)$.

(c) $\Longrightarrow$ (a): Let $\mu \in V^\infty_\Lambda(G; E)$. Let $a_\gamma = \widehat{\mu}(\gamma)$. Then

$$\|a_\gamma\| = \left\| \int \overline{\gamma}\, d\mu \right\| \leq \sup_x |\overline{\gamma(x)}|\, \|\mu\|_\infty .$$

So (a_γ) is bounded. Also $a_\gamma = 0$ for $\gamma \notin \Lambda$. Now if $f_n = \sum_\gamma \widehat{i_n}(\gamma)\gamma$, then

$$\|f_n(x)\| \leq \sum_\gamma |\widehat{i_n}(\gamma)|\, \|a_\gamma\| < \infty.$$

Now we have

$$
\begin{aligned}
(i_n * \mu)(x) &= \int_G i_n(xy^{-1})\, d\mu(y) \\
&= \int_G \sum_\gamma \widehat{i_n}(\gamma)\gamma(xy^{-1})\, d\mu(y) \\
&= \sum_\gamma \widehat{i_n}(\gamma)\gamma(x) \int_G \overline{\gamma(y)}\, d\mu(y) \\
&= \sum_\gamma \widehat{i_n}(\gamma)\gamma(x) a_\gamma \\
&= f_n(x).
\end{aligned}
$$

Thus, $\|f_n(x)\|_\infty \leq \|i_n\|_1 \|\mu\|_\infty$, and the sequence (f_n) is uniformly bounded. So, by hypothesis (c), we have $f_n \to f$ in $L^1(G; E)$ norm. The limit f is the Radon-Nikodým derivative $d\mu/d\lambda$.

(a) $\implies$ (d): Let T be the given operator, and write $S = TQ$. Define a measure $\mu(A) = S(\mathbf{1}_A)$ for $A \in \mathcal{B}(G)$. Approximate by simple functions to get $\int g\, d\mu = S(g)$ for all $g \in L^1(G)$. We claim that $\mu \in V^\infty_\Lambda(G; E)$. First,

$$\|\mu(A)\| = \|S(\mathbf{1}_A)\| \leq \|S\|\, \|\mathbf{1}_A\|_1 = \|S\|\, \lambda(A),$$

so $\mu \in V^\infty(G; E)$. Next, for $\gamma \in \Gamma$, we have $\widehat{\mu}(\gamma) = \int \overline{\gamma}\, d\mu = S(\overline{\gamma})$. Now $\overline{\gamma} \in \Lambda'$ if and only if $\gamma \notin \Lambda$, so $\widehat{\mu}(\gamma) = 0$ if $\gamma \notin \Lambda$. This shows

$\mu \in V_\Lambda^\infty(G; E)$. By hypothesis (a), $f = d\mu/d\lambda \in L^\infty(G; E)$ exists, so $\int g\, d\mu = \int gf\, d\lambda$ for $g \in L^1(G)$. But $\int g\, d\mu = S(g)$, so S is a representable operator.

(d) $\Longrightarrow$ (b): Let $(a_\gamma)_{\gamma \in \Lambda}$ satisfy $\|f_n\|_\infty \leq M$,where

$$f_n = \sum_{\gamma \in \Lambda} \widehat{i_n}(\gamma)\, a_\gamma\, \gamma.$$

We want to define an operator T. For $n \in \mathbb{N}$, define an operator $S_n\colon L^1(G) \to E$ by:

$$S_n(g) = \int_G g\, f_n\, d\lambda.$$

Now S_n is a bounded linear operator:

$$\begin{aligned} \|S_n(g)\| &= \left\| \int_G g(x) f_n(x)\, dx \right\| \\ &\leq \int_G |g(x)|\, \|f_n(x)\|\, dx \\ &\leq \int_G |g(x)|\, dx\, \|f_n\|_\infty \leq \|g\|_1\, M. \end{aligned}$$

So $\|S_n\| \leq M$.

There is a subnet $S_{n(\alpha)}$ of the sequence S_n that converges in the weak* operator topology, say to $S\colon L^1(G) \to E^{**}$. Of course $\|S\| \leq$

M. We claim that S actually maps into E. First, if $\gamma \in \Gamma$, then $\gamma \in L^1(G)$, $\overline{\gamma} \in L^1(G)$, and

$$\begin{aligned} S(\overline{\gamma}) &= \lim_{\alpha} \int \overline{\gamma}\, f_{n(\alpha)}\, d\lambda \\ &= \lim_{\alpha} \sum_{\gamma_1 \in \Lambda} \widehat{i_{n(\alpha)}}(\gamma_1) a_{\gamma_1} \int \gamma_1\, \overline{\gamma}\, d\lambda \\ &= \lim_{\alpha} \widehat{i_{n(\alpha)}}(\gamma) a_\gamma \\ &= a_\gamma \in E. \end{aligned}$$

Now linear combinations of functions $\gamma \in \Gamma$ are dense in $L^1(G)$, so $S(g) \in E$ for all $g \in L^1(G)$.

Next, S vanishes on the subspace $L^1_{\Lambda'}(G) \subseteq L^1(G)$: If $\gamma \in \Lambda'$, then $\overline{\gamma} \notin \Lambda$, so $a_{\overline{\gamma}} = 0$ and (as above) $S(\gamma) = a_{\overline{\gamma}} = 0$. But the linear combinations of functions $\gamma \in \Lambda'$ are dense in $L^1_{\Lambda'}(G)$, so S vanishes on $L^1_{\Lambda'}(G)$. This means that S factors through the quotient map Q. Then by (d), the operator S is a representable operator; so there exists $f \in L^\infty(G; E)$ with

$$S(g) = \int g\, f\, d\lambda \qquad \text{for all } g \in L^1(G).$$

Now $\widehat{f}(\gamma) = a_\gamma$ for all γ as required. ∎

ADDITIONAL COMMENTS

In many cases, the sequences (f_n) of the theorem are, in fact, martingales. This is the point of view of [**6**]. Then part (c) concerns the convergence of a martingale in $L^1(G;E)$-norm. It is known that if a Banach-valued martingale converges in mean, then it must also converge almost everywhere.

Next is a simple characterization of Riesz sets using a "disintegration of measure" argument. (Perhaps this means that Riesz sets can be investigated using martingales.)

PROPOSITION. *Let G be a compact metrizable abelian group, $\Gamma = \widehat{G}$, and $\Lambda \subseteq \Gamma$. The Banach space $E = L^1[0,1]$ has the Λ-Radon-Nikodým property if and only if Λ is a Riesz set.*

PROOF: Suppose Λ is a Riesz set. Let $T\colon L^1(G)/L^1_{\Lambda'}(G) \to L^1[0,1]$ be a bounded operator, and let $Q\colon L^1(G) \to L^1(G)/L^1_{\Lambda'}(G)$ be the quotient map. Disintegration (see, for example, [**11**, Theorem 3.1] or [**7**, Théorèm 1]) yields a family $\{\mu_\omega\}_{\omega\in[0,1]}$ of complex measures μ_ω on G, such that for every $h \in L^1(G)$, we have

$$TQ(h)(\omega) = \int_G h(x)\, d\mu_\omega(x)$$

for almost all $\omega \in [0,1]$. Also, if M is the norm of T, then we have

$$\int_{[0,1]} \|\mu_\omega\|_1 \, d\lambda(\omega) \leq M,$$

so μ_ω has finite variation for almost all ω.

Now if $\gamma \notin \Lambda$, then $\overline{\gamma} \in \Lambda'$, so $TQ(\gamma) = 0$. So we have

$$0 = TQ(\gamma)(\omega) = \int_G \gamma \, d\mu_\omega \quad = \widehat{\mu_\omega}(\gamma)$$

for almost all ω. Combining countably many null sets, we see that for almost every ω, we have $\widehat{\mu_\omega}(\gamma) = 0$ for all $\gamma \notin \Lambda$. So since Λ is a Riesz set, μ_ω is absolutely continuous. Then, when we integrate again to get TQ, we see by [**7**, Proposition 3] that TQ is representable.

For the converse, suppose $L^1[0,1]$ has the Λ-Radon-Nikodým property. Since $L^1(G)$ is isomorphic to $L^1[0,1]$ (or, if G is finite, to a subspace of $L^1[0,1]$) the space $L^1(G)$ also has the Λ-RNP. Let $\mu \in V^1_\Lambda(G, \mathbb{C})$. We must show that μ is absolutely continuous. Consider the "convolution operator" $S\colon L^1(G) \to L^1(G)$ defined by

$$S(h)(y) = (h * \mu)(y) = \int_G h(yx^{-1}) \, d\mu(x).$$

As usual, using Fubini's Theorem, we may compute $\|S(h)\|_1 \leq \|h\|_1 \, \|\mu\|_1 < \infty$, so $S(h) \in L^1(G)$. The Kalton disintegration of this operator is

$$S(h)(y) = \int h(x)\, d\mu_y(x),$$

where μ_y is defined by $\mu_y(A) = \mu(yA^{-1})$.

Now for $\gamma \in \Lambda'$,

$$S(\gamma)(y) = \int \gamma(yx^{-1})\, d\mu(x) = \gamma(y) \int \overline{\gamma(x)}\, d\mu(x) = 0,$$

since $\overline{\gamma} \notin \Lambda$. Therefore the operator S vanishes on $L^1_{\Lambda'}(G)$, so it has the form $S = TQ$, where $Q\colon L^1(G) \to L^1(G)/L^1_{\Lambda'}(G)$ is the quotient map. Since $L^1(G)$ has the Λ-Radon-Nikodým property, we conclude that S is representable. Therefore (by the converse direction of [**7**, Proposition 3]), almost all of the disintegration components μ_y are absolutely continuous. But μ_y is absolutely continuous if and only if μ itself is absolutely continuous, so we have finished the proof that Λ is a Riesz set. ∎

Consider the compact group $G = \mathbb{T}^{\mathbb{N}}$, the product of countably many copies of the circle. (We use the convention $\mathbb{N} = \{1, 2, 3, \cdots\}$.) The dual is $\Gamma = \mathbb{Z}^{(\mathbb{N})}$, the sum of countably many copies of $\mathbb{Z}$. (It may be identified with $\bigcup_{n=0}^{\infty} \mathbb{Z}^n$, where identifications are made by adding zeros to the end of an element.) Let

$$\Lambda_+ = \{ (a_1, a_2, \cdots, a_k, 0, 0, \cdots) : a_1, \cdots, a_{k-1} \in \mathbb{Z}, a_k \in \mathbb{N} \}.$$

This is the positive cone for an ordering of the group Γ. The proof in [**6**] shows that a Banach space E has the analytic Radon-Nikodým property if and only if E has the Λ-Radon-Nikodým property for all sets $\Lambda \subseteq \Lambda_+$ such that

$$\Lambda \cap \mathbb{Z}^n$$

is finite for all n. The condition is, however, not the same as the Λ_+-Radon-Nikodým property: since $\mathbb{Z}^1 \times \{1\} \subseteq \Lambda_+$, it follows that the Λ_+-Radon-Nikodým property implies the $\mathbb{Z}$-Radon-Nikodým property which is equivalent to the Radon-Nikodým property. Now the space $L^1[0,1]$ has the analytic Radon-Nikodým property [**6**]. So L^1-bounded analytic martingales in $L^1[0,1]$ converge. Therefore the sets Λ described above are Riesz sets. Another proof of this fact was given by G. Godefroy [**9**].

There are some questions that seem natural. I will list some of them here.

Can the definition be put in L^1 as well as L^∞? More precisely: If Λ is a Riesz set and E has the Λ-Radon-Nikodým property does it follow that $V^1_\Lambda(G; E) = L^1_\Lambda(G; E)$?

Suppose we are given two situations G_1, Γ_1, Λ_1 and G_2, Γ_2, Λ_2. When does the Λ_1-Radon-Nikodým property imply the Λ_2-Radon-Nikodým property? Some cases are easy. For example: If $G_1 = G_2$, $\Gamma_1 = \Gamma_2$, $\Lambda_1 \subseteq \Lambda_2$, then Λ_2-RNP $\Longrightarrow$ Λ_1-RNP. If G_1 is a quotient group of G_2, so that we may make the identification $\Gamma_1 \subseteq \Gamma_2$, and if $\Lambda_1 = \Lambda_2 \subseteq \Gamma_1$, then Λ_1-RNP $\Longleftrightarrow$ Λ_2-RNP. (This shows that the definition of Λ-Radon-Nikodým property is sensible.)

The next question was suggested to me by G. Pisier. For which sets Λ is the following true: E has the Λ-Radon-Nikodým property if and only if c_0 does not embed in E? Lacunary sets Λ, for example $\Lambda = \{ 2^n : n \in \mathbb{N} \}$, have this property. Presumably more complex sets, such as $\Lambda = \{ 2^n + 2^m : n, m \in \mathbb{N} \}$ will work, too. This corresponds to taking Walsh functions that are the product of two Rademacher functions.

References

1. S. Bochner and A. E. Taylor, *Linear functionals on certain spaces of abstractly valued functions*, Ann. of Math. **39** (1938), 913–944.
2. A. V. Bukhvalov and A. A. Danilevich, *Boundary properties of analytic functions with values in Banach space* (Russian), Mat. Zametki **31** (1982), 103–114; English translation, Math. Notes Acad. Sci. USSR **31** (1982), 104–110.

3. J. Diestel and J. J. Uhl, Jr., "Vector Measures," Mathematical Surveys 15, American Mathematical Society, 1977.

4. P. N. Dowling, *Representable operators and the analytic Radon-Nikodým property in Banach spaces*, Proc. Royal Irish Acad. **85A** (1985), 143–150.

5. P. N. Dowling and G. A. Edgar, *Some characterizations of the analytic Radon-Nikodým property in Banach spaces*, J. Functional Anal. (to appear).

6. G. A. Edgar, *Analytic martingale convergence*, J. Functional Anal. **69** (1986), 258–280.

7. Hicham Fakhoury, *Représentations d'opérateurs à valeurs dans $L^1(X, \Sigma, \mu)$*, Math. Ann. **240** (1979), 203–212.

8. N. Ghoussoub, B. Lindenstrauss, and B. Maurey, *Analytic martingales and plurisubharmonic barriers in complex Banach spaces*, (to appear).

9. G. Godefroy, *On Riesz subsets of abelian discrete groups*, (to appear).

10. W. Hensgen, *Hardy-Räume vektorwertiger Funktionen*, Dissertation, Munich (1986).

11. N. J. Kalton, *The Endomorphisms of L_p $(0 \le p \le 1)$*, Indiana Univ. Math. J. **27** (1978), 353–381.

GENERALIZED MEASURE PRESERVING TRANSFORMATIONS

Ulrich Krengel*
University of Göttingen,
Department of Mathematical Stochastics
Lotzestr. 13, 3400 Göttingen, West-Germany

To the memory of Horst Michel**

0. INTRODUCTION

The notion of generalized measure preserving transformation (gmp-transformation), introduced here, serves to describe movement of matter in situations where the movement of the matter on part of the space may depend on the distribution of the matter on the remaining part of the space. Gmp-transformations are set transformations.

---

* Most of this work was done during a visit at the TU Delft. The travel to the Columbus conference was supported by the DFG.

** Apart from his contributions to ergodic theory and mathematical pyhsics, Horst Michel from Halle (GDR) did a lot for bringing mathematicians from the East and the West together by organizing two very interesting conferences in Vitte/Hiddensee and Georgenthal (see [M1], [M2]). He and his wife died after an automobile accident which occurred on Christmas Day 1987.

ISBN 0-12-231050-0

They induce nonlinear operators T in L_1^+ which generalize the familiar maps $f \to f \circ \tau$ given by measure preserving transformations τ.

Even in a finite space the new notion is more general. In an abstract measure space, gmp-transformations arise naturally by piecing two distinct measure preserving transformations τ_1, τ_2 together, taking τ_1 on part of the space and suitable powers of τ_2 on the complement.

For $f \in L_1^+$, the sequence $f, Tf, T^2f, \dots$ has stationary 1-dimensional marginals. The multidimensional marginals are only asymptotically stationary.

We show that T is nonexpansive in L_p $(1 \leq p \leq \infty)$. It then follows from the theorems of Baillon [B] and of Krengel and Lin [KL] that the averages

$$A_n f = (f + Tf + \dots + T^{n-1} f)/n$$

converge weakly in L_2 for f in L_2 (and in L_1 for finite μ). However, a.e.-convergence need not hold. Using speed limit operators introduced in [KL] and [K1] it is shown that there is no permanent summation method for which the averages of $T^n f$ converge a.e. for all T and f.

1. DEFINITIONS AND EXAMPLES

Let $(\Omega, \mathfrak{A}, \mu)$ be a σ-finite measure space. A gmp-transformation is a map $\varphi : \mathfrak{A} \to \mathfrak{A}$ such that

$$A \subset B \text{ implies } \varphi(A) \subset \varphi(B) \qquad (1.1)$$

and

$$\mu(\varphi(A)) = \mu(A) \quad \text{for all } A \in \mathcal{A} . \tag{1.2}$$

If $\tau : \Omega \to \Omega$ is a measure preserving transformation it induces a gmp-transformation φ_τ by $\varphi_\tau(A) = \tau^{-1}A$. φ_τ has the additional properties $\varphi_\tau(A^c) = (\varphi_\tau(A))^c$ and $\varphi_\tau(A \cup B) = \varphi_\tau(A) \cup \varphi_\tau(B)$. Our φ need not have these properties.

To get acquainted with the new concept we first look at the special case where Ω is a finite set $\{1,2,\ldots,N\}$ and μ the counting measure.

For $1 \leq i \leq N$ let $(\Pi_{i1},\Pi_{i2},\ldots,\Pi_{iN})$ be a permutation of $\{1,\ldots,N\}$. We can think of N people and N places on a beach. Π_{i1} is the place person i likes best, Π_{i2} the place person i likes second best, etc. If $A = \{i_1,\ldots,i_k\}$ with $i_1 < i_2 < \ldots < i_k$ is the set of people going to the beach in this order, person i_1 will go to place $\Pi_{i_1,1}$, person i_2 goes to the place he likes best among the remaining places, etc. Let $\varphi(A)$ be the set of places occupied after i_k made his choice. It is simple to check that φ is a gmp-transformation. The construction may be slightly generalized by introducing an additional permutation $(p_1,\ldots,p_N)$ of $\{1,\ldots,N\}$ determining the order in which the persons can make their choice. p_1 is the number of the person having the first choice, etc. Formally, the condition $i_1 < i_2 < \ldots < i_k$ is replaced by the condition that $p_{j_1} = i_1$, $p_{j_2} = i_2,\ldots,p_{j_k} = i_k$ implies $j_1 < j_2 < \ldots < j_k$.

We say that any such φ comes from a priority rule. A gmp-transformation φ is of the form φ_τ for a measure preserving τ if the first priorities Π_{i1} $(1 \le i \le N)$ of all persons are distinct.

It is not hard to check that for $N \le 3$ any gmp-transformation comes from a priority rule. M. Keane showed by the following example that for $N = 4$ a gmp-transformation need not come from any priority rule:

$\varphi(\{i\}) = 1$ for all i

$\varphi(\{1,2\}) = \varphi(\{2,3\}) = \varphi(\{3,4\}) = \{1,2\}$

$\varphi(\{1,3\}) = \varphi(\{2,4\}) = \varphi(\{1,4\}) = \{1,3\}$

$\varphi(A) = \{1,2,3\}$ for all A with $\mathrm{card}(A) = 3$

(Necessarily $\varphi(\emptyset) = \emptyset$ and $\varphi(\Omega) = \Omega$.)

Assume that φ is given by a priority rule. Clearly, we must have $\Pi_{i1} = 1$ for all i. As only the first n priorities of person p_n matter, the rule must be given by a scheme:

p_1 : 1

p_2 : 1 a

p_3 : 1 b c

p_4 : 1 d e f

where $a = \Pi_{p_2,2}$, $b = \Pi_{p_3,2}$ etc.

No matter what p_4 is, there is one set $\{i,p_4\}$ whose image is $\{1,2\}$ and another $\{j,p_4\}$ whose image is $\{1,3\}$. So $d = 2$ and $d = 3$, a contradiction. Thus φ cannot come from a priority rule.

The next simple conjecture would be that each gmp-transformation φ can be written as a finite product $\varphi_1 \circ \varphi_2 \circ \dots \circ \varphi_n$ where the φ_ν come from priority rules. However, I checked that, with more work, the above φ can again serve as a counterexample. Thus, for $N \geq 4$ the structure of the set of gmp-transformations on $\{1,2,\dots,N\}$ remains to be explored.

To describe a first example in an abstract measure space we now show how one can obtain gmp-transformations by piecing two measure preserving transformations together. This will be a variant of the Chacon-Ornstein filling scheme as explored by Rost; see [R] and [K2]. Actually, the first of the transformations, can be a gmp-transformation.

Let $(\Omega, \mathcal{A}, \mu)$ be a σ-finite measure space, and let ψ be a gmp-transformation and τ an invertible conservative ergodic measure preserving transformation in Ω. Let $\Omega = \Omega' \cup \Omega''$ be a partition of Ω into two disjoint subsets of positive measure. We want to apply ψ on Ω' and then τ on Ω''. It is enough to define $\varphi(A)$ for sets A with $\mu(A) < \infty$. Roughly speaking, $\Omega' \cap A$ is mapped to $\psi(\Omega' \cap A)$ and $\Omega'' \cap A$ to the first available place when mapped with powers of τ.

Formally, we must use an inductive definition. Let $C_0 = \Omega'' \cap A$ and $D_0 = (\psi(\Omega' \cap A))^c$. If C_n and D_n have been defined, put

$$E_{n+1} = \tau C_n \cap D_n^c \qquad C_{n+1} = \tau C_n \cap D_n$$

and

$$D_{n+1} = D_n \setminus E_{n+1}.$$

(E_n is the portion of $\tau^n C_0$ which finds its place at time n,

C_n is the portion which remains to be mapped, and D_n the remaining unoccupied space.) Finally, put

$$\varphi(A) = \psi(\Omega' \cap A) \cup \bigcup_{n=1}^{\infty} E_n. \tag{1.3}$$

Clearly, φ satisfies the monotonicity condition (1.1). We now prove (1.2). Let D be the intersection of the decreasing sequence D_n. First assume $\mu(D) > 0$. Then

$$\Omega = \bigcup_{k=1}^{\infty} \tau^{-k} D \qquad (\text{mod } \mu). \tag{1.4}$$

Put

$$C_{0,k} = \{\omega \in C_0 : \tau^k \omega \in D,\ \tau^i \omega \in D^c\ (i=1,\ldots,k-1)\}.$$

It follows from (1.4) and

$$C_n \subset \tau^n \left(\bigcup_{k=n}^{\infty} C_{0,k} \right)$$

that $\mu(C_n) \to 0$. Now observe that

$$\mu(\Omega'' \cap A) = \mu(C_0) = \mu(C_n) + \sum_{i=1}^{n} \mu(E_i).$$

As the sets E_i are disjoint, (1.3) yields $\mu(\varphi(A)) = \mu(A)$. If $\mu(D) = 0$, then $D_0 = \bigcup_{i=1}^{\infty} E_i$ (mod μ), and $\mu(D_0) \geq \mu(C_0)$ again yields the same assertion.

Iterating the above construction, one can piece finitely many invertible ergodic measure preserving transformations together. Further examples are given in section 4.

2. ASYMPTOTIC STATIONARITY

In this section we define the operator T associated with a gmp-transformation φ and prove asymptotic stationarity of $f, Tf, T^2 f, \ldots$.

For any measurable $f \geq 0$ the function Tf is defined by

$$Tf(\omega) = \sup\{s > 0 : \omega \in \varphi(\{f > s\})\}. \tag{2.1}$$

The supremum of the empty set shall be 0. In other words: Tf is defined by the condition that

$$\{Tf > s\} = \varphi(\{f > s\}) \tag{2.2}$$

for all $s > 0$.

If τ is a measure preserving transformation and $\varphi(A) = \tau^{-1}A$, then $Tf = f \circ \tau$.

If f is of the form $f = \Sigma_{i=1}^{n} \alpha_i 1_{A_i}$ with $A_1 \supset A_2 \supset \ldots \supset A_n$, and $\alpha_i \geq 0$, it is readily checked that a.e.

$$Tf = \Sigma_{i=1}^{n} \alpha_i 1_{\varphi(A_i)}.$$

(We did not require $\varphi(\emptyset) = \emptyset$. Thus Tf may be infinite on the nullset $\varphi(\emptyset)$. We disregard this set in the sequel.)

If $f_1 = f_2$ μ-a.e., then $Tf_1 = Tf_2$ μ-a.e. Thus T maps equivalence classes of functions to equivalence classes. We usually do not distinguish functions and their equivalence classes and use the same letter T in both cases. If we deal with equivalence classes, φ need only be defined mod nullsets.

We say that $f \geq 0$ and $g \geq 0$ have the same distribution if $\mu(f > t) = \mu(g > t)$ for all $t > 0$. (If μ is infinite, this does not imply $\mu(f \leq t) = \mu(g \leq t)$ for all $t > 0$.) By (2.2) and (1.2) $f, Tf, T^2f, \ldots$ have the same distribution. In particular, T maps each space L_p $(1 \leq p \leq \infty)$ into L_p, and we have $\|Tf\|_p = \|f\|_p$.

To prove the asymptotic stationarity, we first consider the simple special case $\mu(\Omega) = 1$.

Theorem 2.1: *Let φ be a gmp-transformation in a probability space $(\Omega, \mathcal{A}, \mu)$. For any $t_1,\dots,t_k > 0$, and for any measurable real valued functions $f_1,\dots,f_k \geq 0$ the sequence*

$$\mu(T^n f_1 > t_1,\dots,T^n f_k > t_k)$$

is increasing. The sequence $(T^n f_1,\dots,T^n f_k)$ converges in distribution.

Proof: The monotonicity of φ implies $\varphi(A \cap B) \subset \varphi(A) \cap \varphi(B)$. Put $A_{ni} = \{T^n f_i > t_i\}$ and $A_n = \{T^n f_1 > t_1,\dots,T^n f_k > t_k\}$. Then

$$\varphi(A_n) = \varphi(\bigcap_i A_{ni}) \subset \bigcap_i \varphi(A_{ni}) = \bigcap_i A_{n+1,i} = A_{n+1}.$$

Therefore, the monotonicity of $\mu(A_n)$ follows from (1.2). As the sequences $T^n f_i$ $(n = 0,1,\dots)$ are stationary for all i, the sequences of distributions of $T^n f_i$ are tight for all i. It follows, that also the sequence of distributions of the vectors $(T^n f_1,\dots,T^n f_k)$ is tight in $\mathbb{R}^k$. The convergence in distribution now follows by induction in k. (If k is fixed and $X(n) = (X_1(n),\dots,X_k(n))$ is a sequence of $\mathbb{R}^k$-valued random variables with tight distributions, the convergence of $P(X_1(n) > t_1,\dots,X_k(n) > t_k)$ for all $t_1,\dots,t_k > 0$ does not imply the convergence in distribution of $X(n)$. E.g., we may have $X(n) \equiv (0,0)$ for odd n and $X(n) \equiv (1,0)$ for even n.) ⌑

A process $(g_0, g_1,\dots)$ is called *asymptotically stationary* if the vectors $(g_n, g_{n+1},\dots,g_{n+k})$ converge in distribution for all k. Clearly, the theorem implies the asymptotic stationarity of the process $(f, Tf, T^2f, \ldots)$.

Remark 2.2: Krengel and Lin [KL] proved the convergence in distribtution of $T^n f$ for a more general class of operators T, namely, for nonexpansive order preserving operators in L_1^+ which decrease the L_∞-norm. (For definitions, see section 3.) For such a more general T the vectors $(T^n f_1, \ldots, T^n f_k)$ need not converge in distribution. To give an example, we apply an idea from [KL], p. 185: Let $\Omega = [0,1[$ and let μ be the Lebesgue-measure λ restricted to Ω. For $0 \le t \le 1$ let τ_t be the identity in Ω, and for $t > 1$ let τ_t be a rotation by some irrational α. For measurable $f \ge 0$ put

$$Tf(\omega) = \lambda(\{t \ge 0 : t \le f(\tau_t^{-1}\,\omega)\}).$$

Take $f_1 = 1_{[0,1/2]}$ and $f_2 = 1 + f_1$. It can be checked that T is order preserving, nonexpansive in L_1^+ and that T decreases the L_∞-norm. However, the vectors $(T^n f_1, T^n f_2)$ do not converge in distribution.

It may be possible, though, that the asymptotic stationarity of the sequence $f, Tf, T^2f, \ldots$ extends to the larger class of operators.

The σ-finite case: In [KL] the concept of convergence in distribution was introduced for σ-finite measures and real valued functions. We briefly discuss the $\mathbb{R}^k$-valued case here.

Let $k \ge 1$ be fixed. $C_{b,0}$ denotes the set of all bounded continuous functions on $\mathbb{E} = \mathbb{R}^k$ which vanish in a neighbourhood of $0 = (0,0,\ldots,0) \in \mathbb{E}$. Let $\mathbb{D}$ be the family of measures γ with $\gamma(U^c) < \infty$ for all neighborhoods U of 0. Thus, $\int h\, d\gamma$ is well defined and finite for $h \in C_{b,0}$ and $\gamma \in \mathbb{D}$. We say that the sequence $\gamma_n \in \mathbb{D}$ *converges in distribution* if $\gamma_n(\mathbb{E})$ converges and $\int h\, d\gamma_n$ converges to a finite limit for all $h \in C_{b,0}$. The numbers $\gamma_n(\mathbb{E})$ or their limit may be infinite.

If f is a measurable $\mathbb{E}$-valued function on a measure space $(\Omega, \mathbb{A}, \mu)$, the distribution of f is the measure γ with $\gamma(B) = \mu(f \in B)$ for Borel-measurable $B \subset \mathbb{E}$. If the coordinates of $f = (f_1,\ldots,f_k)$ belong to some space L_p with $1 \le p < \infty$, the distribution γ of f belongs to $\mathbb{D}$ and $\int h\, d\gamma = \int h \circ f\, d\mu$. Thus, we may say that such a sequence $f^{(n)}$ of $\mathbb{E}$-valued functions converges in distribution if $\int h \circ f^{(n)} d\mu$ converges for all $h \in C_{b,0}$. This notion of convergence in distribution is equivalent to the usual one if μ is finite. (Sometimes convergence in distribution is called weak convergence in probability theory, but we shall need the functional analytic notion of weak convergence below and want to distinguish the two notions.)

As sequence $\gamma_n \in \mathbb{D}$ is called *tight* if there exists for each $\varepsilon > 0$ a number $K(\varepsilon) > 0$ such that

$$\gamma_n(B(K_\varepsilon)^c) < \varepsilon$$

holds for all n, where $B(K) = \{x \in \mathbb{E} : \|x\| \le K\}$ is the ball with Euclidean radius K and center 0.

For $u = (u_1, \dots, u_k) \in \mathbb{E}$ we say that $I \subset \mathbb{E}$ is a quadrant with corner u if I is of the form $H_1 \times H_2 \times \dots \times H_k$ and each H_ν is one of the sets $[u_\nu, \infty[$, $]u_\nu, \infty[$, $]-\infty, u_\nu]$, or $]-\infty, u_\nu[$.

The proof of the following theorem has a lot in common with that of the special case of finite measures and with that of theorem 3.1 in [KL]. We therefore delete it.

Theorem 2.3: *Let γ_n be a sequence of elements of $\mathbb{D}$ for which $\gamma_n(\mathbb{E})$ converges. Then the following assertions are equivalent:*

(i) *γ_n converges in distribution.*

(ii) *The sequence γ_n is tight, and there exists a dense set $D \subset \mathbb{R}$ such that $\gamma_n([u,v[)$ converges if $0 \notin [u,v]$ and all coordinates of u and v belong to D.*

(iii) *The sequence γ_n is tight, and there exists a dense set $D \subset \mathbb{R}$ such that $\gamma_n(I)$ converges for all quadrants I which have positive distance from $0 \in \mathbb{E}$ and for which the corner has all coordinates in D.*

Now we can state the extension of theorem 2.1 to the σ-finite case:

Theorem 2.4: *Let φ be a gmp-transformation in a σ-finite measure space $(\Omega, \mathfrak{A}, \mu)$ and let $f_1, \dots, f_k$ be nonnegative real-valued measurable functions with $\mu(f_i > t) < \infty$ for all $i = 1, \dots, k$ and all $t > 0$. Then, for any $t_1, t_2, \dots, t_k > 0$, the sequence (2.3) is increasing. The sequence*

$(T^n f_1, T^n f_2, \dots, T^n f_k)$ *converges in distribution.*

Again, we delete the proof.

3. AN ERGODIC THEOREM

An operator T in L_p^+ is called L_p-nonexpansive (resp. L_p-norm-decreasing) if $\|Tf - Tg\|_p \le \|f - g\|_p$ (resp. $\|Tf\|_p \le \|f\|_p$) holds for all f,g in the range of definition. T is called order preserving if $f \le g$ implies $Tf \le Tg$. T is called positively homogeneous if $T(\alpha f) = \alpha(Tf)$ holds for all $\alpha > 0$.

If T is the operator associated with a gmp-transformation φ, T is obviously order preserving and positively homogeneous. We now show that T is L_p-nonexpansive for $1 \le p \le \infty$. For measurable $f \ge 0$ we have

$$\int f d\mu = \int_0^\infty \mu(\{f > t\}) dt.$$

By (2.2)

$$\mu(\{Tf > t\}) = \mu(\varphi(\{f > t\})) = \mu(\{f > t\}).$$

Therefore, T is integral preserving, i.e., $\int f d\mu = \int Tf d\mu$ holds for measurable $f \ge 0$. It is simple to see that order preserving integral preserving operators in L_1^+ are L_1-nonexpansive, see [KL], Lemma 2.2. Thus T is L_1-nonexpansive.

To prove the L_∞-nonexpansiveness, it suffices to show that $\|Tf - Tg\|_\infty \le \|f - g\|_\infty$ holds for all $f,g \in L_\infty^+$ with $f \le g$. If $\|f - g\|_\infty = \beta$ we have $0 \le f \le g \le f + \beta$ (mod μ). Hence $\{g > s + \beta\} \subset \{f > s\}$ and by (2.2)

$$\{Tg > s + \beta\} = \varphi(\{g > s + \beta\}) \subset \varphi(\{f > s\}) = \{Tf > s\}$$

mod μ. As $s > 0$ was arbitrary we have $Tf \leq Tg \leq Tf + \beta$ a.e. and therefore $\|Tf - Tg\|_\infty \leq \beta$.

Now the L_p-nonexpansiveness for $1 \leq p \leq \infty$ follows by a special case of a theorem of Browder, for which a simple proof has been sketched in the appendix of [KL]. We thus have

Theorem 3.1: *If φ is a gmp-transformation (or a map of the measure algebra of equivalence classes with the same properties) the operator T defined in section 2 is order preserving, positively homogeneous, integral preserving and L_p-nonexpansive for $1 \leq p \leq \infty$.*

Combining this result with theorem 4.1 in [KL], we arrive at

Theorem 3.2: *If φ is a gmp-transformation, the averages*

$$A_n f = (f + Tf + \ldots + T^{n-1} f)/n$$

converge weakly in L_p for $f \in L_p^+$ $(1 < p < \infty)$. If the measure μ is finite this assertion holds also for $p = 1$.

Actually, the case $1 < p < \infty$ follows already from theorem 3.1 and the ergodic theorem of Baillon [B].

Example 5.2 in [KL] shows that norm convergence in L_p with $1 < p < \infty$ need not hold. In this example, μ is infinite. It seems that for finite μ norm-convergence need not hold either, but the construction of examples is more difficult.

4. ON POINTWISE CONVERGENCE

It has been shown by the author [K1] that the averages $A_n f$ in theorem 3.2 need not converge a.e. The counterexample was build from speed limit operators introduced in [KL]. (The operators in [KL] and [K1] are induced by gmp-transformations, although this was not made explicit, since the notion was not yet formulated at that time.)

Recently, stronger averaging methods than Cesàro-averages have gained considerable attention. We therefore show now that there is no such method which yields almost everywhere convergence for all gmp-transformations.

The basic ideas are taken from section 5 of [KL] and from [K1]. Thus, some arguments need only be sketched here. On the other hand, some arguments are changed, and we must adapt the proof to general averaging methods.

If $A = (\alpha_{nk})$ $(n, k = 0, 1, 2, \ldots)$ is an infinite matrix of real numbers, a sequence $(s_k)_{k=0}^{\infty}$ is called A-convergent to s if $t_n = \Sigma_{k=0}^{\infty} \alpha_{nk}\, s_k$ is well-defined and $\lim_{n\to\infty} t_n = s$. We then write A-lim $s_k = s$. We consider only permanent matrices A, i.e., we require that A is such that for any convergent sequence $s_k \to s$ we have $t_n \to s$. This is the case if and only if A has the following properties

$$\lim_{n\to\infty} \Sigma_{k=0}^{\infty} \alpha_{nk} = 1 \qquad (4.1)$$

$$\text{For all } k \geq 0 \quad \lim_{n\to\infty} \alpha_{nk} = 0 \qquad (4.2)$$

$$\sup\{\Sigma_{k=0}^{\infty} |\alpha_{nk}| \ : n \geq 0\} < \infty \qquad (4.3)$$

(For a proof, see [ZB], p. 57.) Our result is:

Theorem 4.1: *For any permanent* A *there exists a qmp-transformation* Ψ *of a suitable probability space and an indicator function* f *such that* A-lim $T^k f$ *fails to exist a.e., where* T *is the operator induced by* Ψ.

(We switch to the letter Ψ here to facilitate reference to [KL].)

Proof: *Step 1* We begin by constructing some sequences of integers depending on the matrix A. Put $r_1 = 0$. There exists $n_1, m_1 \geq 1$ with

$$\Sigma_{k=m_1+1}^{\infty} |\alpha_{n_1,k}| < \frac{1}{8} \quad \text{and} \quad \Sigma_{k=0}^{m_1} \alpha_{n_1,k} > 1 - \frac{1}{8}. \tag{4.4}$$

If, for some $j \geq 2$ the numbers r_{j-1}, n_{j-1} and m_{j-1} have been determined, we can find (in this order)

$r_j > 2\,m_{j-1}$, $n_j > n_{j-1}$ and $m_j > 2\,r_j$ such that

$$\sum_{k=r_j+1}^{m_j} \alpha_{n_j,k} > 1 - \frac{1}{8} \tag{4.5}$$

and

$$\sum_{k=0}^{r_j} |\alpha_{n_j,k}| + \sum_{k=m_j+1}^{\infty} |\alpha_{n_j,k}| < \frac{1}{8} . \tag{4.6}$$

Step 2: The construction of the speed limit operators in [KL], section 5 depends on the solution of an integral equation. The following more explicit argument seems preferable: Let $\varphi(x)$ be a decreasing strictly positive function in $[0,1[$ and $\varphi(x) = 0$ for $x \geq 1$. If a point moves with speed $\varphi(x)$ to the right when it is in x, and if it starts at a point b_i at time 0, the time needed to reach y with

$b_i \leq y \leq 1$

$$\tau_i(y) = \int_{b_i}^{y} \frac{1}{\varphi(x)}\, dx.$$

For $y > 1$ put $\tau_i(y) = \infty$. The distance travelled up to time t is

$$c_i^t = \sup\{y : \tau_i(y) \leq t\} - b_i.$$

Step 3: We now want to define gmp-transformations ζ_t $(t \geq 0)$ in $[0,1[$ (with Lebesgue measure μ). First we consider a set B which is the union of k disjoint intervals $[a_i, b_i[$ with

$$0 \leq a_1 < b_1 \leq a_2 < b_2 \leq \ldots \leq a_k < b_k \leq 1.$$

The intervals move with the speed permitted for their right-hand endpoints, but when this endpoint catches up with the left-hand endpoint of the interval to the right, no over-taking is allowed. At that moment the two intervals unite to form a bigger interval, whose right-hand endpoint determines the speed.

Formally, put

$$d_k^t = c_k^t \quad \text{and} \quad d_i^t = \mathrm{Min}(c_i^t,\ d_{i+1}^t + (a_{i+1} - b_i))$$

for $i = k-1, k-2, \ldots, 1$. Then the union B^t of the k disjoint intervals $[a_i^t, b_i^t[$ with

$$a_i^t = a_i + d_i^t \ , \quad b_i^t = b_i + d_i^t$$

describes the set, into which B has been transformed by time t. We put

$$\zeta_t(B) = B^t.$$

Clearly, $\mu(\zeta_t(B)) = \mu(B)$. If $\overline{B}$ is obtained from B by deleting one of the intervals, it may be seen as in [KL], p. 188 that $\bar{B}^t \subset B^t$. Thus ζ_t has the properties of a gmp-transformation, except that its domain of definition is only the algebra consisting of the disjoint unions of intervals. Passing to equivalence classes rather than sets, ζ_t can be extended to the full measure algebra by the following lemma.

Lemma 4.2: *Let* J *be an algebra generating* $\mathbb{A}$ *in* $(\Omega, \mathbb{A}, \mu)$, *and let* $\mathbb{A}^\mu$ *be the measure algebra of equivalence classes of* $\mathbb{A}$*-measurable sets. If* ζ *is a map of* J *into* J *which satisfies*

$$A \subset B \Rightarrow \zeta(A) \subset \zeta(B) \quad \text{and} \quad \mu(\zeta(A)) = \mu(A) \qquad (4.7)$$

for all $A, B \in J$, *there is a unique* $\zeta : \mathbb{A}^\mu \to \mathbb{A}^\mu$ *with the same properties.*

Proof: Put $d(A,B) = \mu(A \Delta B)$, show $d(\zeta(A), \zeta(B)) \leq d(A,B)$ and extend ζ by continuity. If $A \subset B$ and A_n, B_n are sets in J with $d(A_n, A) \to 0$, $d(B_n, B) \to 0$ we may assume $A_n \subset B_n$ by passing to $A_n \cap B_n$. ⌑

Step 4: We now construct ψ. Let $(\Omega_n, \mathbb{A}_n, \mu_n)$, $n = 1,2,\ldots$, be copies of $[0,1[$ with Lebesgue-measure, X_n the n-th coordinate map in the product space $(\Omega, \mathbb{A}, \mu)$, and φ_n a speed limit function for Ω_n, to be specified later.

A set of the form $\{\omega : x_n \leq X_n(\omega) < y_n,\ n = 1,\ldots,N\}$ will be called N-interval. Let $\mathbb{H}$ be the family of all sets H which are finite disjoint unions of N-intervals for some n. For any H there exists finitely many disjoint (N-1)-dimensional intervals I_s in $\Omega_2 \times \ldots \times \Omega_N$ with union $\Omega_2 \times \ldots \times \Omega_N$ and numbers

$$0 \leq a_{s1} < b_{s1} \leq a_{s2} < \ldots \leq a_{sk} < b_{sk} \leq 1$$

with

$$H \cap \{(X_2,\ldots,X_N) \in I_s\} = \bigcup_{i=1}^{k} \{(X_2,\ldots,X_N) \in I_s \text{ and } a_{si} \leq X_1 < b_{si}\}.$$

(k can depend on s; if the left-hand side is empty for some s, put $k = 0$.)

Construct a^t_{si} and b^t_{si} as above with φ replaced by φ_1, and $[a_i,b_i[$ replaced by $[a_{si},b_{si}[$. Next put

$$H_{si} = \{(X_2,\ldots,X_N) \in I_s,\ a^1_{si} \leq X_1 < b^1_{si}\}.$$

and let $\psi_1(H)$ be the disjoint union of the sets H_{si} for all s and i. ψ_1 doesn't affect the coordinates $n = 2,\ldots,N$ and it moves the first coordinate with the speed limit φ_1 for one time unit.

Construct ψ_n in the same way using X_n and φ_n instead of X_1 and φ_1. If $n > N$ then $\psi_n(H) = H$, since the interval $[0,1[$ is not moved at all. Hence, if we put

$$\psi(H) = \psi_n \circ \ldots \circ \psi_2 \circ \psi_1(H)$$

this definition is independent of n for $n \geq N$. As $\mathcal{H}$ is an algebra generating $\mathcal{A}$, lemma 4.2 yields the extension of ψ to $\mathcal{A}$. (We do not distinguish sets and equivalence classes here.)

Step 5: Let f be the indicator function of the union B of the sets $\{X_m \leq 2^{-(m+2)}\} = B_m$. Put

$$T_n f = \Sigma_{k=0}^{\infty} \alpha_{nk} T^k f$$

where T is the operator corresponding to the gmp-transformation ψ. It follows from (4.1), (4.3), $\int f d\mu \leq 1/4$ and Fatous lemma that

$$\int \liminf T_n f \, d\mu \leq 1/4.$$

It will therefore be sufficient to show that the speed limits φ_n can be chosen in such a way that

$$\limsup T_n f \geq 3/4 \qquad \text{a.e.}$$

Let h_m be the indicator function of B_m. Then $0 \leq T^k h_m \leq T^k f \leq 1$ for all k and all m. We now make use of the numbers determined in step 1.

Put $\gamma_m = 1/(3 \cdot 2^{m+2})$ and let $\varphi_1(x) = \gamma_1/m_1$ on $[0,4\gamma_1[$. Then $[0,3\gamma_1[$ moves to $[\gamma_1,4\gamma_1[$ by time m_1 under the speed limit φ_1. This implies $T^k h_1 = 1$ on $\{\gamma_1 \leq X_1 < 3\gamma_1\}$ for $k = 0,\dots,m_1$. Hence $T_{n_1} f \geq 3/4$ on this set by (4.4).

Next put $\varphi_1(x) = \gamma_1/(r_2-m_1)$ on $[4\gamma_1,5\gamma_1[$ and $=\gamma_1/(m_2-r_2)$ on $[5\gamma_1,6\gamma_1[$. Then the considered interval moves to $[2\gamma_1, 5\gamma_1[$ by time r_2 and to $[3\gamma_1,6\gamma_1[$ by time m_2. This yields $T_{n_2} f \geq 3/4$ on $\{3\gamma_1 \leq X_1 < 5\gamma_1\}$ by (4.5) and (4.6).

This is continued nine more times. By time m_{11} the interval $[0,3\gamma_1[$ reaches the right-hand end of $[0,1[$ and the maximum of $T_{n_1} f,\dots,T_{n_{11}} f$ is $\geq 3/4$ on $\{\gamma_1 \leq X_1 < 1-\gamma_1\}$.

Now the second coordinate is used. If we put $\varphi_2(x) = \gamma_2/m_{11}$ for $0 \leq x < 4\gamma_2$, the interval $[0,3\gamma_2[$ moves to $[\gamma_2,4\gamma_2[$ by time m_{11}. Let it move to $[2\gamma_2,5\gamma_2[$ by time r_{12} and to $[3\gamma_2,6\gamma_2[$ by time m_{12}. It should now be clear, how to continue. The end of the interval is reached by time m_{33}. Then the maximum of $T_{n_{12}} f, \dots, T_{n_{33}} f$ is $\geq 3/4$ on $\{2\gamma_2 \leq X_2 < 1-\gamma_2\}$, and the third coordinate takes over.

for the k-th coordinate we find finitely many new n_i such that the maximum of the functions $T_{n_i} f \geq 3/4$ on a set of measure $1-2^{-(k+2)}$. This completes the proof. ∎

Note: I would like to use this opportunity to clarify an acknowledgement in the related paper [AK]. The mentioned work by Sine [S] was done independently of [AK].

REFERENCES

[AK] M.A. Akcoglu and U. Krengel: Nonlinear models of diffusion on a finite space. Prob. Th. Rel. Fields 76, 411-420, (1987).

[B] J.-B. Baillon: Comportement asymptotique des itérés de contractions non-linéaires dans les espaces L^p. Comptes Rendus Acad. Sci. Paris 286, 157-159, (1978).

[K1] U. Krengel: An example concerning the nonlinear pointwise ergodic theorem. Israel J. Math. 58, 193-197, (1987).

[K2] U. Krengel: Ergodic Theorems. de Gruyter Studies in Mathematics 6, de Gruyter, Berlin - New York, (1985).

[KL] U. Krengel and M. Lin: Order preserving nonexpansive operators in L_1. Israel J. Math. 58, 170-192 , (1987).

[M1] H. Michel (Editor): Ergodic Theory and Related Topics. Proc. of the Conf. held at Vitte/Hiddensee (GDR), Oct. 19-23, 1981, Mathematical Research 12, Akademie-Verlag, Berlin (1982).

[M2] H. Michel (Editor): Proc. of the Conf. Ergodic Theory and Related Topics II, Georgenthal (Thuringia), GDR,

April 20-25, 1986. Teubner Texte zur Mathematik 94, Teubner, Leipzig, (1987).

[R] H. Rost: Markoff-Ketten bei sich füllenden Löchern im Zustandsraum. Ann. Inst. Fourier, Grenoble, 21, 253-270, (1971).

[ZB] K. Zeller and W. Beekmann: Theorie der Limitierungsverfahren. Ergebn. Math. Grenzgeb. 15, Springer, Berlin - Heidelberg - New York, (1970).

[S] R. C. Sine: A nonlinear Perron-Frobenius theorem. To appear.

TANGENT SEQUENCES OF RANDOM VARIABLES: BASIC INEQUALITIES AND THEIR APPLICATIONS[1]

S. KWAPIEN[2] and W.A. WOYCZYNSKI
Warsaw University and Case Western Reserve University

Summary. Basic inequalities concerning conditionally identically distributed sequences of random variables (tangent sequences) are obtained. The inequalities are applied to obtain comparison theorems for the almost sure convergence of series of tangent sequences. The inequalities for tangent martingale difference sequences are applied to get sharp inequalities for stochastic integrals.

1. Introduction

Let $(\Omega, \mathcal{F}, P)$ be a probability space, and let $\mathcal{F}_0 \subset \mathcal{F}_1 \subset \mathcal{F}_2 \subset \ldots$ be an ascending sequence of sub–σ–fields (a filtration) of $\mathcal{F}$.

Definition 1.1. We shall say that two $(\mathcal{F}_i)$–adapted sequences of random variables (d_i) and (e_i) are *tangent* if, for each $i = 1,2,\ldots, d_i \sim_{\mathcal{F}_{i-1}} e_i$, i.e. for each $c \in \mathbb{R}$, we have $P(d_i < c \mid \mathcal{F}_{i-1}) = P(e_i < c \mid \mathcal{F}_{i-1})$ a.s.

[1]Research partially supported by an NSF Grant.
[2]Visiting at Case Western Reserve University

ISBN 0-12-231050-0

Formally, the above definition has been introduced by the authors (cf. S. Kwapien and W.A. Woyczynski [11]) in connection with a study of stochastic integrals, but it was influenced by an earlier definition of a tangent process proposed by J. Jacod [10], who, in turn, exploited an old idea of K. Ito [9].

It was observed recently by T. McConnell [12] and, independently, P.Hitczenko [8], that the powerful and elegant technique of Burkholder's L–functions, developed to study martingale transforms and subordinated martingale difference sequences, can also be applied to some special tangent sequences. In the present paper we pursue these ideas further.

Following a preliminary Section 2, containing definitions and motivating examples, the paper contains weak type inequalities and results on comparison of moments for various classes of tangent martingale difference sequences (Sections 3 and 4). These are applied to obtain comparison results for tail probabilities of series of tangent sequences, which, in turn, give comparison theorems for the almost sure convergence of such series (Section 5). Section 6 describes some applications of the above results such as best constant estimates for Brownian stochastic integrals, and for the planar Brownian motion.

2. Basic Definitions and Examples.

In view of the way tangency of sequences is used in the following arguments, it is worth noticing that sequences (ξ_i) and (η_i) are tangent if, and only if, for each $(\mathcal{F}_i)$–predictable, bounded sequence (v_i) (i.e. (v_i) is $(\mathcal{F}_{i-1})$–adapted), and for each sequence (φ_i) of Borel measurable, bounded functions, $Ev_i\varphi_i(\xi_i) = Ev_i\varphi_i(\eta_i)$, $i = 1,2,\ldots$.

Hence, in particular, for any stopping time τ, and any tangent sequences (ξ_i) and (η_i),

$$E \sum_{i=1}^{T} v_i \varphi_i(\xi_i) = E \sum_{i=1}^{T} v_i \varphi_i(\eta_i).$$

If sequences (ξ_i) and (η_i) are tangent then, for any $(\mathcal{F}_i)$–predictable sequence (v_i), sequences $(v_i \xi_i)$ and $(v_i \eta_i)$ are tangent, as well.

Definition 2.1. Two $(\mathcal{F}_i)$–adapted sequences (d_i) and (e_i) are said to be *tangentially conditionally independent* (*tci*) if, for each $i = 1,2,\ldots,$ random variables d_i and e_i are $\mathcal{F}_{i-1}$–conditionally independent.

Remark 2.1. If (e_i) is a decoupled tangent sequence to (d_i) then sequences (d_i) and (e_i) are tci. For a definition of decoupled tangent sequences see S. Kwapien and W.A. Woyczynski [9], where they appeared in a crucial way in the theory of stochastic integrals with respect to semimartingales.

Definition 2.2. An $(\mathcal{F}_i)$–adapted sequence (d_i) is said to be *conditionally symmetric* if, for each $i = 1,2,\ldots,$ $d_i \sim_{\mathcal{F}_{i-1}} - d_i$. If (d_i) and (e_i) are tangent sequences and (d_i) is conditionally symmetric then so is (e_i).

Example 2.1. Let (ξ_i) be a sequence of independent random variables, (ξ_i') be its independent copy, and let (v_i) be a sequence of $(\mathcal{F}_i)$–predictable random variables (i.e. v_i is $\mathcal{F}_{i-1}$–measurable for $i = 1,2,\ldots$), where $\mathcal{F}_i = \sigma(\xi_1,\ldots,\xi_i, \xi_1',\ldots,\xi_i')$, $i = 0,1,2,\ldots$. Then, sequences $(d_i) := (v_i \cdot \xi_i)$ and $(e_i) := (v_i \xi_i')$ are tangent tci sequences. Furthermore, if random variables ξ_i, $i = 1,2,\ldots,$ are symmetric then sequences (d_i) and (e_i) are conditionally symmetric.

Example 2.2. Let (ξ_i) be a sequence of independent, identically distributed (iid) symmetric random variables, let (v_i) be a sequence of $(\mathcal{F}_i)$–predictable random variables, where $\mathcal{F}_i = \sigma(\xi_1,\dots,\xi_i)$, i = 0,1,2,..., and let $\epsilon_i = \pm 1$, i = 1,2,... . Then, sequences $(d_i) := (v_i\xi_i)$ and $(e_i) := (\epsilon_i v_i \xi_i)$ are tangent and conditionally symmetric.

Example 2.3. Let (h_i) be the Haar system on [0,1], (α_i) be a sequence in $\mathbb{R}$, $\epsilon_i = \pm 1$, i = 1,2,..., and let $\mathcal{F}_i = \sigma(h_i,\dots,h_i)$, i = 0,1,2,... . Then, sequences $(d_i) := (\alpha_i h_i)$ and $(e_i) := (\epsilon_i \alpha_i h_i)$ are tangent and conditionally symmetric.

The preceding example is a special case of the following, more general,

Example 2.4. (cf. P. Hitczenko [6]). Let (d_i) be an $(\mathcal{F}_i)$–adapted sequence and let $(e_i) := (d_i \circ \phi_i)$, where $\phi_i : (\Omega, \mathcal{F}_i, P) \to (\Omega, \mathcal{F}_i, P)$, i = 1,2,..., are measurable, measure preserving transformations such that $\phi_i(A) = A$ for each $A \in \mathcal{F}_{i-1}$, i = 1,2,... . Then, sequences (d_i) and (e_i) are tangent.

In the remainder of this paper the following notation will be utilized:

$$f_0 = g_0 = 0, \quad f_i := d_1 + \dots + d_i, \quad g_i := e_1 + \dots + e_i, \quad i = 1,2,\dots,$$

$$f^* := \sup_i |f_i| := \sum^* d_i, \quad f := \lim_i f_i,$$

with a similar meaning for g^* and g. Also, for a random variable ξ, the truncation

$$[\xi]^c := \begin{cases} c & \text{if } \xi > c \\ \xi & \text{if } |\xi| \le c \\ -c & \text{if } \xi < -c. \end{cases}$$

In all the inequalities of this paper, sequences of random variables will always have only finitely many non–zero terms, so, we need not concern ourselves with the question of convergence of their series.

3. Weak–type Inequalities for Tangent Martingale Difference Sequences.

THEOREM 3.1. *Let* (d_i) *and* (e_i) *be tangent martingale difference sequences. Then, for any* $a > 0$,

(i) $aP(f^* \geq a) \leq \frac{5}{2} E|g|$;

(ii) (cf. T. McConnell [10]) *if* (d_i) *and* (e_i) *are assumed to be tci, then*

$$aP(f^* \geq a) \leq 2E|g|$$

(*the constant 2 is best possible*);

(iii) *if* (d_i) *and* (e_i) *are assumed to be tci, and to have Gaussian conditional distributions* $\mathcal{L}(d_i | \mathcal{F}_{i-1})$, $i = 1,2,...,$ *then*

$$aP(f^* > a) \leq \kappa E|g|,$$

where $\kappa = (1+3^{-2}+5^{-2}+...)/(1-3^{-2}+5^{-2}...) = 1.346886...$ (*the constant* κ *is best possible*).

We will precede the proof of Theorem 3.1 with an introduction of classes $\mathcal{U}$, $\mathcal{U}^{\text{ind}}$, and $\mathcal{U}^{\text{ind, Gauss}}$, of real functions of two variables, which are similar to the classes of L–functions originally introduced by D.L. Burkholder.

Definition 3.1. A function $u: \mathbb{R}^2 \to \mathbb{R}^+$ is said to be in class $\mathcal{U}$ (resp. $\mathcal{U}^{\text{ind}}$, resp. $\mathcal{U}^{\text{ind, Gauss}}$) if the following three conditions are satisfied:

(a) for each (resp. independent, resp. independent Gaussian) equidistributed random variables ξ and η with $E\xi = E\eta = 0$, and for all $x,y \in \mathbb{R}$, $Eu(x + \xi, y+\eta) \geq u(x,y)$;

(b) for each $x \in \mathbb{R}$ such that $|x| \geq 1$ and all $y \in \mathbb{R}$, $u(x,y) \leq |y|$;

(c) for all $x,y \in \mathbb{R}$, $u(x,y) \leq |y| + u(0,0)$.

It is clear that $\mathcal{U} \subset \mathcal{U}^{\text{ind}} \subset \mathcal{U}^{\text{ind, Gauss}}$, and that the class $\mathcal{U}$ is non–empty (take $u(x,y) = |y|$). It is also easy to see that the condition (a) implies that if $u \in \mathcal{U}$ then u is biconvex in "skew coordinates" i.e. the following two functions of $t \in \mathbb{R}$

$$u(x + t, y + t) \text{ and } u(x + t, y - t) \tag{3.1}$$

are convex for each $x,y \in \mathbb{R}$. Furthermore, if $u,v \in \mathcal{U}$ (resp. $\mathcal{U}^{\text{ind}}$, resp. $\mathcal{U}^{\text{ind, Gauss}}$) then $\max(u,v) \in \mathcal{U}$(resp. $\mathcal{U}^{\text{ind}}$, resp. $\mathcal{U}^{\text{ind, Gauss}}$), so, there exists a maximal function $u_{\max} \in \mathcal{U}$ (resp. $u_{\max}^{\text{ind}} \in \mathcal{U}^{\text{ind}}$, resp. $u_{\max}^{\text{ind, Gauss}} \in \mathcal{U}^{\text{ind, Gauss}}$). We do not know the explicit formula for $u_{\max}$(resp. $u_{\max}^{\text{ind}}$). Its knowledge, or at least the information about $u_{\max}(0,0)$ (resp. $u_{\max}^{\text{ind}}(0,0)$, $u_{\max}^{\text{ind,Gauss}}(0,0)$), is important for what follows. The best constants in Theorem 3.1 (i) (resp. (ii), resp. (iii)) are equal to $(u_{\max}(0,0))^{-1}$ (resp. $(u_{\max}^{\text{ind}}(0,0))^{-1}$, resp. $u_{\max}^{\text{ind,Gauss}}(0,0))^{-1}$). We do know, however that $u_{\max}(0,0) \leq 7/15$ ($< 1/2$) and $u_{\max}^{\text{ind}}(0,0) = 1/2$.

The "largest" function in $\mathcal{U}$ we could identify is described in

PROPOSITION 3.1. *Let* $D = \{(x,y){:} = (9+(4x)^2)^{1/2} + |4y| \leq 5\} \subset \mathbb{R}^2$, and

$$u_0(x,y) := \begin{cases} \frac{2}{5}(y^2 - x^2 + 1) & \text{if } (x,y) \in D, \\ |y| & \text{if } (x,y) \notin D, \end{cases}$$

Then $u_0 \in \mathcal{U}$.

Proof. Indeed, if $(x,y) \notin D$ then, for any equidistributed ξ,η with $E\eta = 0$, $Eu_o(x + \xi, y + \eta) \geq E|y + \eta| \geq |y| = u_o(x,y)$ since $u_o(s,t) \geq |t|$ for all $(s,t) \in \mathbb{R}^2$.

If $(x,y) \in D$, and ξ and η are as above, then let $\tilde{\xi} = \varphi(\xi)$, and $\tilde{\eta} = \varphi(\eta)$, where $\varphi(s) := [y + s]^2 - y$, with $[.]^2$ denoting the truncation defined in Section 2.

First, we will prove that

$$(\xi,\eta) \neq (\tilde{\xi},\tilde{\eta}) \Rightarrow (x + \xi, y + \eta) \notin D \text{ and } (x + \tilde{\xi}, y + \tilde{\eta}) \notin D \tag{3.2}$$

If $\eta \neq \tilde{\eta}$ then $|y+\eta| > 2$, and $|y+\tilde{\eta}| = 2$. Since $D \subset [-1,1] \times [-1/2, 1/2]$, we get that both $(x+\xi, y+\eta)$ and $(x+\tilde{\xi}, y+\tilde{\eta})$ are in D^c.

If $\xi \neq \tilde{\xi}$ then $|y+\xi| - |x-y| > 1$, because $|x-y| < 1$ for $(x,y) \in D$. Thus $(x+\xi, y+\eta) \in D^c$. Similarily, $|x+\tilde{\xi}| = |y+\tilde{\xi} + x-y| \geq 2-1 = 1$, so that $(x+\tilde{\xi}, y+\tilde{\eta}) \in D^c$ and (3.2) has been proven.

Next, by (3.2), we have that

$$Eu_o(x+\xi, y+\eta) =$$

$$E(u_o(x+\xi, y+\eta) - u_o(x+\tilde{\xi}, y+\tilde{\eta})) + Eu_o(x+\tilde{\xi}, y+\tilde{\eta}) =$$

$$E(|y+\eta| - |y+\tilde{\eta}|) + Eu_o(x+\tilde{\xi}, y+\tilde{\eta}) \geq \tag{3.3}$$

$$E(|y+\eta| - |y+\tilde{\eta}|) + \frac{2}{5} E((y+\tilde{\eta})^2 - (x+\tilde{\xi})^2 + 1) =$$

$$E(|y+\eta| - |y+\tilde{\eta}|) + u_o(x,y) + \frac{2}{5}(2y - 2x)E\tilde{\eta} ,$$

where the inequality takes place because $u_o(s,t) \geq (2/5)(t^2 - s^2 + 1)$ if $|t| \leq 2$, and

$|y+\tilde{\eta}| \le 2$, and the last equality holds true because $\tilde{\xi}$ and $\tilde{\eta}$ are equidistributed.

Finally, since, by the definition of $\tilde{\eta}$, $|(y+\eta)-(y+\tilde{\eta})| \le |y+\eta| - |y+\tilde{\eta}|$, and since $(4/5)|x-y| < 1$ for $(x,y) \in D$, we also have that

$$E(|y+\eta| - |y+\tilde{\eta}|) \ge (4/5)(y-x)\, E((y+\eta)-(y+\tilde{\eta})) = -(4/5)(y-x)E\tilde{\eta}\,,$$

which, together with (3.3) completes the proof of property (a) for u_o. Obviously, u_o also satisfies properties (b) and (c). Q.E.D.

Proof of Theorem 3.1 (i). *Firstly,* notice that it suffices to prove that for any tangent martingale difference sequences (d_i) and (e_i)

$P(|f| \ge 1) \le 5/2\, E|g|$.

Indeed, if we put $d_i' = I(\tau \ge i)d_i a^{-1}$, $e_i' = I(\tau \ge i)e_i a^{-1}$, $i = 1,2,\ldots$, where $\tau = \min\{k: |f_k| \ge a\}$, then (d_i') and (e_i') are tangent martingale difference sequences, $P(f^* \ge a) = P(|f'| \ge 1)$, and $E|g'| \le E|g|a^{-1}$.

Let $u \in \mathcal{U}$ and let (d_n) and (e_n) be tangent martingale difference sequences. Then

$$P(|f_n| \ge 1) \le P(|g_n| - u(f_n,g_n) + u(0,0) \ge u(0,0)) \le$$

$$\le \frac{1}{u(0,0)} E(|g_n| - u(f_n,g_n) + u(0,0)) \le \frac{1}{u(0,0)} E|g_n|,$$

where the first inequality follows from property (b) (Definition 3.1), the second from (c) and the Chebyshev's inequality, and the last from the fact that, by property (a) the sequence $Eu(f_n,g_n)$, $n = 0,1,2,\ldots$, is increasing. Since, by Proposition 3.1, we can take $u = u_o$, the proof of Theorem 3.1(i) is complete. Q.E.D.

Proof of Theorem 3.1. (ii) (Sketch). T.McConnell [12] and P. Hitczenko [8] observed, independently, that if a function u is biconvex in "skew coordinates" (cf. 3.1)) then u satisfies condition (a) from Definition 3.1 for $\mathcal{U}^{ind}$, and thus, a proof of Theorem 3.1 (ii) can be carried out mimicking the proof of Theorem 3.1 (i) with the function u_0 replaced by the Burkholder function $u^B \in \mathcal{U}^{ind}$, where

$$u^B(x,y) := \begin{cases} \frac{1}{2}(y^2 - x^2 + 1) & \text{if } |x| + |y| \leq 1, \\ |y| & \text{if } |x| + |y| > 1. \end{cases}$$

Q.E.D.

Proof of Theorem 3.1 (iii). Again, with proof of Theorem 3.1(i) as a model, in the present case it is easy to see that the condition (a) (Definition 3.1) for u in $\mathcal{U}^{ind,\ Gauss}$ is equivalent to the condition that u is pre–subharmonic (i.e. u coincides with a subharmonic function on $\mathbb{R}^2$ except for a set of Lebesque measure zero, where u is smaller). Therefore, $u_{max}^{ind,Gauss}$ coincides with the function u^D which, in the strip $|x| \leq 1$, is equal to the solution of a Dirichlet Problem which satisfies the boundedness assumption (c) of Definition 3.1, and boundary conditions $u(\pm 1,y) = |y|$, and which for $|x| > 1$, is given by $u^D(x,y) = |y|$. The solution can be readily written in the form

$$u^D(x,y) = E|Y^{x,y}(\tau(x,y))|, \quad \text{for } |x| \leq 1, \tag{3.11}$$

where $(X^{x,y}(t), Y^{x,y}(t))$ is a two–dimensional Brownian motion starting at (x,y) in the strip, and $\tau(x,y)$ is the first exit time of this process from the strip.

Since $u^D(0,0) = \sqrt{2/\pi}\ E(\sqrt{\Theta}) = \kappa^{-1}$ (cf. B. Davis [5]), where Θ is the first exit time from $[-1,1]$ of the standard Brownian motion in $\mathbb{R}$, the proof of Theorem 3.1 (iii) is complete Q.E.D.

Remark 3.1. The same constant κ and the same function u^D had been used by B. Davis [5] to find the best constant in the Kolmogorov's inequality for the Hilbert transform on the unit circle (cf. also Section 6, Remark 6.2).

4. Comparison of Moments of Tangent Martingale Difference Sequences

THEOREM 4.1. *Let* $1 < p < \infty$ *and let* (d_i) *and* (e_i) *be tangent martingale difference sequences. Then,*

(i) $E|f|^p \le (p^*-1)^{2p} E|g|^p$,

where $p^* := \max(p, p/p-1)$ (cf. P. Hitczenko [11])

(ii) *if* (d_i) *and* (e_i) *are assumed to be tci then*

$$E|f|^p \le (p^*-1)^p \, E|g|^p$$

(cf. T. McConnell [12], and P. Hitczenko [8]);

(iii) *if* (d_i) *and* (e_i) *are assumed to be tci and to have Gaussian conditional distributions* $\mathscr{L}(d_i \mid \mathscr{F}_{i-1})$, $i = 1,2,\ldots$, *then*

$$E|f|^p \le \cot^p(\pi/2p^*) E|g|^p,$$

and the constant $\cot^p(\pi/2p^*)$ *is best possible.*

In this section, we again exploit ideas stemming from D.L. Burkholder's use of L–functions and introduce.

Definition 4.1. Let $1 < p < \infty$ and $C > 0$. A function $u: \mathbb{R}^2 \to \mathbb{R}$ is said to be in class $\mathscr{U}_{p,C}$ (resp. $\mathscr{U}^{ind}_{p,C}$, resp. $\mathscr{U}^{ind,Gauss}_{p,C}$) if the following three conditions are satisfied:

(a) for each (resp. independent, resp. independent, Gaussian) equi-distributed random variables ξ and η with $E\xi = 0$ and $E|\xi|^p < \infty$, and for all $x,y \in \mathbb{R}$, $Eu(x+\xi, y+\eta) \ge u(x,y)$;

(b) for each $x,y \in \mathbb{R}$, $u(x,y) \le C|x|^p - |y|^p$;

(c) for some $K > 0$ and each $x,y \in \mathbb{R}$, $|u(x,y)| \le K(|x|^p + |y|^p)$, so that, in particular, $u(0,0) = 0$.

Proof of Theorem 4.1. It is easy to see that if the class $\mathcal{U}_{p,C}$ (resp. $\mathcal{U}^{ind}_{p,C}$, resp. $\mathcal{U}^{ind,Gauss}_{p,C}$) is non-empty then the inequality

$$E|f|^p \le C\, E|g|^p$$

holds true for any tangent (resp. tangent tci, resp. tangent tci with Gaussian conditional distributions) martingale difference sequence (d_i) and (e_i). Indeed, by property (b) of Definition 4.1,

$$C\, E|g_i|^p - E|f_i|^p \ge E\, u(g_i,f_i),\ i = 1,2,\dots,$$

and $E\, u(g_i,f_i) \ge 0$ by property (c), because $Eu(g_i,f_i)$, $i = 1,2,\dots$, is a nondecreasing sequence, in view of property (a). Thus, the remainder of the proof will consist in showing that the classes $\mathcal{U}^{ind}_{p,(p^*-1)^p}$, and $\mathcal{U}^{ind,Gauss}_{p,\cot^p(\pi/2p^*)}$ are nonempty. It will be convenient to begin with completion of the proof of

(ii) Here, by the previously mentioned observation of T. McConnell, and P. Hitczenko, the function $u^B_p(x,y)\!:= -u(x+y, x-y)$, where u is the function introduced by D.L. Burkholder ([2], Formula 7), belongs to $\mathcal{U}^{ind}_{p,(p^*-1)^p}$.

(i) An argument of P. Hitczenko [8] reduces this part to part (ii) as follows: if (d_i) and (e_i) are tangent martingale difference sequences then we can construct (possibly enlarging the underlying probability space) a sequence (a_i) of random variables which is a decoupled tangent sequence for both (d_i) and (e_i). Thus, we obtain (i) applying twice Theorem 4.1 (ii).

(iii) As in the proof of Theorem 3.1 (iii), we observe that the condition (a) for $u \in \mathcal{U}^{ind,Gauss}_{p,c}$ means that u is pre-subharmonic. Therefore, $\mathcal{U}^{ind,Gauss}_{p,C}$ is nonempty if, and only if, the function $C|x|^p - |y|^p$ has a subharmonic minorant. Here, proceeding in a fashion influenced by the work of S.K. Pichorides [11] on best constants in L^p-estimates for Hilbert transform on the unit circle, we can also constructively show that the class $\mathcal{U}^{ind,Gauss}_{p,\cot^p(\pi/2p^*)}$ is non-empty (see Appendix). Q.E.D.

Remark 4.1. The constant $\cot^p(\pi/2p^*)$ is best possible in Theorem 4.1 (iii). We do not know if the constant $C = p^* - 1$ is the smallest constant for which $U^{ind}_{p,C}$ is nonempty, and we do not know the best constant in Theorem 4.1 (ii). It would be also interesting to find explicit formulae for functions in $\mathcal{U}_{p,C}$ so that we would not have to rely on repeated application of Theorem 4.1 (ii) in the proof of Theorem 4.1 (i). It could help to improve the constant in Theorem 4.1 (i).

5. Comparison of Tail Probabilities and of the Almost Sure Convergence of Series of Tangent Sequences.

The following result is a modified version of the decoupling inequality in Theorem 2.1 of S. Kwapien and W.A. Woyczynski [11]. The present version is more useful in some applications. Its proof is similar to the proof of the above mentioned

Theorem 2.1, the only novelty being an application of the weak–type inequality from Section 4.

THEOREM 5.1. *Let* (d_i) *and* (e_i) *be two tangent sequences of random variables. Then for each* a,b,c, > 0 *with* $c \geq 2b$

$$P(f^* \geq a) \leq 6\frac{b}{a} + 3\,\frac{a+c}{a}\,P(g^* \geq b) + 2P(\Sigma^* E([e_i]^c \mid \mathcal{F}_{i-1}) \geq b) \tag{5.1}$$

Proof. Define stopping times $\sigma := \inf\{k: |e_k| \geq c\}$,

$$\rho := \inf\{k: |\sum_{i=1}^{k} [e_i]^c| \geq b\}, \quad \lambda := \inf\{k: |\sum_{i=1}^{k+1} E([e_i]^c \mid \mathcal{F}_{i-1})| \geq b\}.$$

Then

$$P(f^* \geq a) \leq P(d^* \geq c) + P(\Sigma^* ([d_i]^c - E([d_i]^c \mid \mathcal{F}_{i-1})) \geq \tfrac{5}{6}a)$$
$$+ P(\Sigma^* E([d_i]^c \mid \mathcal{F}_{i-1}) \geq \tfrac{1}{6}a),$$

and, under the above notation, the three terms on the right–hand side can be estimated as follows:

$$P(d^* \geq c) \leq E\sum_{i=1}^{\sigma} I(|d_i| \geq c) + P(\sigma < \infty) =$$
$$E\sum_{i=1}^{\sigma} I(|e_i| \geq c) + P(\sigma < \infty) = 2P(e^* \geq c) \leq 2P(g^* \geq \tfrac{c}{2}) \leq 2P(g^* \geq b).$$

Next,

$$P(\Sigma^*([d_i]^c - E([d_i]^c | \mathcal{F}_{i-1})) \geq \tfrac{5}{6} a) \leq$$

$$P(\rho < \infty) + P(\lambda < \infty) + P(\Sigma^* I(i \leq \min(\rho,\lambda))([d_i]^c - E([d_i]^c | \mathcal{F}_{i-1})) \geq \tfrac{5}{6} a) \leq$$

$$P(\Sigma^*[e_i]^c \geq b) + P(\Sigma^* E([e_i]^c | \mathcal{F}_{i-1}) \geq b) + \tfrac{5}{2}\tfrac{6}{5a} E\left|\sum_{i=1}^{\min(\rho,\lambda)} ([e_i]^c - E[e_i]^c | \mathcal{F}_{i-1}))\right| \leq$$

$$P(g^* \geq b) + P(\Sigma^* E([e_i]^c | \mathcal{F}_{i-1}) \geq b) + \tfrac{3}{a}((2b+c)P(\Sigma^*[e_i]^c \geq b) + 2bP(\Sigma^*[e_i]^c \leq b)) =$$

$$\tfrac{6b}{a} + P(g^* \geq b) + P(\Sigma^* E([e_i]^c | \mathcal{F}_{i-1}) \geq b) + \tfrac{3c}{a} P(\Sigma^*[e_i]^c \geq b) \leq$$

$$\leq 6\tfrac{b}{a} + P(\Sigma^* E([e_i]^c | \mathcal{F}_{i-1}) \geq b) + \tfrac{3c+a}{a} P(g^* \geq b),$$

where the second inequality follows by Theorem 3.1 (i). Therefore

$$P(f^* \geq a) \leq$$

$$6\tfrac{b}{a} + 3\tfrac{a+c}{a} P(g^* \geq b) + P(\Sigma^* E([e_i]^c | \mathcal{F}_{i-1}) > b) + P(\Sigma^* E([e_i]^c | \mathcal{F}_{i-1}) > \tfrac{1}{6} a).$$

If $6b/a > 1$ then the inequality (5.1) is trivial. If $6b/a \leq 1$ then $a/6 \geq b$ which, in view of the above, also gives (5.1) Q.E.D.

In the case of either conditionally symmetric or nonnegative sequences (d_i) and (e_i), Theorem 5.1 can be strengthened as follows.

THEOREM 5.2. *Let* (d_i) *and* (e_i) *be two tangent sequences of random variables and let* $a,b > 0$. *Then*

(i) *if* (d_i) *and* (e_i) *are conditionally symmetric, and* $b \le a$, *then*

$$P(f^* \ge a) \le 3(\tfrac{b}{a} + P(g^* \ge b));$$

(ii) *if* (d_i) *and* (e_i) *are nonnegative, then*

$$P(f^* \ge a) \le \tfrac{b}{a} + 2P(g^* \ge b)\ .$$

Proof. (i) Let $\sigma := \inf\{k: \sum_{i=1}^{k} [e_i]^{2b} \ge b\}$. Then

$$P(f^* \ge a) \le P(d^* \ge 2b) + P(\Sigma^* [d_i]^{2b} \ge a)\ .$$

The first term is estimated (as before) from above by $2P(g^* \ge b)$, and for the second term, in view of Doob's Maximal Inequality and condition $b \le a$, we have that

$$P(\Sigma^* [d_i]^{2b} \ge a) \le P(\sigma < \infty) + \tfrac{1}{a}(E(\sum_{i=1}^{\sigma} [d_i]^{2b})^2)^{1/2} =$$

$$P(\Sigma^* [e_i]^{2b} \ge b) + \tfrac{1}{a}(E(\sum_{i=1}^{\sigma} [e_i]^{2b})^2)^{1/2} \le$$

$$P(g^* \ge b) + \frac{3b}{a},$$

because $|\Sigma_{i=1}^{\sigma} [e_i]^{2b}| \le 3b$ in view of the definition of σ.

(ii) Let $\lambda := \inf\{k: \sum_{i=1}^{k} \min(e_i,a) \geq b\}$. Then

$$P(f^* \geq a) \leq \frac{1}{a} E \min(\Sigma(\min(d_i,a),a) \leq$$

$$P(\lambda < \infty) + \frac{1}{a} E \sum_{i=1}^{\lambda} \min(d_i,a) = P(\lambda < \infty) + \frac{1}{a} E \sum_{i=1}^{\lambda} \min(e_i,a) \leq$$

$$P(\lambda < \infty) + \frac{b+a}{a} P(\lambda < \infty) + \frac{b}{a} P(\lambda = \infty) =$$

$$\frac{b}{a} + 2P(\lambda < \infty) \leq \frac{b}{a} + 2P(g^* \geq b). \text{ Q.E.D.}$$

Remark 5.1. Let $\varphi: \mathbb{R}^+ \to \mathbb{R}^+$ be an increasing and continuous function such that $\varphi(2x) \leq C\,\varphi(x)$, for all $x \geq 0$ and some $C > 0$. Then, there exists a constant $L = L(C)$ such that for any sequences (d_i) and (e_i) satisfying the assumptions of Theorem 5.2, we have that $E\varphi(f^*) \leq LE\varphi(g)$. This result is due to P. Hitczenko [7].

The following corollary is obtained immediately from Theorems 5.1 and 5.2.

COROLLARY 5.1. *If* (d_i) *and* (e_i) *are tangent sequences of random variables and satisfy one of the following three conditions*:

(i) (d_i) *and* (e_i) *are conditionally symmetric*;

(ii) (d_i) *and* (e_i) *are nonnegative*;

(iii) (d_i) *and* (e_i) *are uniformly bounded martingale difference sequences*;

then, the series $\Sigma\, d_i$ *converges a.s. if, and only if, the series* $\Sigma\, e_i$ *converges a.s.*

Remark 5.2. If (d_i) and (e_i) are tangent sequences and if, additionally, (e_i) is a CI–sequence (i.e. (e_i) is $\mathcal{G}$–conditionally independent where $\mathcal{G}:= \sigma\{\mathcal{L}(e_n | \mathcal{F}_{n-1}), n = 1,2,...\}$, cf. S. Kwapien and W.A. Woyczynski [11]), then if the series Σe_i converges a.s. then the series Σd_i converges a.s. This follows immediately from Theorem 2.2 of the above mentioned paper.

Example 5.1. In general, if the sequences (d_i) and (e_i) are just assumed to be tangent, the a.s. convergence of the series $\Sigma\ d_i$ need not imply the a.s. convergence of the series $\Sigma\ e_n$. To see this take (ξ_i) to be a sequence of independent random variables, (ξ_i') be its independent copy and assume that $\xi_i \to 0$ a.s. as $i \to \infty$ but the series $\Sigma(\xi_i - \xi_i')$ diverges a.s. Now define

$$d_{2i+1} = \xi_i,\ d_{2i+2} = -\xi_i,\ i = 0,1,2,...,\ \text{and}$$

$$e_{2i+1} = \xi_i',\ e_{2i+2} = -\xi_i,\ i = 0,1,2,... .$$

Both sequences are $(\mathcal{F}_i)$–adapted and tangent for $\mathcal{F}_k := \sigma(d_i, e_i, i = 1,2,...,k)$, $k = 1,2,...,$ and, moreover, (e_i) has property CI with respect to $\mathcal{G} = \sigma((\xi_i))$. In view of the construction, the partial sums of $\Sigma\ d_i$ are either 0 or ξ_i so that the series Σd_i converges. On the other hand, the series $\Sigma e_i = \Sigma(\xi_i - \xi_i')$ diverges.

6. Applications and Generalizations

Results of previous sections permit finding best constants in inequalities for Brownian stochastic integrals and for the planar Brownian motion.

THEOREM 6.1. *Let* $(X(t), Y(t))$, $t \in T = [0,t_\infty]$, *be a standard planar Brownian motion and let* $F(t)$, $t \in T$, *be a predictable process with respect to* $\mathcal{F}(t) := \sigma(X(s), Y(s): s \le t)$, $t \in T$. *Then, for any* $a > 0$, *and* $1 < p < \infty$,

$$a\,P(\sup_{t\in T}|\int_0^t F(t)dX(t)| > a) \le \kappa\, E|\int_T F(t)dY(t)| \quad \text{if} \quad E(\int_T F^2(t)dt)^{1/2} < \infty,$$

and

$$E|\int_T F(t)dX(t)|^p \le \cot^p \frac{\pi}{2p^*}\, E|\int_T F(t)dY(t)|^p \quad \text{if} \quad E(\int_T F^2(t)dt)^{p/2} < \infty,$$

where $\kappa = 1.346886...$ *is the constant from Theorem* 3.1 (ii) *and, as before,* $p^* = \max(p,p/(p-1))$.

Proof. If F is a predictable step process for the partition $0 = t_0 < t_1 <...< t_n = t_\infty$, then the sequences $F(t_{i-1})(X(t_i) - X(t_{i-1}))$ and $F(t_{i-1})(Y(t_i) - Y(t_{i-1}))$, $i = 1,2,...,n$, are $(\mathcal{F}(t_i))$–tangent tci martingale difference sequences with Gaussian conditional distributions, and the inequalities of Theorem 6.1 for them follow directly from Theorems 3.1 (iii) and 4.1 (iii). The general case is then obtained by approximation. Q.E.D.

If F is an $(\sigma(X(s), s \le t), t \in T)$–predictable process then F is independent of the process $Y(t)$, $t \in T$, and the integral $\int_T F(s)Y(s)$ is equidistributed with the random variable $(\int_T F^2(s)ds)^{1/2}Y_1$. Hence, we arrive at the following

COROLLARY 6.1 . *Let* X(t), t ∈ T, *be a standard one–dimensional Brownian motion, and let* F(t), t ∈ T, *be a process predictable with respect to* X. *Then, for each* $a > 0$ *and* $1 < p < \infty$,

$$aP(\sup_{t \in T} |\int_0^t F(s)dX(s)| > a) \leq \sqrt{2/\pi}\, \kappa\, E(\int_T F^2(s)ds)^{1/2},$$

$$E\varphi((\int_T F^2(s)ds)^{1/2}) \leq \kappa\, E|\int_T F(s)dX(s)|, \text{ if } E(\int_T F^2(s)ds)^{1/2} < \infty,$$

and

$$(\cot^p \tfrac{\pi}{2p^*}\, \Gamma(\tfrac{p+1}{2})\sqrt{2p/\pi})^{-1} E(\int_T F^2(s)ds)^{p/2} \leq$$

$$E|\int_T F(s)dX(s)|^p \leq$$

$$\cot^p \tfrac{\pi}{2p^*}\, \Gamma(\tfrac{p+1}{2})\sqrt{2p/\pi}\; E(\int_T F^2(s)ds)^{p/2},$$

if $E(\int_T F^2(s)ds)^{p/2} < \infty$, *where* $\varphi(t) := \sqrt{2/\pi} \int_{1/t}^{\infty} \exp(-u^2/2)du$, $t > 0$, $\kappa =$ 1.346886... *is the constant from Theorem* 3.1 (ii), *and, as before,* $p^* = \max(p, p/(p-1))$.

Choosing τ to be a stopping time with respect to $\mathcal{F}(t)$, t ∈ T, and setting $F(t) = I(t \leq \tau)$ in Theorem 6.1, we immediately obtain

COROLLARY 6.2. *Let* $(X(t), Y(t)), t \in T$, *be a standard planar Brownian motion and let* τ *be a stopping time for* (X,Y). *Then*

$$aP(|X_\tau| \geq a) \leq \kappa E|Y_\tau| \text{ if } E\sqrt{\tau} < \infty,$$

and

$$E|X_\tau|^p \leq \cot^p \frac{\pi}{2p^*} E|Y_\tau|^p \text{ if } E\sqrt{\tau^p} < \infty,$$

where κ *and* p^* *are as before.*

Remark 6.1. Inequalities of the above type are well known or, at least, can be deduced from known results (cf. D.L. Burkholder [1], and B. Davis [5], [6]). However, the constants obtained here seem to us to be better than the known ones.

Remark 6.2. In light of Corollary 6.2, the results of S.K. Pichorides [13] and B. Davis [5] on the *Hilbert transform* on the unit circle come as no surprise. Indeed, if u and v are conjugate harmonic functions on the unit disc with $u(0,0) = v(0,0) = 0$, then the boundary values of u and v are equidistributed with $X(\tau)$ and $Y(\tau)$ for some stopping time τ.

Let us also point out another straightforward application of Theorems 3.1 (i) and 4.1 (i) to tangent sequences arising in Example 2.4.

THEOREM 6.2. *Let* $(\Omega, \mathcal{F}, P;(\mathcal{F}_i))$ *be a filtered probability space and let* ϕ_i: $(\Omega,\mathcal{F}_i,P) \to (\Omega, \mathcal{F}_i, P)$ *be measurable, measure–preserving transformation such that* $\phi_i A = A$ *whenever* $A \in \mathcal{F}_{i-1}$, $i = 1,2,\ldots$. *Let us introduce operators*

$$\mathrm{Th}:=\sum_{i=1}^{\infty}\Delta_i h\circ\phi_i,\quad \textit{and}\quad T^*h:=\Sigma^*\Delta_i h\circ\phi_i$$

for $h\in L_1(\Omega,\mathscr{F},P)$, *where* $\Delta_i h=E(h\,|\,\mathscr{F}_i)-E(h\,|\,\mathscr{F}_{i-1})$, $i=1,2,\ldots$. *Then*

(i) *for any* $a>0$, $aP(T^*h\geq a)\leq(5/2)\,E|h|$,

(ii) *for any* p, $1<p<\infty$, T *is a bounded operator in* $L_p(\Omega,\mathscr{F},P)$ *with* $\|T\|_p\leq(p^*-1)^2$. (cf. P. Hitczenko [6])

Remark 6.3. In the case of *UMD Banach–space–valued tangent tci sequences*, P. Hitczenko [8] and T. McConnell [12] proved that inequalities analogous to Theorems 3.1 (ii) and 4.1 (ii) hold true. Their proofs simply adopted D.L. Burkholder's L–function developed originally for martingale transforms in UMD Banach spaces.

Remark 6.4. The concept of *tangent sequences* can be generalized to random variables which are *not necessarily defined on the same probability space*. Namely, if $(\Omega',\mathscr{F}',P';(\mathscr{F}_i'))$ and $(\Omega'',\mathscr{F}'',P'';(\mathscr{F}_i''))$ are two filtered probability spaces then an $(\mathscr{F}_i')$–adapted sequence (d_i'), and an $(\mathscr{F}_i'')$–adapted sequence (d_i''), are said to be tangent if the sequences $(\mathscr{L}(d_i'\,|\,\mathscr{F}_{i-1}'))$ and $(\mathscr{L}(d_i''\,|\,\mathscr{F}_{i-1}''))$ of conditional probability distributions (which themselves are random variables with values in the space of all probability distributions on $\mathbb{R}^1$) are equidistributed. All the inequalities of this paper obtained for tangent sequences defined on the same space carry over to the above context. This follows from the fact that, given two tangent sequences as defined above, one can find another filtered probability space $(\tilde\Omega,\tilde{\mathscr{F}},\tilde P;(\tilde{\mathscr{F}}_i))$ and $(\tilde{\mathscr{F}}_i)$–tangent sequences $(\tilde d_i')$ and $(\tilde d_i'')$ on $\tilde\Omega$ such that double sequences $(d_i',\mathscr{L}(d_i'\,|\,\mathscr{F}_{i-1}'))$ and $(\tilde d_i',\mathscr{L}(\tilde d_i'\,|\,\tilde{\mathscr{F}}_{i-1}))$, are equidistributed, and the same holds true for double sequences $(d_i'',\mathscr{L}(d_i''\,|\,\mathscr{F}_{i-1}''))$ and $(\tilde d_i'',\mathscr{L}(\tilde d_i''\,|\,\tilde{\mathscr{F}}_{i-1}''))$.

Theorem 3.1 (i) can be generalized in a way that encompasses both the case of *tangent* and of *subordinated* (cf. D.L. Burkholder [3]) martingale difference sequences. Both of these are examples of conditionally subordinated sequences defined below. The method also works for Hilbert space valued random variables.

Definition 6.1. Let (d_i) and (e_i) be two Hilbert space valued, $(\mathcal{F}_i)$–adapted, sequences of random variables. (d_i) is said to be *conditionally subordinated* to (e_i) if, for each increasing $\psi:\mathbb{R}^+ \to \mathbb{R}^+$,

$$E(\psi(|d_i|)|\mathcal{F}_{i-1}) \le E(\psi(|e_i|)|\mathcal{F}_{i-1}).$$

THEOREM 6.3. *Let* (d_i) *and* (e_i) *be Hilbert–space–valued* $(\mathcal{F}_i)$*–martingale difference sequences, and let* (d_i) *be conditionally subordinated to* (e_i)*. Then, for any* $a > 0$,

$$aP(f^* \ge a) \le 3\,E|g|.$$

Proof. The proof is very similar to the proof of Theorem 3.1 (i) except that instead of the function u_0 we use a function $u: H \times H \to \mathbb{R}^+$ defined by the formula

$$u(x,y) = \begin{cases} 1/3(|y|^2 - |x|^2 + 1) & \text{if } (x,y) \in D, \\ |y| & \text{if } (x,y) \in D^{c}, \end{cases}$$

where $D = \{(x,y): (5+4|x|^2)^{1/2} + 2|y| \le 3\}$. It turns out that u satisfies the following property: for any H–valued random variables ξ and η such that $E\xi = E\eta = 0$, and such that for any nondecreasing function $\psi: \mathbb{R}^+ \to \mathbb{R}^+$, $E\psi(|\xi|) \le E\psi(|\eta|)$, we have that $Eu(x + \xi, y + \eta) \ge u(x,y)$. Clearly, this property is all we need to conclude the proof, and to see that u enjoys this property we proceeds follows.

If $(x,y) \in D^c$ then, as in the case of u_0, we have that $Eu(x + \xi, y + \eta) \geq E|y + \eta| \geq |y| = u(x,y)$ because $u(\bar{x},\bar{y}) \geq |\bar{y}|$ for all $\bar{x},\bar{y} \in H$.

Let $(x,y) \in D$. For a $z \in H$ let us define $\tilde{z} := \sqrt{5}\, z/\max(\sqrt{5},|z|)$. If $z \neq \tilde{z}$ then $|z|, |\tilde{z}| \geq \sqrt{5}$. Since $D \subset \{(\bar{x},\bar{y}): |\bar{x}| \leq 1, |\bar{y}| \leq (3 - \sqrt{5})/2\}$, we obtain that if $(\xi,\eta) \neq (\tilde{\xi},\tilde{\eta})$ then $(x + \xi,y + \eta) \in D^c$ and $(x + \tilde{\xi},y + \tilde{\eta}) \in D^c$. Therefore

$$Eu(x + \xi,y + \eta) =$$

$$E(u(x + \xi,y + \eta) - u(x + \tilde{\xi},y + \tilde{\eta})) + Eu(x + \tilde{\xi},y + \tilde{\eta}) =$$

$$E(|y + \eta| - |y + \tilde{\eta}|) + u(x,y) + \tfrac{2}{3}((y, E\tilde{\eta}) - (x,E\tilde{\xi})) + \tfrac{1}{3}(E|\tilde{\eta}|^2 - E|\tilde{\xi}|^2) \geq$$

$$u(x,y) + E(|y+\eta| - |y+\tilde{\eta}|) - \tfrac{2}{3}(|x| + |y|)\max(|E\tilde{\xi}|,|E\tilde{\eta}|) \geq$$

$$u(x,y) + E(|y+\eta| - |y+\tilde{\eta}|) - \tfrac{2}{3} E|\eta-\tilde{\eta}| \geq u(x,y),$$

where the second equality follows from the fact that $u(\bar{x},\bar{y}) \geq (1/3)(|\bar{y}|^2 - |\bar{x}|^2 + 1)$ for $|\bar{y}| \leq (3 + \sqrt{5})/2$, and from the fact that $|y+\tilde{\eta}| \leq (3 + \sqrt{5})/2$. The first inequality follows from the inequality $E|\tilde{\eta}|^2 \geq E|\tilde{\xi}|^2$ which, in turn, is implied by the assumptions on η and ξ and by standard Hilbert space estimates. The second inequality uses two facts: that $|x| + |y| \leq 1$ for $(x,y) \in D$, and that $|E\tilde{\xi}| = E|(\xi-\tilde{\xi})| \leq E|\eta-\tilde{\eta}|$ and $|E\tilde{\eta}| \leq E|\eta-\tilde{\eta}|$ in view of assumptions on ξ and η. Finally, the last inequality follows from inequality $|y+z| - |\bar{y}+\tilde{z}| \geq (2/3)|z-\tilde{z}|$ valid for any $z, \bar{y} \in H$ such that $|\bar{y}| \leq (3 - \sqrt{5})/2$. Q.E.D.

7. **Appendix: Subharmonic minorants of** $\Phi(x,y) = C|x|^p - |y|^p$, $1 < p < \infty$.

The question that remains to be answered is: for which constants C the function Φ admits a subharmonic minorant, and, in particular, what is smallest such C? If $\Phi(x,y) = C|x|^p - |y|^p$ admits a subharmonic minorant then a maximal subharmonic minorant U for Φ exists and is a p–homogeneous function. Therefore, in polar coordinates $(r,\theta) \in \mathbb{R}^+ \times [-\pi,\pi]$, $U(x,y) = u(r,\theta) = r^p g(\theta)$ and, moreover, g satisfies the following two conditions:

(i) $p^2 g + g'' \geq 0$ on $[-\pi,\pi]$ (in the distributional sense),

(ii) $g(\theta) \leq C|\cos\theta|^p - |\sin\theta|^p =: h(\theta)$.

Case $p > 2$. From general considerations on maximal subharmonic minorants it follows that U exists if, and only if, we can find $\theta_0 \in (0,\pi/2)$, $B \in \mathbb{R}^+$ and $\epsilon > 0$ such that

(iii) $-B \cos p(\theta-\pi/2) \leq h(\theta)$ for $|\theta-\pi/2| \leq \theta_0 + \epsilon$,

(iv) $-B \cos p\,\theta_0 = h(\theta_0 + \pi/2)$.

Moreover, in such a case function g is given by the formula

$$\text{(v)} \qquad g(\theta) = \begin{cases} h(\theta) & \text{if } |\theta+\pi/2| \text{ or } |\theta-\pi/2| > \theta_0, \\ -B \cos p(\theta+\pi/2) & \text{if } |\theta+\pi/2| > \theta_0, \\ -B \cos p(\theta-\pi/2) & \text{if } |\theta-\pi/2| < \theta_0. \end{cases}$$

Here we proceed as follows: (i) and (ii) are fulfilled if, and only if,

$$f(\theta) := \frac{C|\sin\theta|^p + B\cos p\theta}{|\cos\theta|^p} \geq 1 \quad \text{for } \theta \in (-\theta_0 - \epsilon, \theta_0 + \epsilon),$$

and $f(\theta_0) = 1$. Since, for $|\theta| < \pi/2$, $f'(\theta) = p(C \sin^{p-1}\theta - B \sin(p-1)\theta)/\cos^{p+1}\theta$, and since, on $(0,\pi/(p-2))$, $(\sin(p-1)\Theta/\sin^{p-1}\Theta)' = -(p-1)\sin(p-2)\Theta/\sin^{p}\Theta < 0$ so that $\sin(p-1)\Theta/\sin^{p-1}\Theta$ is decreasing from $+\infty$ to 0 on $(0,\pi/(p-1))$, we get that for each $B, C > 0$ there exists exactly one $\Theta_0 \in (0,\pi/(p-1))$ for which $f'(\Theta_0) = 0$. At this point, f attains its minimal value on $(0, \pi/(p-1))$. Since $f(\Theta_0) = 1$, we obtain two conditions

$$\frac{C \sin^{p}\Theta_0 + B \cos p\Theta_0}{\cos^{p}\Theta_0} = 1, \quad \text{and} \quad C \sin^{p-1}\Theta_0 - B \sin(p-1)\Theta_0 = 0.$$

Solving these equations for C and B we get that

$$C = \cot^{p-1}\Theta_0 \tan(p-1)\Theta_0, \quad \text{and} \quad B = \cos^{p-1}\Theta_0/\cos(p-1)\Theta_0.$$

Hence, the smallest C for which U exists is the minimum of function $\varphi(\Theta)$: $= \cot^{p-1}\Theta \tan(p-1)\Theta$ on $(0,\pi/2(p-1))$. Since

$$\varphi'(\Theta) = -\frac{\cot^{p-2}\Theta \cos p\Theta \sin(p-2)\Theta}{\sin^{2}\Theta \cos^{2}(p-1)\Theta},$$

we get that the minimal $C = C_p = \cot^{p}(\pi/2p)$ is obtained for $\Theta_0 = \pi/2p$. In this case $B = B_p = \cos^{p-1}(\pi/2p)/\sin(\pi/2p)$. For these C_p and B_p, in view of (v), we obtain that (cf. Section 4) for $p > 2$

$$U_p(x,y) = \begin{cases} C_p|x|^p - |y|^p & \text{if } |y| < |x| \cot(\pi/2p), \\ B_p \ \mathrm{Re}\ (-iz)^p & \text{if } y \geq |x| \cot(\pi/2p), \\ B_p \ \mathrm{Re}\ (iz)^p & \text{if } y \leq -|x| \cot(\pi/2p), \end{cases} \tag{7.1}$$

where $z = x + iy$ and z^p is the homomorphic branch in $\mathbb{C}\backslash\mathbb{R}^-$ with $1^p = 1$. Notice, that there is no problem with the subharmonicity of U_p at $(0,0)$ since, by (i), $\int_{-\pi}^{\pi} g(\Theta)d\Theta \geq 0$.

Case $1 < p < 2$. Once again, by general considerations, it follows that U exists if, and only if, we can find $\Theta_0 \in (0,\pi/2)$, $B \in \mathbb{R}^+$, and $\epsilon > 0$, such that

($\overline{\text{iii}}$) $B \cos p\Theta \leq h(\Theta)$ for $\Theta \in (-\Theta_0 - \epsilon, \Theta_0 + \epsilon)$, and

($\overline{\text{iv}}$) $B \cos p\Theta_0 = h(\Theta_0)$,

and, in this case, g is given by the formula

$$(\overline{\text{v}}) \qquad g(\Theta) = \begin{cases} h(\Theta) \text{ if } & \Theta \in (-\pi + \Theta_0, -\Theta_0] \text{ or } \Theta \in (\Theta_0, \pi-\Theta_0] \\ B \cos p\Theta & \text{if } \Theta \in (-\Theta_0, \Theta_0] \\ B \cos p(\Theta-\pi) & \text{if } \Theta \in (\pi-\Theta_0, \pi] \\ B \cos p(\Theta+\pi) & \text{if } \Theta \in [-\pi, -\pi + \Theta_0]. \end{cases}$$

Once again, (iii)' and (iv)' are equivalent to conditions

$$\bar{f}(\Theta) := \frac{C \cos^p\Theta - B \cos p\Theta}{\sin^p\Theta} \geq 1 \text{ for } (-\Theta_0 - \epsilon, \Theta_0 + \epsilon),$$

and $\bar{f}(\Theta_0) = 1$. Since, for $\Theta \in (0,\pi/2)$, $\bar{f}'(\Theta) = p(B \cos(p-1)\Theta - C \cos^{p-1}\Theta)$ $\cos^{p-1}\Theta/\sin^{p+1}\Theta$, and $(\cos(p-1)\Theta/\cos^{p-1}\Theta)' = (p-1) \sin(2-p)\Theta/\cos^p\Theta > 0$, so that $\cos(p-1)\theta/\cos^{p-1}\theta$ increases on $(0,\pi/2)$ from 1 to $+\infty$, we get that for each $C > B > 0$ there exists exactly one $\theta_0 \in (0,\pi/2)$ such that $\bar{f}'(\theta_0) = 0$. At this point $\bar{f}$ attains its minimal value on the interval $(0,\pi/2)$. Since $\bar{f}(\theta_0) = 1$ we thus have two conditions

$$\frac{C \cos^p\theta_0 - B \cos p\theta_0}{\sin^p\theta_0} = 1, \quad \text{and} \quad B \cos(p-1)\theta_0 - C \cos^{p-1}\theta = 0.$$

Hence

$$C = (\tan \theta_0)^{p-1}\cot(p-1)\theta_0, \quad \text{and} \quad B = \sin^{p-1}\theta_0/\sin(p-1)\theta_0.$$

The minimum of $\tan^{p-1}\theta \cot(p-1)\theta$ on $(0,\pi/2)$ is obtained when $\theta_0 = \pi/2p$ and is equal to $C_p = \tan^p\pi/2p = \cot^p(\pi/2p^*)$. Also $B_p = \sin^{p-1}(\pi/2p)/\cos(\pi/2p) = \cos^{p-1}(\pi/2p^*)/\sin(\pi/2p^*)$. Finally, in view of $(\overline{\text{v}})$, for such C_p and B_p the maximal subharmonic minorant of Φ for $1 < p < 2$ is given by the formula

$$U_p(x,y) = \begin{cases} C_p|x|^p - |y|^p & \text{if } |x| \le |y| \tan(\pi/2p), \\ B_p \operatorname{Re} z^p & \text{if } x > |y| \tan(\pi/2p), \\ B_p \operatorname{Re}(-z)^p & \text{if } x < -|y| \tan(\pi/2p). \end{cases} \tag{7.2}$$

So, U_p defined by (7.1) for $2 < p < \infty$, and by (7.2) for $1 < p < 2$, is the maximal subharmonic minorant of Φ, and C_p is the smallest value of C for which such a minorant exists.

Acknowledgement. The authors would like to thank Burgess Davis and the referee for a careful reading of the manuscript, and for pointing out to us that the material in the appendix (which we decided to keep anyway to make the paper self–contained) is, essentially, already contained in a paper by M. Essen [4].

References

[1] D.L. Burkholder, Distribution function inequalities for martingales, *Annals of Probability* 1 (1983), 19–42.

[2] D.L. Burkholder, An elementary proof of an inequality of R.E.A.C. Paley, *Bull. London Math. Soc.* 17 (1985), 474–478.

[3] D.L. Burkholder, Differential subordination of harmonic functions and martingales, preprint.

[4] M. Essen, A superharmonic proof of the M. Riesz conjugate function theorem, Ark. Math. 22 (1984), 241–249.

[5] B. Davis, On weak type (1,1) inequality for conjugate functions, *Proc. Amer. Math. Soc.* 44 (1974), 307–311.

[6] B. Davis, Applications of the conformal invariance of Brownian motion, *Proc. Symp. Pure Math.* (AMS) 35 (1979), 303–310.

[7] P. Hitczenko, Comparison of moments for tangent variables, *Probab. Theory and Rel. Fields* (to appear).

[8] P. Hitczenko, On tangent sequences of UMD–space valued random vectors, *Studia Math.* (submitted).

[9] K. Ito, Differential equations determining a Markov process, *Journ. Pan–Japan Math. Coll.* No. 1077 (1942).

[10] J. Jacod, Une generalisation des semimartingales: les processus admettant un processus a accroissement independants tangent, *Springer Lecture Notes in Math.* 1059 (1984), 91–118.

[11] S. Kwapien and W.A. Woyczynski, Semimartingale integrals via decoupling inequalities and tangent processes, C.W.R.U. preprint #86–56 (1986).

[12] T.R. McConnell, Decoupling and stochastic integration in UMD Banach space, *Prob. Theory and Math. Stat.* (to appear).

[13] S.K. Pichorides, On the best values of the constants in the theorems of M. Riesz, Zygmund and Kolmogorov, *Studia Math.* 44 (1972), 165–179.

Institute of Mathematics
Warsaw University
Warsaw, Poland 00901

Department of Mathematics and Statistics
Case Western Reserve University
Cleveland, Ohio 44106

On a pointwise ergodic theorem for multiparameter groups

F. J. Martín-Reyes
Universidad de Málaga, Facultad de Ciencias, Departamento de Matemáticas
29071-Málaga, Spain

1. Introduction.

Let $\mathbf{Z}$ be the set of the integers number, let d be a positive integer and let $\mathbf{Z}^d$ be the cartesian product of $\mathbf{Z}$ by itself, d times. If $u = (u_1,..., u_d)$ and $v = (v_1,..., v_d)$ are elements of $\mathbf{Z}^d$ we say that $u \le v$ if $u_i \le v_i$ for every i, and, in this case, we write $[u, v]$ to indicate the set $\{w \in \mathbf{Z}^d : u \le w \le v\}$. Finally, for any integer n, we denote by $\bar{n}$ to the element of $\mathbf{Z}^d$ with each coordinate equals to n.

Let $(X, \mathcal{F}, \nu)$ be a finite measure space and let $\{T_u : u \in \mathbf{Z}^d\}$ be a group of null-preserving transformations from X to X. Associated with the group and for every measurable function f and any non-negative integer n we define

$$A_n f(x) = (n+1)^{-d} \sum_{\bar{0} \le u \le \bar{n}} f(T_u x)$$

and the maximal operator

$$Mf = \sup_{n \ge 0} A_n |f|$$

It is known that if the transformations preserve the measure then $A_n f$ converges a.e. for every $f \in L_1(d\nu)$. But if ν is not preserved by the transformations, do the averages $A_n f$ converge a.e. for every $f \in L_1(d\nu)$? Of course, the same question makes sense for $L_p(d\nu)$ with $1 < p < \infty$. Therefore, we are interested in characterizing the finite measures ν such that $A_n f$ converges a.e. for every $f \in L_p(d\nu)$, $1 \le p < \infty$, (this problem for $d = 1$ and semigroups was studied in [9] and [10]). We will prove (Theorem 3.1) that ν is such a measure if and only if there exists a finite measure γ equivalent to ν such that the operators A_n are uniformly bounded from $L_p(d\nu)$ to weak-$L_p(d\gamma)$. As a corollary, we obtain that if

ISBN 0-12-231050-0

$$\sup_{n\geq 0} \| A_n \|_{L_p(d\nu)} < \infty$$

then the averages $A_n f$ converge a.e. for every $f \in L_p(d\nu)$ (see [1] and[10] for $p > 1$ and $d = 1$; in both cases the authors consider semigroups). In the case $p = 1$ we obtain a nicer condition: the averages $A_n f$ converge a.e. for every $f \in L_1(d\nu)$ if and only if there exists a measure γ equivalent to ν such that

$$\sup_{n\geq 0} (n+1)^{-d} \sum_{\bar{0}\leq u \leq \bar{n}} \gamma(T_{-u} E) \leq \nu(E)$$

for any measurable set E. Observe that this condition is closely related to those in [4] and [11].

In order to solve the question, we begin by considering measures $\nu = Vd\mu$ (not necesarily finite) where V is a positive measurable function and μ is a finite measure which is preserved by the transformations. In particular we prove that the averages $A_n f$ converge a.e. for every $f \in L_p(d\nu)$ if and only if

$$\inf_{\bar{0}\leq u} V(T_u x) > 0 \quad (p = 1)$$

$$MV^{1-q}(x) < \infty \quad \text{a.e.} \quad (1 < p < \infty, \ p + q = pq)$$

In section 3 we apply these results to obtain necessary and sufficient conditions in the case of general measures (Theorem 3.1).

In what follows we will consider two sets as equals if they agree up to a set of measure zero and if $1 < p < \infty$ then q will be its conjugate exponent, i.e. the number such that $q + p = pq$.

2. The measure preserving case.

In this section we will assume $\nu = Vd\mu$ where V is a positive measurable function and μ is a finite measure which is preserved by all the transformations T_u. Observe that we do not assume that ν is finite.

Theorem 2.1. Let $1 \leq p < \infty$. The following are equivalent:

(a) The sequence $A_n f$ converges a.e. for all f in $L_p(Vd\mu)$

(b) $Mf(x) < \infty$ a.e. for all f in $L_p(Vd\mu)$

(c) There exists a positive measurable function W such that

$$\int_{\{x:Mf(x)>\lambda\}} W d\mu \leq \lambda^{-p} \int_X |f|^p V d\mu$$

for all $\lambda > 0$ and all f in $L_p(Vd\mu)$

(d) There exists a positive measurable function W such that

$$\sup_{n\geq 0} \int_{\{x:|A_n f(x)|>\lambda\}} W d\mu \leq \lambda^{-p} \int_X |f|^p V d\mu$$

for all $\lambda > 0$ and all f in $L_p(Vd\mu)$

(e) $\inf_{\bar{0}\leq u} V(T_u x) > 0$ a.e. if $p = 1$; $MV^{1-q}(x) < \infty$ a.e. if $1 < p < \infty$

(f) There exists a positive measurable function W such that

$$\sup_{n\geq 0} \int_X |A_n f|^p W d\mu \leq \int_X |f|^p V d\mu$$

for all f in $L_p(Vd\mu)$

Proof.

It is clear that (a) ⇒ (b), (c) ⇒ (d), and (f) ⇒ (d). On the other hand, (b) implies the continuity in measure of the operator M from $L_p(Vd\mu)$ to $L_0(d\mu)$ where $L_0(d\mu)$ is the space of all measurable functions provided with the topology of the convergence in measure (see Theorem 1.1.1. in [6], page 10). Then, (c) follows from (b) by Nikishin's theorem (see [5], pp. 536-537). To finish the proof of the theorem it will suffice to prove (d) ⇒ (e), (e) ⇒ (f) and (e) ⇒ (a).

In order to prove (d) ⇒ (e) we will need the following lemmas.

Lemma 2.2. Let S be a measurable transformation from X into itself such that S preserves μ. Then there exists a countable family $\{B_i\}^{\infty}_{i=0}$ of measurable sets such that

(a) $X = \cup^{\infty}_{i=0} B_i$

(b) $B_i \cap B_j = \varnothing$ if $i \neq j$

(c) $S^{-1} B_i \cap B_i = \varnothing$ if $i \geq 1$

(d) $S^{-1} E = E$ for every measurable subset $E \subset B_0$

The proof of this lemma can be seen in [8] (Lemma 2.9).

Lemma 2.3. Let n be a non negative integer. Then there exists a countable family

$\{B_i\}^{\infty}_{i=0}$ of measurable sets such that

(i) $X = \cup^{\infty}_{i=0} B_i$

(ii) $B_i \cap B_j = \emptyset$ if $i \neq j$

(iii) For every B_i and all $u \in [-\bar{n}, \bar{n}]$ one of the following conditions holds:

(a) $T_u B_i \cap B_i = \emptyset$

(b) $T_u E = E$ for every measurable subset $E \subset B_i$

(iv) For every B_i there exists a set $D_i \subset [\bar{0}, \bar{n}]$ such that $\{T_u B_i : u \in D_i\}$ is a family of disjoint sets and $\cup_{u \in D_i} T_u E = \cup_{\bar{0} \le u \le \bar{n}} T_u E$ for any measurable subset E of B_i. Furthermore, if $F_i = \{u \in [-\bar{n}, \bar{n}] : T_u E = E$ for any subset E of $B_i\}$ and $D_i + F_i = \{u + v : u \in D_i, v \in F_i\}$ then

$$[\bar{0}, \bar{n}] \subset D_i + F_i \subset [-\bar{n}, 2\bar{n}]$$

Proof of Lemma 2.3. Let $u \in [-\bar{n}, 2\bar{n}]$ and let $\{B_{i,u}\}^{\infty}_{i=0}$ be the decomposition determined by (2.2) and the transformation T_u. Then

$$X = \bigcap_{-\bar{n} \le u \le \bar{n}} \left(\bigcup_{i \ge 0} B_{i,u} \right) = \bigcup_{c \in \mathcal{C}} B_c$$

where $\mathcal{C}$ is the set of maps from $[-\bar{n}, \bar{n}]$ to $\mathbf{N}$ (the set of the natural numbers) and $B_c = \cap_{-\bar{n} \le u \le \bar{n}} B_{c(u),u}$. It is clear that $\{B_c : c \in \mathcal{C}\}$ is a countable family of sets with the properties (i) and (ii). We will now see that (iii) holds. Let us fix B_c and $u \in [-\bar{n}, \bar{n}]$. If $c(u) = 0$ and $E \subset B_c$ we have $T_{-u} E = E$ and then $T_u E = E$ because $A \subset B_c \subset B_{c(u),u}$ (see Lemma 2.2). If $c(u) \ge 1$ we have $T_{-u}(B_{c(u),u}) \cap B_{c(u),u} = \emptyset$ ((d) in Lemma 2.2) and thus $T_u B_c \cap B_c = \emptyset$.

By the property (iii), it is clear that if $u \in [\bar{0}, \bar{n}]$ and $v \in [\bar{0}, \bar{n}]$ then $T_u B_i \cap T_v B_i = \emptyset$ or $T_u E = T_v E$ for any measurable set $E \subset B_i$. It follows from this that there exists a set $D_i \subset [\bar{0}, \bar{n}]$ such that $\{T_u B_i : u \in D_i\}$ is a family of disjoint sets and

$$\bigcup_{u \in D_i} T_u E = \bigcup_{\bar{0} \le u \le \bar{n}} T_u E$$

for any measurable set $E \subset B_i$. It is clear that $D_i + F_i \subset [-\bar{n}, 2\bar{n}]$. Then to finish the proof of the lemma we only have to show $[\bar{0}, \bar{n}] \subset D_i + F_i$. Let $v \in [\bar{0}, \bar{n}]$. Since $T_v B_i \subset \cup_{\bar{0} \le u \le \bar{n}} T_u B_i = \cup_{u \in D_i} T_u B$ we have that there exists $u \in D_i$ such that $T_v B_i \cap T_u B_i \neq \emptyset$ and then $T_{v-u} B_i \cap B_i \neq \emptyset$. Since $v - u \in [-\bar{n}, \bar{n}]$ property (ii) implies that $T_{v-u} E = E$ for any measurable subset E of B_i and therefore $v - u \in F_i$. This finishes the proof of the lemma.

(d) ⇒ (e) (p =1)

We may assume $W \le 1$.

Let n be a non negative integer and let $\{B_i\}^{\infty}_{i=0}$, $\{D_i\}^{\infty}_{i=0}$ $\{F_i\}^{\infty}_{i=0}$ be the sets determined by lemma (2.3) for the number n. Let E be any subset of B_i and let $R = \bigcup_{-2\bar{n}\le u\le -\bar{n}} T_u E$ (observe that $R = T_{-2\bar{n}} (\bigcup_{u\in D_i} T_u E)$. For every $x \in R$ we have

$$A_n \chi_E \ge (3n+1)^{-d} (\#F_i)$$

where $\#F_i$ stands for the number of elements of F_i. Thus, by (d)

$$\int_R W d\mu \le (3n+1)^d (\#F_i)^{-1} \int_E V d\mu \tag{2.4}$$

On the other hand, keeping in mind the inclusion $[\bar{0}, \bar{n}] \subset D_i + F_i$, the definitions of F_i and the property of D_i, we have

$$\int_E \sum_{-2\bar{n}\le u\le -\bar{n}} W(T_u x) d\mu \le \int_E \sum_{u\in D_i+F_i} W(T_{u-2\bar{n}} x) d\mu \tag{2.5}$$

$$= \int_X W \sum_{u\in D_i+F_i} \chi_{T_{u-2\bar{n}}E}\, d\mu$$

$$= \int_X (\#F_i)\, W \sum_{u\in D_i} \chi_{T_{u-2\bar{n}}E}\, d\mu$$

$$= (\#F_i) \int_R W d\mu$$

Therefore (2.4) and (2.5) give

$$\int_E \sum_{-2\bar{n}\le u\le -\bar{n}} W(T_u x) d\mu \le (3n+1)^d \int_E V(T_u x) d\mu$$

Since this inequality holds for any measurable subset $E \subset B_i$ and any B_i we have

$$(3n+1)^{-d} \sum_{-2\bar{n}\le u\le -\bar{n}} W(T_u x) \le V(x) \text{ a.e.}$$

If we let n tend to ∞ we have (see for instance [7], Theorem 2.8)

$$0 < \mathcal{W}(x) \le V(x) \text{ a.e.}$$

where $\mathcal{W}$ is an invariant function with respect to each T_u. Therefore

$$0 < \inf_{\bar{0}\le u} V(T_u x) \text{ a.e..}$$

(d) ⇒ (e) $(p > 1)$

As in case $p = 1$, we may assume $W \le 1$.

Let n be a non negative integer and let $\{B_i\}^{\infty}_{i=0}$, $\{D_i\}^{\infty}_{i=0}$, $\{F_i\}^{\infty}_{i=0}$ be the sets determined by Lemma (2.3) for the number n. For every integer s we consider

$$H_{is} = \{x \in B_i : 2^s \le (3n+1)^{-d} \sum_{u\in D_i+F_i} V^{1-q}(T_u x) < 2^{s+1}\}$$

Let E be any subset of H_{is} and consider the set $R = \cup_{-2\bar{n}\le u\le -\bar{n}} T_u E$ and $R' = \cup_{\bar{0}\le u\le \bar{n}} T_u E$. It is clear that $R' = \cup_{u\in D_i} T_u E$ (Lemma 2.3) and $R = T_{-2\bar{n}} R' = \cup_{u\in D_i} T_{u-2\bar{n}} E$. If $x \in R$ and $f = V^{1-q}\chi_{R'}$ we have since $E \subset H_{is}$

$$A_{3n} f(x) \ge 2^s$$

Then, by (d)

$$\int_R W d\mu \le 2^{-sp} \int_{R'} |f|^p V d\mu = 2^{-sp} \int_{R'} V^{1-q} d\mu \qquad (2.6)$$

On the other hand, keeping in mind the inclusion $[\bar{0}, \bar{n}] \subset D_i + F_i$ (Lemma 2.3) the definition of F_i and the properties of D_i, we have

$$\int_R \sum_{-2\bar{n}\le u\le -\bar{n}} W(T_u x)\, d\mu \le \int_E \sum_{u\in D_i+F_i} W(T_{u-2n} x) \qquad (2.7)$$

$$= (\#F_i) \int_E \sum_{u\in D_i} W(T_{u-2n} x) = (\#F_i) \int_R W\, d\mu$$

Then, (2.6) and (2.7) give

$$\int_E \sum_{-2\bar{n}\le u\le -\bar{n}} W(T_u x) \le 2^{-sp} \int_E (\#F_i) \sum_{u\in D_i} V^{1-q}(T_u x) d\mu = 2^{-sp} \int_E \sum_{u\in D_i+F_i} V^{1-q}(T_u x)\, d\mu$$

Since $E \subset H_{is}$ and $[\bar{0}, \bar{n}] \subset D_i + F_i$ the last term is smaller than or equal to

$$2^p \int_E (3n+1)^{-dp} \Big[\sum_{\bar{0}\le u\le \bar{n}} V^{1-q}(T_u x)\Big]^{1-p} d\mu$$

and therefore we have obtained

$$\int_E \sum_{-2\bar{n}\le u\le -\bar{n}} W(T_u x)\, d\mu \le 2^p \int_E (3n+1)^{-dp} \Big[\sum_{\bar{0}\le u\le \bar{n}} V^{1-q}(T_u x)\Big]^{1-p} d\mu$$

Since this inequality holds for any measurable set E included in H_{is}, for any H_{is} and all B_i we have

$$(3n+1)^{-d} \sum_{-2\bar{n}\le u\le -\bar{n}} W(T_u x) \Big[(3n+1)^{-d} \sum_{\bar{0}\le u\le \bar{n}} V^{1-q}(T_u x)\Big]^{p-1} \le 2^p \text{ a.e.}$$

Now, as in case $p = 1$, we have that

$$(3n+1)^{-d} \sum_{-2\bar{n}\le u\le -\bar{n}} W(T_u x)$$

converges a.e. to a positive function. Then

$$\limsup_{n\to\infty} A_n V^{1-q}(x) < \infty \quad \text{a.e.}$$

and therefore $MV^{1-q}(x) < \infty$ a.e..

<u>(e) ⇒ (f)</u> ($p = 1$)

Let's take $W(x) = \inf_{\bar{0}\le u} V(T_u x)$. Since W is an invariant function with respect to each transformation T_u we have

$$\int_X |A_n f|\, W d\mu \le \int_X A_n |f|\, W d\mu = \int_X |f|\, W d\mu \le \int_X |f|\, V d\mu$$

<u>(e) ⇒ (f)</u> ($p > 1$)

Let's take $W = [MV^{1-q}]^{1-p}$. By Hölder's inequality and the definition of W,

$$|A_n f|^p \le A_n(|f|^p V) |A_n V^{1-q}|^{p-1}$$
$$\le A_n(|f|^p V)\, W^{-1}$$

Therefore, since the transformations preserve the measure μ,

$$\int_X |A_n f|^p W d\mu \le \int_X A_n(|f|^p V)\, d\mu = \int_X |f|^p V d\mu$$

<u>(e) ⇒ (a)</u> ($p = 1$)

It follows from (e) that $X = \cup^{\infty}_{s=1} X_s$ where $X_s = \{x : \sup_{\bar{0}\le u} V^{-1}(T_u x) < s\}$. It is clear that X_s is T_u-invariant for every $u \in \mathbf{Z}^d$ and $L_1(X_s, Vd\mu) \subset L_1(X_s, d\mu)$. Then, (a) follows from this inclusion because the averages $A_n f$ converge a.e. for all f in $L_1(X_s, d\mu)$ (classical result).

(e) ⇒ (a) $(p > 1)$

Let N be any positive integer and let $\sigma_N = \min(V^{1-q}, N)$. Then, by (e), we have $X = \cup_{s=1}^{\infty} X_s$ where $X_s = \{x : \sup_N \lim_{n\to\infty} A_n \sigma_N < s\}$. It is clear that each X_s is T_u-invariant for every $u \in \mathbf{Z}^d$. We will now see that $L_p(X_s, Vd\mu) \subset L_1(X_s, d\mu)$ and then the implication follows as the case $p = 1$.

By Hölder's inequality we have

$$\int_{X_s} |f|\, d\mu \leq \left(\int_{X_s} |f|^p\, Vd\mu\right)^{\frac{1}{p}} \left(\int_{X_s} V^{1-q}\, d\mu\right)^{\frac{1}{q}}$$

Thus, if we prove that $V^{1-q} \in L_1(X_s, d\mu)$, we have as a consequence $L_p(X_s, Vd\mu) \subset L_1(X_s, d\mu)$. Now, observe that

$$\int_{X_s} \sigma_N\, d\mu = \int_{X_s} \lim_{n\to\infty} A_n\, \sigma_N\, d\mu \leq s\mu(X_s)$$

and therefore by monotone convergence theorem we have $V^{1-q} \in L_1(X_s, d\mu)$.

Remark. The proof of (e) ⇒ (a) shows that the conditions in Theorem 2.1 are equivalent to the existence of a countable family X_s of subsets of X such that

(i) $X = \cup_{s=1}^{\infty} X_s$

(ii) For every s, X_s is invariant with respect to all transformations T_u

(iii) For every s,

$$V^{-1} \in L_\infty(X_s, d\mu) \quad \text{if } p = 1$$
$$V^{1-q} \in L_1(X_s, d\mu) \quad \text{if } 1 < p < \infty$$

Therefore, if one of the transformations T_u is ergodic then the conditions in Theorem 2.1 are equivalent to $V^{-1} \in L_\infty(X, d\mu)$ if $p = 1$ and $V^{1-q} \in L_1(X, d\mu)$ if $1 < p < \infty$.

3. The general case.

In this section we will characterize the finite measures ν such that the averages $A_n f$ converge a.e. for every $f \in L_p(d\nu)$.

Theorem 3.1 Let $(X, \mathcal{F}, \nu)$ be a finite measure space and let $\{T_u : u \in \mathbf{Z}^d\}$ a group of null-preserving transformations from X to X. Let $1 \leq p < \infty$. The following are equivalent:

(a) The sequence $\{A_n f\}$ converge a.e. for all f in $L_p(d\nu)$

(b) $Mf(x) < \infty$ a.e. for all f in $L_p(d\nu)$

(c) There exists a finite measure γ equivalent to ν such that

$$\gamma(\{x : Mf(x) > \lambda\}) \leq \lambda^{-p} \int_X |f|^p d\nu$$

for all $\lambda > 0$ and any function $f \in L_p(d\nu)$

(d) There exists a finite measure γ equivalent to ν such that

$$\sup_{n \geq 0} \gamma(\{x : |A_n f(x)| > \lambda\}) \leq \lambda^{-p} \int_X |f|^p d\nu$$

for all $\lambda > 0$ and any function f in $L_p(d\nu)$

(e) There exists a finite measure γ equivalent to ν such that

$$\sup_{n \geq 0} \int_X |A_n f|^p d\gamma \leq \int_X |f|^p d\nu$$

for all functions f in $L_p(d\nu)$.

If $p = 1$ then (a), (b), (c), (d) and (e) are equivalent to

(f) There exists a finite measure γ equivalent to ν such that

$$\sup_{n \geq 0} (n+1)^{-d} \sum_{0 \leq u \leq \bar{n}} \gamma(T_{-u} E) \leq \nu(E)$$

for all measurable sets E.

Proof.

The proofs of (a) $\Rightarrow$ (b) $\Rightarrow$ (c) $\Rightarrow$ (d) and (e) $\Rightarrow$ (d) are the same as the corresponding in Theorem 2.1. To prove (d) $\Rightarrow$ (e) and (d) $\Rightarrow$ (a) it will suffice, by Theorem 2.1, to see that (d) implies the existence of a finite measure μ equivalent to ν which is preserved by the transformations T_u. In order to do this, we first observe that, by Marcinkiewicz's interpolation theorem, there exists a constant $C > 0$ such that

$$\sup_{n\geq 0} \int_X |A_n f|^{2p} d\gamma \leq C \int_X |f|^{2p} d\nu \tag{3.2}$$

for every $f \in L^{2p}(d\nu)$. Then if E is a measurable set we have, by Hölder's inequality and (3.2),

$$\int_X A_n \chi_E \, d\gamma \leq C(\nu(E))^{\frac{1}{2p}} [\gamma(X)]^{\frac{1}{p'}}$$

where $2p + p' = 2pp'$. Therefore, the sequence $\{\int_X A_n \chi_E \, d\gamma\}$ is bounded. We now consider a Banach's limit, L, and we define for every measurable set E,

$$\mu(E) = L(\{\int_X A_n \chi_E \, d\gamma\})$$

It is clear that μ is finitely additive. On the other hand, (3.3) implies

$$\mu(E) \leq C(\nu(E))^{\frac{1}{2p}} (\gamma(X))^{\frac{1}{p'}}$$

Thus, if $\{E_n\}$ is a decreasing sequence of sets, $E_1 \supset E_2 \supset \ldots$, with $\cap_{n\geq 1} E_n = \emptyset$ we have $\lim \nu(E_n) = 0$ and then $\lim \mu(E_n) = 0$. Therefore μ is a countably additive finite measure. We will now see that μ is preserved by each T_u. To avoid difficulties with the notation we will only see it for $u = (1, 0,\ldots, 0)$. Then

$$\left|\int_X A_n \chi_E(x)\, d\gamma - \int_X A_n \chi_{T_{-u}E}(x)\, d\gamma\right|$$

$$= \left|(n+1)^{-d}\left[\sum_{\{\bar{0}\leq v\leq \bar{n};\ v_1=0\}} \int_X \chi_E(T_v x)\, d\gamma - \sum_{\{\bar{0}\leq v\leq \bar{n};\ v_1=n+1\}} \int_X \chi_E(T_v x)\, d\gamma\right]\right|$$

$$\leq 2\frac{(n+1)^{d-1}}{(n+1)^d} \gamma(X) = 2(n+1)^{-1} \gamma(X)$$

and therefore $\lim \int_X (A_n \chi_E - A_n \chi_{T_{-u}E})\, d\gamma = 0$ and as a consequence of this, we have

$$\mu(E) = \mu(T_{-u}E)$$

To finish the implications (d) ⇒ (e) and (d) ⇒ (a) we will now see that μ is equivalent to γ and therefore μ will be equivalent to ν. It is clear that $\mu \ll \gamma$. Conversely let E be a set with $\mu(E) = 0$. Since μ is preserved by each T_u then $\mu(\cup_{u \in Z^d} T_u E) = 0$. But it is clear by the definition of μ that $\mu(\cup_{u \in Z^d} T_u E) = \gamma(\cup_{u \in Z^d} T_u E)$. Then $\gamma(\cup_{u \in Z^d} T_u E) = 0$ and thus $\gamma(E) = 0$ as we wished to prove.

We will now assume $p = 1$ and we will see that (e) is equivalent to (f). Observe first that (f) is (e) for characteristic functions. Therefore (f) follows from (e). On the other hand, if (f) holds we have (e) for characteristic functions and then (e) follows by standard arguments.

Corollary 3.4. Let $(X, \mathcal{F}, \nu)$ be a finite measure space and let $\{T_u : u \in \mathbf{Z}^d\}$ be a group of null preserving transformations from X to X. Let $1 \le p < \infty$. If the operators A_n are uniformly bounded from $L_p(d\nu)$ in $L_p(d\nu)$ then the averages $A_n f$ converge a.e. for all f in $L_p(d\nu)$.

Proof. It is a consequence of the equivalence of (a) and (e) in Theorem 3.1.

Final remarks.

(a) The uniform boundedness of the averages is not a necessary condition for the a.e. convergence of the averages. Furthermore, as the following example shows, it is possible to get a.e. convergence of the averages in such a way that the averages are bounded operators on L_p but the a.e. limit is not necessarily in L_p (an example for d=1 can be found in [3]).

<u>Example</u>. Let $X = \cup^{\infty}_{n=1} X_n$ where $X_n = \{(i, j, n): 1 \le i \le n, 1 \le j \le n\}$ and consider the measure space $(X, \mathcal{F}, \nu)$ where $\mathcal{F}$ is the power set of X and ν is the measure determined by $\nu(\{(i,j,n)\}) = n^{-2} 2^{-i-j-n}$. It is easy to see that ν is a finite measure. Let $\{T_u: u \in \mathbf{Z}^2\}$ be the group of transformations defined by

$$T_{(1,0)}(i,j,n)=(i+1,j,n) \quad (i<n), \quad T_{(1,0)}(n,j,n)=(1,j,n)$$
$$T_{(0,1)}(i,j,n)=(i,j+1,n) \quad (j<n), \quad T_{(0,1)}(i,n,n)=(i,1,n)$$

On one hand, the averages are bounded operators on L_p since $\nu(T_{(-1,0)}E) \le 2\nu(E)$ and $\nu(T_{(0,-1)}E) \le 2\nu(E)$. On the other hand, the averages $A_n f$ converge a. e. for all f in $L_p(d\nu)$ since the transformations are periodic.

Now, we will show a function f in $L_p(d\nu)$ for which the a.e. limit of $A_n f$ is not in $L_p(d\nu)$.

Let f be the function defined by

$$f(i,j,n) = 2^{2n/p} \quad \text{if } i=j=n$$
$$f(i,j,n) = 0 \qquad \text{in other case}$$

The function f is clearly in $L_p(d\nu)$ and the a.e. limit of the averages is the function F given by $F(i,j,n)=2^{2n/p}n^{-2}$. Then

$$\|F\|^p_{p,d\nu} = \sum_{n=1}^{\infty}\sum_{i=1}^{n}\sum_{j=1}^{n} 2^{2n}n^{-2p-2}2^{-i-j-n} \geq \frac{1}{4}\sum_{n=1}^{\infty} 2^{n}n^{-2p-2} = \infty$$

(b) After the example in remark (a), we already know that the a.e. convergence of the averages does not imply the uniform boundedness of the averages. But if the sequence A_nf converges a.e. to a function in $L_p(d\nu)$ for every f in $L_p(d\nu)$, are the averages uniformly bounded in $L_p(d\nu)$? The answer is negative. An example for d=1 and p=1 can be found in [11].

(c) In proving Theorem 3.1 we have shown the necessity of a finite measure μ equivalent to ν and invariant for all the transformations T_u. That is not true if we work with a semigroup instead of a group as the non invertible transformation T(a)=a, T(b)=a on X={a,b} shows. A less trivial example on X=[0,1) can be seen in [2]. The difficulty in both examples is with the dissipative part. This difficulty is handled in the one dimensional case (see [1] and [10]) by showing that the dissipative part is absorbed in the conservative part.

The existence of the finite measure equivalent to ν and invariant for all the transformations T_u is not sufficient for A_nf to converge a.e. for every function f in $L_p(d\nu)$, $1 \leq p < \infty$, as Theorem (2.1) shows. However for $p=\infty$ we have the following result which is included for the sake of completeness.

Theorem 3.5. Let $(X,\mathcal{F},\nu)$ be a σ-finite measure space and let $\{T_u: u \in \mathbf{Z}^d\}$ be a group of null-preserving transformations from X to X. The sequence A_nf converges a.e. for every f in $L_\infty(d\nu)$ if and only if there exists a finite measure μ equivalent to ν and invariant for all the transformations T_u.

Proof.

Assume A_nf converges a.e. for every f in $L_\infty(d\nu)$. We may assume ν to be finite. Then $\mu(E) = \lim \int A_n\chi_E d\nu$ defines a finite measure $\mu \ll \nu$ and invariant with respect to all the transformations T_u. The support X_μ of μ is an absorbing set for every T_u, but since each T_u

is invertible , X_μ is an invariant set. Hence $0 = \mu(X-X_\mu) = \lim \int A_n\chi_{X-X\mu}\, dv = v(X-X_\mu)$ and μ is equivalent to v. The converse is classical.

Acknowledgment. I would like to thank Professor Akcoglu and the University of Toronto for their support and hospitality while working on these questions. Also I wish to thank the referee for his helpful comments and Professor Assani for pointing out to me the example in [3] and for the helpfull discussions we had related to the results in this paper.

References

1. I. Assani, Quelques résultats sur les operateurs positifs a moyennes bornées dans L_p, *Ann. Scien. Clermont - Ferrand, nº 85, Probabilités, Fasc.* **3,** (1985), 65-72.
2. I. Assani, An equivalent measure for some nonsingular transformations and applications, preprint.
3. Y. N. Dowker, A note on the ergodic theorems, *Bull. Amer. Math. Soc.* 55, (1949), 379-383.
4. N. Dunford and D.S. Miller, On the ergodic theorem, *Trans. Amer. Math. Soc.* 60, (1946), 538-549.
5. J. García-Cuerva and J.L. Rubio de Francia, Weighted norm inequalities and related topics, *North-Holland* (1985).
6. M. de Guzmán, Real variable methods in Fourier Analysis, *North-Holland* (1981).
7. U. Krengel, Ergodic theorems, *Walter de Gruyter* (1985).
8. F.J. Martín-Reyes, Inequalities for the maximal function and convergence of the averages in weighted L_p-spaces, *Trans. Amer. Math. Soc.* 296, (1986), 61-82.
9. F.J. Martín-Reyes and A. de la Torre, Weighted weak type inequalities for the ergodic maximal function and the pointwise ergodic theorem. *Studia Math.* 87, (1987), 33-46.
10. F.J. Martín-Reyes and A. de la Torre, On the almost everywhere convergence of the ergodic averages, to appear in *Ergodic Theory and Dynamical Systems*.
11. C. Ryll-Nardzewski, On the ergodic theorems, I (Generalized ergodic theorems), Studia Math. 12, (1951), 65-73.

THE TWO-PARAMETER STRONG LAW FOR PARTIALLY EXCHANGEABLE ARRAYS

by

Terry R. McConnell[1] and Eric Rieders[2]

Syracuse University S.U.N.Y. Brockport

ABSTRACT: We obtain necessary and sufficient conditions on an RCE array to have unrestricted convergence in the two-parameter strong law of large numbers. We also investigate some related results in ergodic theory.

AMS(1980) Subject Classification. Primary 28D05; Secondary 60G60

Key Words and Phrases: Partial exchangeability, RCE arrays, unrestricted convergence.

[1]Supported in part by NSF grant DMS87-00802

[2]Some of the results of this paper will appear in the Syracuse University Ph.D. dissertation of the second author.

ISBN 0-12-231050-0

1. Introduction and Statement of Results.

In this paper we prove a two-parameter strong law of large numbers which asserts the existence of the almost sure limit of $(1/nm)\sum_{i=1}^{n}\sum_{j=1}^{m}W_{i,j}$ as $n\wedge m$ tends to infinity under suitable hypotheses on an array of random variables $W_{i,j}$. Wiener [16] seems to have been the first to consider such questions in the context of ergodic theory, but related results in harmonic analysis were known earlier [5]. Here we will obtain necessary and sufficient conditions for the two-parameter strong law in the case of a row and column exchangeable array, and also discuss some related results in ergodic theory.

Convergence as $n\wedge m$ tends to infinity is often called unrestricted convergence. Other two-parameter modes of convergence, e.g. sectorial convergence in which one requires the ratios n/m and m/n to remain bounded, have also been studied but will not be considered here.

An array $W = \{ W_{i,j} \}$ of random variables indexed by $\mathbb{Z}_+^2$ is called row and column exchangeable (RCE) if the joint distribution of the array is invariant under interchange of rows or columns. The array is said to be dissociated if the sub-families of random variables indexed by two subsets A and B of $\mathbb{Z}_+^2$ are independent whenever A and B are such that $(i,j)\in A$ implies $(i,k)\notin B$ and $(k,j)\notin B$ for all k. Strong laws for dissociated arrays were first studied in [10]. (Also see [2] and [14].) We shall label an array RCED if it is both RCE and dissociated.

The following result, due independently to Aldous[1] and Hoover

[4] (resp. Aldous [1]) describes the structure of RCE (resp. RCED) arrays.

THEOREM 1.1. *Let* $X = \{X_i\}_{i=0}^{\infty}$, $Y = \{Y_j\}_{j=0}^{\infty}$, *and* $Z = \{Z_{i,j}\}_{(i,j)\in\mathbb{Z}_+^2}$ *be independent families of i.i.d.* U(0,1) *random variables. Also let* U *be a* U(0,1) *random variable independent of* X, Y *and* Z. *An array* W *is* RCE *if and only if there exists a Borel function* f *on* $[0,1]^4$ *such that*

$$\{W_{i,j}\} \stackrel{d}{=} \{ f(U, X_i, Y_j, Z_{i,j}) \}. \tag{1.1}$$

The array is RCED if and only if there exists a Borel function g *on* $[0,1]^3$ *such that*

$$\{W_{i,j}\} \stackrel{d}{=} \{ g(X_i, Y_j, Z_{i,j}) \}. \tag{1.2}$$

This result may be viewed as an analogue of DeFinetti's Theorem since it shows that RCE arrays are mixtures of RCED arrays, just as exchangeable sequences are mixtures of i.i.d. sequences. The functions f and g are not unique and it may be difficult to find *any* such functions, depending on what information is used to describe the array W. For example, if X^k, Y^k, and Z^k are independent copies of the X, Y, and Z, and $\{g_k\}_{k=1}^{\infty}$ is bounded in $L^1([0,1]^3)$, then

$$W_{i,j} = \sum_{k=1}^{\infty} 2^{-k} g_k(X_i^k, Y_j^k, Z_{i,j}^k)$$

is again RCED, hence has a representation of the form (1.2) for some function g. However, it is not clear how to construct the function g from the g_k.

In order to state our main results we need to introduce some notation. Given an integrable function g on $[0,1]^3$ let

$$\lambda(x) = \int_{[0,1]^2} |g(x,y,z)|\,dydz,$$

$$\mu(y) = \int_{[0,1]^2} |g(x,y,z)|\,dxdz, \text{ and}$$

$$\tilde{\delta}(g) = \int_{[0,1]^3} |g(x,y,z)| \; [\; 1 + \log_+ \frac{\|g\|_1 |g(x,y,z)|}{\lambda(x)\mu(y)}]dxdydz.$$

We state our main result for RCED arrays, leaving to the reader the obvious reformulation needed for the case of an RCE array. (See Corollary 1.1.)

THEOREM 1.2. Let W be an RCED array, represented as in (1.2) above. Then we have

$$(1.3) \qquad \lim_{n \wedge m \to \infty} (1/nm) \sum_{i=1}^{n} \sum_{j=1}^{m} W_{i,j} \quad \text{exists, a.s.,}$$

if and only if $\tilde{\delta}(g) < \infty$. Whenever the limit in (1.3) exists it equals $EW_{1,1}$.

In the special case of an i.i.d array, i.e., $W_{i,j} = h(Z_{i,j})$ for some Borel function h on [0,1], the condition $\tilde{\delta}(h) < \infty$ reduces to the Zygmund condition, $h \in L\mathrm{Log}_+L$. In this case, (1.3) is due to Smythe [15]. In case $W_{i,j} = h(X_i, Y_j)$, (1.3) was derived by McConnell [7] as a straightforward consequence of work of Rosinski and Woyczynski [13]. They introduced the condition $\delta(h) < \infty$ in connection with double stochastic integration relative to the symmetric Cauchy process. (The quantity δ is defined in exactly the same way as $\tilde{\delta}$ but with no z-integration in the expressions for $\lambda(x)$ and $\mu(y)$.) The connection with double stochastic integration breaks down for general RCE arrays.

The functional $\tilde{\delta}$ is scalar homogeneous but is not a norm. It is however a quasi-norm, i.e., satisfies $\tilde{\delta}(f+g) \leq \beta\tilde{\delta}(f) + \beta\tilde{\delta}(g)$ for some constant $\beta > 1$ [8]. It is proved in [9] that the space of Borel functions f such that $\tilde{\delta}(f) < \infty$ is complete, and that dyadic step functions are dense in this space. (These proofs are given for the functional δ of functions of two variables, but they carry over to the present situation without any essential change.) Perhaps surprisingly, this space is not normable, i.e., there exists no norm on the space of Borel functions equivalent to $\tilde{\delta}$. On the other hand, the α-convexification, $\delta(|f|^{\alpha})^{1/\alpha}$ for $1<\alpha<\infty$, <u>is</u> equivalent to a norm. For a proof of the latter fact see [12, section 7]. The former seems not to have been noticed before, so we sketch a proof here.

Suppose there were a norm $\| \ \|$ on $L^0([0,1]^3)$ and a constant $0 < c < \infty$ such that

$$(1.4) \qquad (1/c)\ \|f\| \ \leq \ \tilde{\delta}(f) \ \leq \ c\|f\|$$

holds for all Borel measurable f. Choose functions $\lambda_i \geq 0$, $i=1,2,\dots,N$ supported in disjoint subintervals of [0,1] such that $\int_0^1 \lambda_i(x)dx = 1$.

Let $f(x,y,z) = N^{-1}\sum_{i=1}^{N}\lambda_i(x)\lambda_i(y)$. Then $\|f\|_1 = 1$, and one computes easily that $\delta(f) = 1 + \log N$. On the other hand, $\delta(\lambda_i(.)\lambda_i(.)) = 1$ for each i, and it follows from (1.4) that $\delta(f) \leq c^2$ since $\| \ \|$ is a norm. Thus $1 + \log N \leq c^2$, which is a contradiction for large enough N.

As mentioned above, it is not always possible to find a representing function g. Thus it is useful to reformulate Theorem 1.2 in a way which does not explicitly involve the function g. To do this, introduce the marginal tail fields

$$\mathcal{T}_1 = \bigcap_{n=1}^{\infty}\sigma\{W_{i,j}:\ i\geq n,\ 0\leq j<\infty\} \text{ and } \mathcal{T}_2 = \bigcap_{m=1}^{\infty}\sigma\{W_{i,j}:\ j\geq m,\ 0\leq i<\infty\},$$

and the tail field

$$\mathcal{T} = \bigcap_{n=1}^{\infty}\sigma\{W_{i,j}:\ i\wedge j \geq n\}.$$

COROLLARY 1.1 <u>Let</u> $W_{i,j}$ <u>be</u> <u>an</u> RCE <u>array</u>. <u>Then</u> $(1/nm)\sum_{i=1}^{n}\sum_{j=1}^{m}W_{i,j}$ <u>converges</u> a.s. <u>as</u> $n\wedge m$ <u>tends</u> <u>to</u> <u>infinity</u> <u>if</u> <u>and</u> <u>only</u> <u>if</u>

$$E[\,|W_{1,1}|\,(1 + \log_+ \frac{|W_{1,1}|}{(E|W_{1,1}|\,|\mathcal{F}_1)(E|W_{1,1}|\,|\mathcal{F}_2)})\,|\mathcal{F}\,] < \infty, \text{ a.s.}$$

The limit, if it exists, is almost surely equal to $E(W_{1,1}|\mathcal{F})$.

Turning now to a related result in ergodic theory, let τ and σ be measure preserving transformations of Lebesgue spaces. We may assume these to be copies of the unit circle, identified with $[0,1)$, and the invariant measure to be Lebesgue measure.

THEOREM 1.3. *Let* f *be a Borel function on* $[0,1)^2$. *Then*

$$(1.5) \qquad \lim_{n\wedge m\to\infty} (1/nm) \sum_{i=1}^{n} \sum_{j=1}^{m} f(\tau^i x, \sigma^j y) \text{ exists for a.e. } (x,y)$$

whenever $\delta(f) < \infty$. *If* τ *and* σ *are ergodic the limit equals* $\int_{[0,1)^2} f$ *almost everywhere*.

This improves upon the best previously known sufficient condition, $f \in L\log_+L$ [16], since it is easy to find functions f satisfying $\delta(f) < \infty$ but not belonging to $L\log_+L$. For example, take $f(x,y) = h(x)h(y)$ where $h \in L^1([0,1))$, but $h \notin L\log_+L$. Unfortunately, the condition $\delta(f) < \infty$ is not necessary for (1.5) in general, even for nonnegative f. We will give an example to illustrate this in section 2. (v.i. Example 2.1.)

Related results hold for the strong maximal function of harmonic analysis [9] and for two-parameter U-statistics ([8] and [11].) It is also possible to formulate and prove two-parameter generalizations of the Marcinkiewicz-Zygmund strong laws of large numbers for certain arrays [11].

2. *Proofs*.

Throughout this section we will use the following notation:

$$S_{n,m}(f) = (1/nm) \sum_{i=1}^{n}\sum_{j=1}^{m} f(X_i, Y_j, Z_{i,j}),$$

$${}^{*}S^{*}(f) = \sup_{n,m} |S_{n,m}(f)|,$$

$${}^{*}f^{*} = \sup_{i,j} |f(X_i, Y_j, Z_{i,j})|/(ij), \text{ and}$$

$$E = \{f: \tilde{\delta}(f) < \infty\}.$$

A number of well known 1-parameter results will be used repeatedly, and sometimes implicitly. These are summarized in the following lemma.

LEMMA 2.1 _Let_ ζ_j, $j=1,2,\ldots$ _be_ i.i.d. _real-valued random variables. Then the following statements are all equivalent:_

(2.1) $\quad E|\zeta_1| < \infty,$

(2.2) $\quad \sup_j |\zeta_j|/j < \infty$, a.s.,

(2.3) $\quad \sup_m (1/m) |\sum_{j=1}^{m} \zeta_j| < \infty$, a.s.,

(2.4) $\quad \sup_m (1/m) \sum_{j=1}^{m} |\zeta_j| < \infty$, a.s.,

(2.5) $\quad (1/m) \sum_{j=1}^{m} \zeta_j$ _converges_ a.s. _as_ $m \to \infty$, _and_

(2.6) $\quad \sum_{j=1}^{\infty} \zeta_j^2/j^2 < \infty$, a.s.

This result will often be applied with ζ_j of the form $\zeta_j = g(X_j)$ for some sequence X_j of i.i.d. $U(0,1)$ random variables.

Proof of Theorem 1.2: First note that it is sufficient to prove

(2.7) $\quad {}^{*}S^{*}(f) < \infty$, a.s., if and only if $\tilde{\delta}(f) < \infty$, and

(2.8) $S_{n,m}(f) \to \int_{[0,1]^3} f(x,y,z)dxdydz$, a.s., if f is a step function.

(A step function is a linear combination of indicator functions of cubes.) Indeed, it is straightforward to show that (2.7) implies the quantitative statement that for each $\varepsilon>0$ there is an η such that $\tilde{\delta}(f) < \eta$ implies $P({}^*S^*(f) > \varepsilon) < \varepsilon$. (See, for example, the argument on page 48 of [8].) This, together with (2.8) and the fact that step functions are dense in E easily yields the a.s convergence in (2.8) for general f satisfying $\tilde{\delta}(f) < \infty$. On the other hand, if $S_{n,m}(f)$ converges a.s. then with probability one there exist $N=N(\omega)$ and $M=M(\omega)$ such that

$$(2.9) \qquad \sup\{|S_{n,m}(f)| : n\geq N,\ m\geq M\} < \infty.$$

It follows easily from (2.9) and a one-parameter result ((2.1) $\Leftrightarrow$ (2.3)) that $f(x,.,.) \in L^1$ for a.e. x and $f(.,y,.) \in L^1$ for a.e. y. For example, if $A = \{x: f(x,.,.) \notin L^1\}$ had positive measure, then with probability one there would exist $n>N$ such that $X_n \in A$. However, the resulting unboundedness of the averages $(1/m)\sum_{j=1}^{m} f(X_n,Y_j,Z_{n,j})$ for the least such n would contradict (2.9). It follows, again from the one-parameter results, that ${}^*S^*(f)$ is finite almost surely. Thus, by (2.7), $\tilde{\delta}(f)$ is finite.

Since (2.8) is an easy consequence of the one-parameter strong law, it is sufficient to prove (2.7). To begin with the implication $\Leftarrow$, assume f satisfies $\tilde{\delta}(f) < \infty$. Since $\tilde{\delta}$ is scalar homogeneous and $\tilde{\delta}(f) = \tilde{\delta}(|f|)$ we may assume $\|f\|_1 = 1$ and $f\geq 0$. We shall prove

$$(2.10) \qquad E_Z\int_0^1 \sup_i f(X_i,y,Z_{i,1})/i \; dy < \infty, \text{ a.s.},$$

where E_Z denotes conditional expectation given the $\{X_i\}$ and $\{Y_j\}$. We would then be finished since (2.10) implies

$$\sup_m (1/m) \sum_{j=1}^{m} \sup_i f(X_i,Y_j,Z_{i,j})/i < \infty, \text{ a.s.},$$

$$\Rightarrow \qquad \sup_i (1/i) \sup_m (1/m) \sum_{j=1}^{m} f(X_i,Y_j,Z_{i,j}) < \infty, \text{ a.s.},$$

$$\Rightarrow \quad E_Z\int_0^1 \sup_m (1/m) \sum_{j=1}^m f(x,Y_j,Z_{1,j})dx < \infty, \text{ a.s.},$$

$$\Rightarrow \quad \sup_n (1/n) \sum_{i=1}^n \sup_m (1/m) \sum_{j=1}^m f(X_i,Y_j,Z_{i,j}) < \infty, \text{ a.s},$$

$$\Rightarrow \quad {}^*S^*(f) < \infty, \text{ a.s.}$$

All these implications except for the second and last, which are trivial, use Fubini's Theorem together with the equivalence of (2.1)-(2.3). For example, to obtain the first implication above use (2.1) $\Rightarrow$ (2.3) with $\zeta_j = \sup_i f(X_i,Y_j,Z_{i,j})/i$, which are i.i.d, conditional on $\{X_i\}$.

To prove (2.10) it is enough to show

$$\sum_{i=1}^{\infty} E_Z\int_0^1 (1/i)\ f(X_i,y,Z_{i,1})I(\ f(X_i,y,Z_{i,1}) > \mu(y)i)dy < \infty, \text{ a.s.},$$

since $E_Z\int_0^1 \mu(y)dy = \|f\|_1 \leq \tilde{\delta}(f) < \infty$. By Kolmogorov's Three Series Theorem it suffices to show

$$\text{(2.11)} \quad \sum_{i=1}^{\infty} P(\lambda(X_i) > i) < \infty, \text{ and}$$

$$\text{(2.12)} \quad \sum_{i=1}^{\infty} E_X[\ E_Z\int_0^1 (1/i)f(X_i,y,Z_{i,1})\ I(f(X_i,y,Z_{i,1}) > \mu(y)i)dy\ ]\wedge 1 < \infty.$$

Statement (2.11) is an immediate consequence of $\|f\|_1 < \infty$. After replacing X_i by X_1, $Z_{i,1}$ by $Z_{1,1}$, and the expectation signs by integrals, the expression in (2.12) may be estimated from above by

$$\int_{[0,1]^3} f(x,y,z) \sum_{i=1}^{\infty} (1/i)\ I(\ \lambda(x) \leq i < f(x,y,z)/\mu(y)).$$

But the last quantity is bounded by $2\tilde{\delta}(f)$. To see this, use the elementary inequality

$$\text{(2.13)} \quad \log_+(B/A\vee 1) - 2 \leq \sum_{i:\ A\leq j<B} (1/i) \leq 2\log_+(B/A\vee 1),$$

valid for all A,B >0, to perform the summation.

Turning to the implication $\Rightarrow$ in (2.7), we shall prove

$$\text{(2.14)} \quad {}^*f^* < \infty \Rightarrow \tilde{\delta}(f) < \infty.$$

This is sufficient, since ${}^*S^*(f) \geq {}^*f^*/4$. We may again assume $\|f\|_1 = 1$ and $f \geq 0$.

Let $\{\varepsilon_i\}$ and $\{\tilde{\varepsilon}_i\}$ be independent copies of the Rademacher sequence which are also independent of the $\{X_i\}$, $\{Y_j\}$ and the $\{Z_{i,j}\}$. The following string of implications follows by combining the assertions of Lemma 2.1 with Fubini's Theorem as above:

$$
\begin{aligned}
{}^*f^* < \infty, \text{ a.s.} &\Leftrightarrow E_{X,Z} \sup_j f(X_1, Y_j, Z_{1,j})/j < \infty, \text{ a.s.}, \\
&\Leftrightarrow \sum_{i=1}^{\infty} (\sup_j f(X_i, Y_j, Z_{i,j})/(ij))^2 < \infty, \text{ a.s.}, \\
&\Rightarrow \sup_j (1/j)(\sum_{i=1}^{\infty} f^2(X_i, Y_j, Z_{i,j})/i^2)^{1/2} < \infty, \text{ a.s.}, \\
&\Leftrightarrow E_{Y,Z}(\sum_{i=1}^{\infty} f^2(X_i, Y_1, Z_{i,1})/i^2)^{1/2} < \infty, \text{ a.s.}, \\
&\Leftrightarrow \sum_{i=1}^{\infty}\sum_{j=1}^{\infty} f^2(X_i, Y_j, Z_{i,j})/(ij)^2 < \infty, \text{ a.s.}
\end{aligned}
$$

It follows from a result of Bonami (see, e.g. [6]) that the latter is equivalent to almost sure convergence of $\sum_{i,j} \tilde{\varepsilon}_i \varepsilon_j f(X_i, Y_j, Z_{i,j})/(ij)$ where convergence means convergence of the partial sums over the sets $\{(i,j): i \leq n, j \leq m\}$ as $n \wedge m$ tends to infinity. This, in turn, is equivalent to

(2.15) $\quad \sum_{j=1}^{\infty} \varepsilon_j f(x, Y_j, Z_{1,j})/j$ converges in $L^1(dx \otimes dP_Z)$, a.s.

(See the argument at the top of p. 1574 in [7].) It follows from Fubini's Theorem and Kahane's contraction principle, applied conditionally given the $\{Y_j\}$, that

$$\sum_{j=1}^{\infty} \varepsilon_j (1/j) f(x, Y_j, Z_{1,j}) I(\mu(Y_j) \leq j) \text{ converges in } L^1(dx \otimes dP_Z), \text{ a.s.}$$

Next we will apply a fundamental result in the theory of probability on Banach spaces due to Hoffman-Jorgensen [3].

LEMMA 2.2. Let B be a Banach space and ξ_j a sequence of independent B-valued random variables such that

$$(2.16) \qquad \sum_{j=1}^{\infty} \xi_j \text{ converges in B-norm almost surely, and}$$

$$(2.17) \qquad E\sup_j \|\xi_j\|_B < \infty.$$

Then $E\| \sum_{j=1}^{\infty} \xi_j \|_B < \infty$.

We may apply this result with $B = L^1(dx \otimes dP_Z)$ and $\xi_j = \varepsilon_j(1/j)f(x,Y_j,Z_{1,j})I(\mu(Y_j) \leq j)$. Then, as noted above, (2.16) holds. Moreover, $\sup_j \|\xi_j\|_B \leq 1$, a.s., so (2.17) also holds. It follows that

$$E\int_0^1 | \sum_{j=1}^{\infty} \varepsilon_j(1/j)f(x,Y_j,Z_{1,j})I(\mu(Y_j) \leq j)|dx < \infty.$$

Now by Fubini's Theorem and Levy's inequality for symmetric real-valued random variables,

$$(2.18) \qquad \begin{aligned} &\int_0^1 E \sup_j (1/j)\, f(x,Y_j,Z_{1,j})I(\mu(Y_j) \leq j)dx \\ &\leq 2\int_0^1 E \sup_n |\sum_{j=1}^{n} \xi_j|dx \leq 4\int_0^1 E|\sum_{j=1}^{\infty} \xi_j|dx < \infty. \end{aligned}$$

Let $A_j(x) = \{f(x,Y_j,Z_{1,j}) > 3j\lambda(x)\}$ and $B_j(x) = \{f(x,Y_k,Z_{1,k}) \leq 3\lambda(x)$ for $k=1,2,\ldots,j-1\}$ (set $B_1(x) \equiv \Omega$). Then $A_j(x)$ and $B_j(x)$ are independent, $A_j(x)\cap B_j(x)$ are disjoint, and

$$P(B_j(x)) \geq P(\sup_k (1/k)f(x,Y_k,Z_{1,k}) \leq 3\lambda(x)) \geq 2/3$$

by Doob's inequality. Therefore the first expression in (2.18) dominates

$$\sum_{j=1}^{\infty}\int_0^1 (E|\xi_j|;\, A_j(x)\cap B_j(x))dx = \sum_{j=1}^{\infty}\int_0^1 (E|\xi_j|;A_j(x))P(B_j(x))dx$$

$$\geq (2/3)\sum_{j=1}^{\infty}\int_0^1 E(|\xi_j|;A_j(x))dx =$$

$$(2/3)\int_{[0,1]^3} f(x,y,z) \sum_{j=1}^{\infty} (1/j) I(\mu(y) \leq j < f(x,y,z)/3\lambda(x))dxdydz.$$

The finiteness of $\tilde{\delta}(f)$ then follows easily from (2.13), and the proof of Theorem 1.2 is complete.

Proof of Theorem 1.3. For Borel functions g on [0,1) define $\tau^* g(x) = \sup_n (1/n) \left| \sum_{i=1}^{n} g(\tau^i x) \right|$ and similarly define $\sigma^* g$. Also extend these to functions of two variables by allowing τ and σ to operate on the first and second variables respectively.

Suppose the variables ζ_i of Lemma 2.1 have the form $g(X_i)$ for some Borel function g and sequence X_i of i.i.d. U(0,1) random variables. Then each of the statements (2.1)-(2.6) implies

(2.19) $\tau^* g < \infty$, a.s., and $\sigma^* g < \infty$, a.s.

This follows from Birkhoff's Ergodic Theorem.

As in the case of Theorem 1.2, it is sufficient to prove the implication $\delta(f) < \infty \Rightarrow \sigma^*\tau^*(f) < \infty$, a.e., for nonnegative f. Then using Fubini's Theorem, Theorem 1.2, (2.1)-(2.3), and (2.19) we have the following string of implications:

$\delta(f) < \infty \Rightarrow {}^*f^* < \infty$, a.s., $\Rightarrow \int (\sup_j f(x,Y_j)/j)dx < \infty$, a.s.,

$\Rightarrow \tau^*(\sup_j f(\cdot,Y_j)/j) < \infty$, a.s., $\Rightarrow \sup_j \tau^*(f(\cdot,Y_j))/j < \infty$, a.s.,

$\Rightarrow \int_0^1 \tau^*(f(\cdot,y))dy < \infty$, a.s., $\Rightarrow \sigma^*\tau^*(f) < \infty$, a.s.,

and the proof is complete.

Next we provide an example of τ, σ and $f \geq 0$ such that $\delta(f) = \infty$, but such that the limit in (1.5) exists.

EXAMPLE 2.1. Let $\tau = \sigma$ be given by the rotation

$$\tau(x) = x + \alpha \pmod 1,$$

for some fixed irrational number α. Take $f(x,y) = h(x-y)$ where $h \geq 0$ has period 1 and satisfies

$$\int_0^1 h(x)dx < \infty \quad \text{and} \quad \int_0^1 h(x)\log_+ h(x)dx = \infty.$$

It is then not difficult to show that $\delta(f) = \infty$.

Now for any $n \geq m \geq 1$ we have

$$(1/nm) \sum_{i=1}^{n} \sum_{j=1}^{m} f(\tau^i x, \sigma^j y) = (1/nm) \sum_{i=1}^{n} \sum_{j=1}^{m} h(x-y + (i-j)\alpha)$$

$$\leq (1/nm) \sum_{k=-m}^{n} mh(x-y-k\alpha) = (1/n) \sum_{k=-m}^{n} h(x-y-k\alpha).$$

A similar estimate holds if $m > n$. Thus by the one-parameter maximal ergodic theorem

$$(2.20) \qquad |\{(x,y): \sup_{n,m} (1/nm) \sum_{i=1}^{n} \sum_{j=1}^{m} f(\tau^i x, \sigma^j y) > \lambda\}| \leq 2\|h\|_1/\lambda,$$

where $|\ |$ denotes 2-dimensional Lebesgue measure. It follows easily from this inequality that one has convergence in (1.5). For example, choose step functions $h_n \to h$ in L^1 and let $f_n(x,y) = h_n(x-y)$. Then the f_n are each bounded and hence $\delta(f_n) < \infty$. By Theorem 1.3 convergence in (1.5) holds for the f_n. A standard approximation argument using (2.20) completes the proof.

Finally, we note that by combining the methods and results of this paper one can prove a two-parameter analogue of Lemma 2.1.

COROLLARY 2.1. _The following are equivalent for any_ RCE _array_ $W_{i,j}$:

$$\sup_{i,j} |W_{i,j}|/ij < \infty, \text{ a.s.},$$

$$\sup_{n,m} (1/nm) \left|\sum_{i=1}^{n} \sum_{j=1}^{m} W_{i,j}\right| < \infty, \text{ a.s.},$$

$$\sup_{n} (1/n) \sum_{i=1}^{n} \sup_{m} (1/m) \sum_{j=1}^{m} |W_{i,j}| < \infty, \text{ a.s.},$$

$$\lim_{n \wedge m \to \infty} (1/nm) \sum_{i=1}^{n} \sum_{j=1}^{m} W_{i,j} \ \underline{\text{exists}} \text{ a.s.},$$

$$\sum_{i,j} W_{i,j}^2/(ij)^2 < \infty, \text{a.s.}, \ \underline{\text{and}}$$

$$\sum_{i,j} \varepsilon_i \tilde{\varepsilon}_j W_{i,j}/(ij) \text{ converges, a.s.,}$$

where ε_i *and* $\tilde{\varepsilon}_j$ *are Rademacher sequences independent of the* $W_{i,j}$, *and the mode of convergence is as described above.*

References

[1] D.J. Aldous, Representations for partially exchangeable arrays of random variables, *J. Multivariate Anal.* 11(1981), 581-598.

[2] G.K. Eagleson and N.C. Weber, Limit theorems for weakly exchangeable arrays, *Math. Proc. Camb. Phil. Soc.* 84(1978), 123-130.

[3] J. Hoffman-Jorgensen, Sums of independent Banach space valued random variables, *Studia Math.* 52(1974), 159-186.

[4] D.N. Hoover, Relations on probability spaces and arrays of random variables, preprint(1979), Princeton University.

[5] B. Jessen, J. Marcinkiewicz, and A. Zygmund, Note on differentiability of multiple integrals, *Fund. Math.* 25(1935), 217-234.

[6] W. Krakowiak and J. Szulga, Random multilinear forms, *Ann. Probab.* 14(1985), 955-973.

[7] T.R. McConnell, A two-parameter maximal ergodic theorem with dependence, *Ann. Probab.* 15(1987), 1569-1585.

[8] —————— , Two-parameter strong laws and maximal inequalities for U-statistics, *Proc. Royal Soc. Edinburgh* 107A(1987), 133-151.

[9] —————— , On the strong maximal function and rearrangements, *Studia Math.*, to appear.

[10] W.G. McGinley and R. Sibson, Dissociated random variables, Math. Proc. Camb. Phil. Soc. 77(1975), 185-188.

[11] E. Rieders, Ph.D. Dissertation, Syracuse University, 1988.

[12] J. Rosinski, On stochastic integral representation of stable processes with sample paths in Banach spaces, J. Multivariate Anal. 20(1986), 277-302.

[13] J. Rosinski and W.A. Woyczynski, On Ito stochastic integration with respect to p-stable motion: inner clock, integrability of sample paths, and multiple integrals, Ann. Probab. 14(1986), 271-286.

[14] B.W. Silverman, Limit theorems for dissociated random variables, Adv. Appl. Prob. 8(1976), 806-819.

[15] R.T. Smythe, Sums of independent random variables on partially ordered sets, Ann. Probab. 2(1974), 906-917.

[16] N. Wiener, The ergodic theorem, Duke Math. J. 5(1939), 1-18.

MULTI-PARAMETER WEIGHTED ERGODIC THEOREMS FROM THEIR SINGLE PARAMETER VERSION

by James H. Olsen

North Dakota State University

Let $(X,\mathcal{F},\mu)$ be a probability space, T a linear operator of $L_p(X,\mathcal{F},\mu)=L_p$, $1\le p<\infty$. If T takes positive functions to positive functions, we say that T is POSITIVE. If there is an operator S of L_p such that $|Tf| \le S|f|$, for all $f\in L_p$ we say that T is DOMINATED by S. We note that if T is dominated by S, S is necessarily positive. If there exists a constant C such that

$$\int \Big|\sup_n \frac{1}{n} \sum_{k=0}^{n-1} T^k f\Big|^p d\mu \le C \int |f|^p du \text{ for all } f\in L_p ,$$

we say that T ADMITS A DOMINATED ESTIMATE ON L_p.

Let $\bar{a} = (a(k);\ k=1,2,...)$ be a complex sequence. If

$$\lim_{n\to\infty} \frac{1}{n} \sum_{k=0}^{n-1} a(k)T^k f \text{ exists a.s. for all } f\in L_p,$$

ISBN 0-12-231050-0

we will say that $(\bar{a},T)$ is BIRKHOFF. If $\bar{a}$ is bounded, we will put

$$\|\bar{a}\|_{\infty} = \sup_k |a(k)|.$$

In this note, we point out that the general approach used by Sucheston (8) to imply multi-parameter ergodic theorems from their single parameter counterparts can be used to deduce multi-parameter weighted ergodic theorems from their single parameter versions as well. As a corollary, we note that this implies multi-parameter versions of ergodic theorems in which the convergence is taken along subsequences which have a density. For each $m \in N$, define $I^m = I_1 \times I_2 \times \ldots \times I_m$, with $I_k = N$, $k=1,2,\ldots,m$. Let $\leq$ be the partial order on I^m given by $s \leq t$ if $s_k \leq t_k$ for $K=1,2,\ldots,n$, when $s=(s_1,s_2,\ldots s_n)$, $t=(t_1,t_2,\ldots t_n)$. Let

Let $L(1), L(2),\ldots,L(m)$ be Orlicz spaces with Δ_2 and let $T(k,n)$, $k=1,2,\ldots,n$, $n \in N$ be positive linear operators from $L(k)$ to $L(1)$. We will need a slight variation of the following theorem, which is Theorem 1.2 of [8].

<u>Theorem 1</u>. Suppose that for $k=1,2,\ldots,n$ and $\underline{\bar{X}} \in L(k)$, $\underline{\bar{X}} \geq 0$, a) $\lim_n T(k,n)\underline{\bar{X}} = T(k,\infty)$ exists a.s.

b) $\sup_n T(k,n)\underline{\bar{X}} \in L(k-1)$ for $k \geq 2$.

c) $T(1,\infty)$ maps $L(1)$ to $L(1)$.

Then $\lim_{I^m} U_{\underline{s}}\overline{\underline{X}} = T(1,\infty)T(2,\infty)\ldots T(m,\infty)\ \overline{\underline{X}}$ exists a.s.

where $U_{\underline{s}}\overline{\underline{X}} = T(1,s_1)T(2,s_2)\ldots T(m,s_m)\ \overline{\underline{X}}$.

We will be interested in the case $L(1)=L(2)=\ldots=L(m)=L_p$, $1<p<\infty$. Such L_p spaces are Orlicz spaces with Δ_2 (see the introduction to (3)). We will want to replace the requirement that each $T(k,n)$ will be positive with the requirement that each $T(k,n)$ is dominated by a positive operator.

The proof of Theorem 1.2 of [8], our Theorem 1, is an induction based on Proposition 1.1 of the same paper. The proof of that result remains valid with minor changes if we assume that if $S(k,n)$ is a positive operator that dominates $T(k,n)$ for $k=1,2,\ldots,m$, $n\in N$, then

$$\sup_n S(k,n)\overline{\underline{X}} \in L(k-1),\ k\geq 2,\ \overline{\underline{X}} \in L_m,$$

and $\lim_n S(k,n)\overline{\underline{X}} = S(k,\infty)\overline{\underline{X}}$ exists a.s.

as well as $S(1,\infty)$ maps $L(1)$ to $L(1)$. We still need [$\lim_n T(k,n)\overline{\underline{X}}=T(k,\infty)\overline{\underline{X}}$] exists a.s., but $\sup_n T(k,n)\overline{\underline{X}}\in L(k-1)$ for $k\geq 2$ is redundant. We omit the verification, but formally restate Theorem 1 taking these facts into consideration. For a related result, see Theorem 6.1, p.488 [3].

<u>Theorem 1'</u>. Suppose that for every $k=1,2,\ldots,m$, $n\in N$,

and $\overline{X} \in L(k)$, $T(k,n)$ is dominated by $S(k,n)$ and

a) $\lim_n T(k,n)\overline{X} = T(k,\infty)$ and

$\lim_n S(k,n)\overline{X} = S(k,\infty)$ exist a.s.,

b) $\sup_n S(k,n)\overline{X} \in L(k-1)$ for $k \geq 2$.

c) $S(1,\infty)$ maps $L(1)$ to $L(1)$.

Then $\lim_m U_{\underline{s}_I}\overline{X} = T(1,\infty)\ldots T(m,\infty)\overline{X}$ exists a.s. for $\overline{X} \in L(m)$.

Hence, $U_s X = T(1,s_1)T(2,s_2)\ldots T(m,s_m)\overline{X}$ and

$s=(s_1,s_2,\ldots,s_m)$.

<u>Theorem 2</u>. Let $T_1,\ldots,T_n$ be linear operators of L_p, $1<p<\infty$, such that each T_i is dominated by the bounded linear operator S_i of L_p. Let $\overline{a}_i$ be bounded sequences of complex numbers such that for each i, $(\overline{a}_i,T_i)$ is Birkhoff. Suppose further that each S_i (and hence T_i) admits a dominated estimate on L_p. Then for each $f \in L_p$, we have

$$\lim_{m_1,\ldots,m_n} \frac{1}{m_1\cdots m_n} \sum_{k_1=0}^{m_1-1} \cdots \sum_{k_n=0}^{m_n 1} a_1(k_1)\cdots a_n(k_n) T_1^{k_1} \ldots T_n^{k_n} f$$

exists a.s. The limit taken as the $m_1,m_2,\ldots,m_n$ tend to infinity independently of each other.

PROOF: Since

$$\left|\frac{1}{m}\sum_{k=0}^{m-1} a_i(k)T_i^k f\right| \leq \|\overline{a}_i\|_\infty \frac{1}{m}\sum_{k=0}^{m-1} S_i^k|f|,$$

$$\sup_m \left|\frac{1}{m} \sum_{k=0}^{m-1} a_i(k) T_i^k f\right| \in L_p,$$

since each S_i admits a dominated estimate. Further, $\lim_{m\to\infty} \frac{1}{m} \sum_{k=0}^{m-1} a_i(k) T_i^k f$ exists a.s. since $(\bar{a}_i, T_i)$ is Birkhoff, and

$$\lim_{m\to\infty} \|\bar{a}_i\|_\infty \frac{1}{m} \sum_{k=0}^{m-1} S_i^k |f| \text{ exists a.s.}$$

since $p>1$, S_i is bounded, positive and admits a dominated estimate. Therefore, the theorem of Sucheston applies, with $T(k,n) = \frac{1}{n} \sum_{j=0}^{n-1} a_k(j) T_k^j$, and the result follows.

This result implies the multi-parameter version of the theorem that $(\bar{a}, T)$ is Birkhoff if T is Dunford-Schwartz and $\bar{a}$ is a bounded Besicovitch sequence (see 6). It also implies a multi-parameter version of the weighted theorem of Bellow-Losert (1, Theorem 5.2, 5.4 and Corollary 5.3).

If $\bar{n} = \{n(k): k=1,2,...\}$ is an increasing sequences of positive integers that has a positive density, i.e.,

$$\sup_N \frac{|i: i \in \bar{n} \cap \{1,...,N\}|}{N} > 0,$$

then the averages $\frac{1}{N} \sum_{k=0}^{N-1} T^{n(k)} f$

can be written $\frac{1}{M} \sum_{k=0}^{M-1} a_k T^k f$

$$\text{where} \quad a_k = \begin{cases} \frac{n(N)}{N} & \text{if } k \in \bar{n} \\ 0 & \text{if } k \notin \bar{n} \end{cases}$$

Then the above result implies the following:

COROLLARY: Let $T_1, \ldots, T_n$ be as in Theorem 2. Suppose for each T_i we have $\lim_{N \to \infty} \frac{1}{N} \sum_{k=1}^{N} T_1^{n_1(k)} f$

exists a.s. for the sequence $\bar{n}_i$ of positive density. Then

$$\lim_{m_1, \ldots, m_n} \frac{1}{m_1 \cdots m_n} \sum_{k_1=1}^{m_1} \cdots \sum_{k_n=1}^{m_n} T_1^{n_1(k_1)} \cdots T_n^{n_n(k_n)} f$$

exists a.s. The limit is taken as the m_i tend to infinity independently of each other.

This corollary implies, for example, multi-parameter versions of the case $\bar{n}$ is a uniform sequence (see 2) and T is a positive contraction of L_p (see 5), or a Lamperti contraction (see 4 and 7).

Theorem 2 remains true if we allow p=1 and require that

$$\lim_{n\to\infty} \frac{1}{n} \sum_{k=0}^{n-1} S_i^k|f| \text{ exists a.s.}$$

REFERENCES

1. A. Bellow and V. Losert, THE WEIGHTED POINTWISE ERGODIC THEOREM ALONG SUBSEQUENCES, TAMS, 288, 307-46.

2. A. Brunel and M. Keane, ERGODIC THEOREMS FOR OPERATOR SEQUENCES, ZW 12, 231-40.

3. N.E. Frangos and Louis Sucheston, ON MULTIPARAMETER ERGODIC AND MARTINGALE THEOREMS IN INFINITE MEASURE SPACES, Probab. Th. Rel. Fields 71, 477-490 (1986).

4. C. Kan, ERGODIC PROPERTIES OF LAMPERTI OPERATORS, Can. J. Math. 30 (1978) 1206-1214.

5. J.H. Olsen, ACKOGULU'S ERGODIC THEOREM FOR UNIFORM SEQUENCES, Can. JM 32, 880-884.

6. __________, THE INDIVIDUAL WEIGHTED ERGODIC THEOREM FOR BOUNDED BESICOVITCH SEQUENCES, Can. Bull. Math. 25(1982)468-71.

7. __________, THE INDIVIDUAL ERGODIC THEOREM FOR LAMPERTI CONTRACTIONS, C.R. Math. Rep. Acad. Sci. Canada 3 (1981), 113-118.

8. Louis Sucheston, ON ONE-PARAMETER PROOFS OF ALMOST SURE CONVERGENCE OF MULTIPARAMETER PROCESSES, ZW 63(1983), 43-49

WEAKLY ERGODIC PRODUCTS OF (RANDOM) NON-NEGATIVE MATRICES.

Steven Orey*

University of Minnesota, Department of Mathematics,

Minneapolis, MN

0. Introduction. Given a sequence (P_n) of non-negative $N \times N$-matrices we study the products $P_{0,n} = P_0 P_1 \cdots P_{n-1}$ in the weakly ergodic situation. In Section 1 we recall some facts in the special case $P = P_0 = P_1 = \ldots$ with P primitive, so that Perron-Frobenius theory applies. These facts motivate our work in Sections 2 and 3 where the general weakly ergodic case is treated. In Sections 4 and 5 we consider the situation in which (P_n) is a stationary random sequence which is weakly ergodic with probability one.

In case all the P_n are sub-stochastic there is a natural probabilistic interpretation at hand, for one may consider P_n as the transition probability matrix at time n of a sub-Markov chain, that is a Markov chain which at each step survives with probability less than or equal to one. Then some of our results can be interpreted as information about the asymptotic behavior of the Markov chain conditioned

*) Work partially supported by NSF Grant MCS83-01080.

ISBN 0-12-231050-0

to survive up to time n.

Proceeding to some notation we define $X = \{1,2,\dots N\}$

and

$$P_{k,n} := P_k P_{k+1} \cdots P_{n-1} \ , \ 0 \leq k \leq n;$$

$$P_{n,n} = I \text{ (identity)}, \ n=0,1,\dots; \tag{0.1}$$

$$u_n(i) := \sum_k P_{0,n}(i,k) \ , \ n = 0,1,\dots; \tag{0.2}$$

$$u_n := \max_i u_n(i) \ . \tag{0.3}$$

Writing $y = (P_n)$, the matrix P_m is the m-th coordinate of the sequence y and this may be indicated by writing it as P_m^y. Let θ be the left-shift on sequences, so $\theta y = (P_{n+1})$, and $P_m^{\theta y} := P_{m+1}^y$. As usual θ^k is the k-th iterate of θ. We extend our notation to other functions of y. For example $u_n(i)$ is a function of y and may be written $u_n^y(i)$. We have for instance

$$u_n^{\theta y}(i) = \sum_k P_{0,n}^{\theta y}(i,k) = \sum_k P_{1,n+1}^y(i,k)$$

or, omitting the superscript y we will write simply

$$u_n^{\theta}(i) = \sum_k P_{0,n}^{\theta}(i,k) = \sum_k P_{1,n+1}(i,k)$$

Introduce the stochastic matrices

$$ {}^{n}\hat{P}_{k,m}(i,j) = P_{k,m}(i,j)u_{n-m}^{\theta^m}(j)/u_{n-k}^{\theta^k}(i) \ , $$
$$ 0 \le k \le m \le n \ , \ (i,j) \in X \times X, \ n = 0,1,\ldots \ . \qquad (0.4) $$

When all P_n are sub-stochastic this gives the conditional probability that the corresponding sub-Markov chain is at j at time m, given that it is at i at time k and that it is still alive at time n. Under the weak ergodicity assumption introduced in Section 2 it is shown that

$$ \lim_{n\to\infty} {}^{n}\hat{P}_{k,m}(i,j) = \hat{P}_{k,m}(i,j) \ , \ 0 \le k \le m $$

exists; the limit is stochastic and $\hat{P}_{k,n} = \hat{P}_{k,m}\hat{P}_{m,n}$ for $k \le m \le n$. In some instances there exists a probability distribution $\hat{\pi}$ on x such that as $n \to \infty$

$$ \frac{1}{n+1} \sum_{m=0}^{n} {}^{n}\hat{P}_{0,m}(i,k) \longrightarrow \hat{\pi}(k) \ , \quad (i,k) \in X \times X \ . \qquad (0.5) $$

In Section 4 we show that if (P_n) is a stationary random sequence which is weakly ergodic with probability one then (0.5) holds with probability one. We make use of ergodic theorems for random Markov chains developed in Nowrotzki [6], Cogburn [1].

It is shown in Section 2 that under the assumption of weak ergodicity the ratios u_n/u_{n-1}^{θ} converge to a finite positive limit L . Turning again to the random situation it is shown in Section 5 that with probability one

$$\frac{1}{n} \sum_{k=0}^{n-1} \log L^{\theta^k} \longrightarrow \gamma$$

where γ is the dominant Lyapunov exponent.

Our work in Sections 3 and 4 uses the Birkhoff contraction coefficient and was greatly helped by the excellent exposition given in Seneta [7].

Finally we explain some notational conventions. If $n \to \infty$, we write $f(n) \sim g(n)$ to signify $f(n)/g(u) \longrightarrow 1$. We now explain the notation

$$f = \overline{O}(g) \tag{0.6}$$

where f and g are functions of the same variables (one of which might be the sequence variable y), and $g \geq 0$. Then (0.6) means that there exist positive constants ϵ_0 and K_0 such that

$$|f| \leq K_0 g \quad \text{whenever} \quad g < \epsilon_0 \; ; \tag{0.7}$$

i.e. f is of order less than or equal to g uniformly.

<u>1. Powers of a matrix</u>. To motivate our work we consider briefly the situation when $P_n = P$, $n = 0,1,\ldots$ so

$P_{0,n} = P^n$, and P is primitive. In this case (1.10) and (1.12) were noted in [3], [8] and the latter authors consider extensions to certain infinite matrices P .

Recall that P primitive means that for some n $P^n(i,j) > 0$ for all i,j . In that case the Perron-Frobenius theorem applies, see e.g. [7]. P has a positive eigenvalue λ , with all other eigenvalues having smaller norm, and there exist associated left and right eigenvectors α and h , unique up to constant multiple; $\alpha(i)h(i) > 0$ for all $i \in X$. Let α and h be chosen so that $\sum \alpha(i)h(i) = 1$. For all sufficiently big n $P^n(i,j) > 0$, $((i,j) \in X \times X)$, and then

$$P^n(i,j) = \lambda^n h(i)\alpha(j)[1 + e_n(i,j)] \qquad (1.1)$$

defines $e_n(i,j)$. Then

$$e_n(i,j) \longrightarrow 0 \text{ exponentially fast as } n \to \infty . \qquad (1.2)$$

Writing

$$P^n(i,X) := \sum_k P^n(i,k) \ , \ \alpha_n(X) := \sum_k \alpha_n(k)$$

we have

$$u_n(i) = P^n(i,X) \sim \lambda^n h(i)\alpha(X) . \qquad (1.3)$$

For $0 \leq m \leq n$

$$ {}^{n}\hat{P}_{0,m}(i,j) = \frac{P^{m}(i,j)u^{\theta^{m}}_{n-m}(j)}{u_{n}(i)} = \frac{P^{m}(i,j)P^{n-m}(j,X)}{P^{n}(i,X)} \tag{1.4} $$

Clearly

$$ \hat{P}(i,j) := \lim_{n} {}^{n}\hat{P}_{0,1}(i,j) = \frac{P(i,j)h(j)}{h(i)\lambda} \tag{1.5} $$

and as $n \to \infty$

$$ {}^{n}\hat{P}_{0,m}(i,j) \sim \frac{P^{m}(i,j)h(j)}{h(i)\lambda^{m}} =: \hat{P}^{m}(i,j) \tag{1.6} $$

More specifically, by (1.2), there exists $K > 0$ and $s < 1$ such that

$$ |({}^{n}\hat{P}_{0,m}(i,j)/\hat{P}^{m}(i,j)) - 1| < Ks^{n-m}; \tag{1.7} $$

(for small m the fraction may take on the form $0/0$ and we interpret that as 1). Observe that $\hat{P}$ is a stochastic matrix.

Let

$$ \hat{\pi}(i) = \alpha(i)h(i) \tag{1.8} $$

and by the definition of α and h

$$\hat{\pi} = \hat{\pi}\hat{P} \tag{1.9}$$

Clearly $\hat{P}$ is also a primitive matrix, so $\hat{\pi}$ is the unique invariant probability. One obtains

$$\frac{1}{n+1} \sum_{m=0}^{n} {}^{n}\hat{P}_{0,m}(i,j) \longrightarrow \hat{\pi}(j) \tag{1.10}$$

either directly from (1.1), (1.2) and (1.4) or by using (1.7) and the fact that $\hat{P}^{m}(i,j) \longrightarrow \hat{\pi}(j)$ as $m \longrightarrow \infty$.

In connection with (1.10) note that also

$$\lim_{m\to\infty} \lim_{n\to\infty} {}^{n}\hat{P}_{0,m}(i,j) \longrightarrow \hat{\pi}(j) \ ; \tag{1.11}$$

but on the other hand

$${}^{n}\hat{P}_{0,n}(i,j) \longrightarrow \frac{\alpha(j)}{\alpha(X)} \ . \tag{1.12}$$

When P is substochastic we have a probabilistic interpretation (see introduction). According to (1.9) and (1.10), n^{-1} times the expected number of visits paid to j up to time n by the sub-Markov chain conditioned to survive up to time n converges to $\hat{\pi}(j)$; and $\hat{\pi}$ is the unique invariant distribution for the sub-Markov chain conditioned to survive forever. Furthermore (1.5) shows that $\hat{P}$ is obtained from P by a familiar similarity transformation.

2. Weak ergodicity. The notation will be that of the introduction. The theory of weak ergodicity is set forth in [7] under the assumption

$$P_k \text{ has no vanishing rows or columns, } k = 0,1\ldots \qquad (2.1)$$

We will assume (2.1). (This assumption could be avoided by making minor changes in the theory.) Now define

$$\alpha_n(j) = \sum_i P_{0,n}(i,j) \Big/ \sum_{i,k} P_{0,n}(i,k) \ , \quad j \in X \ . \qquad (2.2)$$

Call (P_n) weakly ergodic if (2.1) holds, and as $n \to \infty$,

$$P_{k,n}(i,j) \sim u_{n-k}^{\theta^k}(i)\, \alpha_{n-k}^{\theta^k}(j) \ ,$$
$$(i,j) \in X \times X \ , \ k = 0,1,\ldots \ . \qquad (2.3)$$

(In the context of Section 1, (1.1) implies (2.3).) Define the error term $e_n(i,j)$ by

$$P_{0,n}(i,j) = u_n(i)\alpha_n(j)(1 + e_n(i,j)) \ . \qquad (2.4)$$

Note that $-1 \leq e_n(i,j) < \infty$. The basic device for the development of the theory of weak ergodicity is the Birkhoff contraction coefficient $b(T)$, introduced originally by

Birkhoff [1]. For a detailed exposition of properties of the Birkhoff contraction coefficient in our present context we refer to Seneta [7]. For T a non-negative matrix satisfying (2.1) one has

$$b(T) = [1-(\varphi(T))^{1/2}]/[1+(\varphi(T))^{1/2}] \qquad (2.5)$$

where $\varphi(T) := 0$ if T has any zero entries and otherwise

$$\varphi(T) := \min_{i,j,k,l} \left[\frac{T(i,k)\ T(j,l)}{T(i,l)\ T(j,k)}\right] .$$

The key fact is that for S also a non-negative matrix satisfying (2.1) the inequality

$$b(ST) \leq b(S)b(T) \qquad (2.7)$$

obtains; see [7,§3.1] . The second important fact is expressed by

$$\varphi(P_{0,n}) \leq [1 + e_n(i,j)] \leq (\varphi(P_{0,n}))^{-1} \qquad (2.8)$$

as shown in the proof of [7, Lemma 3.3, p. 84]. From this we obtain that

$$e_n := \max_{i,j} |e_n(i,j)| \qquad (2.9)$$

satisfies

$$e_n = \bar{O}(b(P_{0,n})) \tag{2.10}$$

where we use the notation explained in (0.6) .

Theorem 2.1. Let (P_n) be weakly ergodic. Then there exists a function $h: X \longrightarrow (0,\infty)$ such that as $n \longrightarrow \infty$

$$\frac{u_n(i)}{u_n(j)} \longrightarrow \frac{h(i)}{h(j)}, \quad (i,j) \in X \times X \tag{2.11}$$

and more specifically

$$\frac{u_n(i)}{u_n(j)} = \frac{h(i)}{h(j)} [1 + \bar{O}(e_n)] . \tag{2.12}$$

Proof: Note

$$u_{n+m}(i) = \sum_k P_{0,n}(i,k) u_m^{\theta^n}(k)$$

and use (2.4) to obtain

$$u_{n+m}(i) = u_n(i) \sum_k \alpha_n(k) [1 + e_n(i,k)] u_m^{\theta^n}(k)$$

and therefore

$$u_n(i) \sum_k \alpha_n(k) u_m^{\theta^n}(k)[1-e_n] \leq u_{n+m}(i)$$

$$\leq u_n(i) \sum_k \alpha_n(k) u_m^{\theta^n}(k)[1+e_n]$$

By weak ergodicity there exists n_0 such that $e_n < 1$ for $n \geq n_0$, and then for $(i,j) \in X \times X$

$$\frac{u_n(i)[1-e_n]}{u_n(j)[1+e_n]} \leq \frac{u_{n+m}(i)}{u_{n+m}(j)} \leq \frac{u_n(i)[1+e_n]}{u_n(j)[1-e_n]}, \quad n \geq n_0, \; m \geq 0, \tag{2.13}$$

which in turn implies

$$\frac{u_n(i)}{u_n(j)}\left[\frac{1-e_n}{1+e_n} - 1\right] \leq \frac{u_{n+m}(i)}{u_{n+m}(j)} - \frac{u_n(i)}{u_n(j)} \leq \frac{u_n(i)}{u_n(j)}\left[\frac{1+e_n}{1-e_n} - 1\right], \quad n \leq n_0, \; m \geq 0 \tag{2.14}$$

Since the extremes in (2.13) do not depend on m we see that

$$1 \leq \sup_m \max_{i,j} \frac{u_m(i)}{u_m(j)} := K < \infty. \tag{2.15}$$

However K will depend on the sequence (P_n). The expressions in brackets in (2.14) are $\overline{O}(e_n)$ terms and using also (2.15) we see that $(u_n(i)/u_n(j))$ is a Cauchy sequence converging to a limit which must, by (2.15) lie in the interval $[K^{-1}, K]$. For fixed $k \in X$ one can write

$$\frac{u_n(i)}{u_n(j)} = \frac{u_n(i)}{u_n(k)} \left[\frac{u_n(j)}{u_n(k)}\right]^{-1}$$

to conclude the limit in (2.11) has the desired form with

$$h(i) = \lim_n \frac{u_n(i)}{u_n(k)}$$

one possible choice for h . Letting $m \to \infty$ in (2.14) we obtain

$$\frac{h(i)}{h(j)} - \frac{u_n(i)}{u_n(j)} = \frac{u_n(i)}{u_n(j)} \bar{O}(e_n)$$

and hence (2.12). Q.E.D.

The function h in Theorem 2.1 is determined only up to a positive multiplicative constant. We now define h to be the unique function satisfying (2.11) and also

$$\max_i h(i) = 1 . \tag{2.16}$$

<u>Corollary 2.2</u> For (P_n) weakly ergodic there exists $L > 0$ such that

$$\frac{u_n(i)}{u^{\theta}_{n-1}(j)} = \frac{Lh(i)}{h^{\theta}(j)} [1 + \bar{O}(e^{\theta}_{n-1})] , \quad (i,j)\in X\times X , \tag{2.17}$$

and

$$\sum_k P_0(i,k)h^\theta(k) = Lh(i) \quad , \quad i\in X \ . \qquad (2.18)$$

Proof. By (2.12) we have

$$u_n(i) = \sum_k P_0(i,k)\ u^\theta_{n-1}(k) =$$

$$\sum_k P_0(i,k)\ u^\theta_{n-1}(j)[u^\theta_{n-1}(k)/u^\theta_{n-1}(j)]$$

$$= u^\theta_{n-1}(j) \sum_k P_0(i,k)[h^\theta(k)/h^\theta(j)](1 + \bar{o}(e^\theta_{n-1}))$$

and therefore

$$u_n(i)/u^\theta_{n-1}(j) = [v(i)/h^\theta(j)](1 + \bar{o}(e^\theta_{n-1})) \qquad (2.19)$$

where

$$v(i) := \sum_k P_0(i,k)h^\theta(k) \qquad (2.20)$$

Write

$$u_n(i)/u_n(\ell) = (u_n(i)/u^\theta_{n-1}(j)) \cdot (u_n(\ell)/u^\theta_{n-1}(j))^{-1}$$

and let $n\to\infty$ to obtain

$$\frac{h(i)}{h(\ell)} = \frac{v(i)}{h^{\theta}(j)} \cdot \frac{h^{\theta}(j)}{v(\ell)} = \frac{v(i)}{v(\ell)}$$

and hence

$$\frac{v(i)}{h(i)} = \frac{v(\ell)}{h(\ell)} =: L \, , \quad (i,\ell) \in X \times X \, . \tag{2.21}$$

Relations (2.19), (2.20) and (2.21) imply (2.17), (2.18) .

Q.E.D.

Henceforth L will denote the number defined in (2.17).

Corollary 2.3. With (P_n) , L as in Corollary 2.2 and $1 \leq m \leq n$, $(i,j) \in X \times X$,

$$\frac{u_n(i)}{u^{\theta}_{n-m}(j)} = \frac{h(i) L \cdot L^{\theta} \dots L^{\theta^{m-1}}}{h^{\theta^m}(j)} [1 + \bar{O}(e^{\theta}_{n-1})] \cdot$$

$$\cdot [1 + \bar{O}(e^{\theta^2}_{n-1})] \dots [1 + \bar{O}(e^{\theta^m}_{n-m})] \tag{2.22}$$

Proof. Write

$$\frac{u_n(i)}{u^{\theta}_{n-m}(j)} = \frac{u_n(i)}{u^{\theta}_{n-1}(i_1)} \frac{u^{\theta}_{n-1}(i_1)}{u^{\theta^2}_{n-2}(i_2)} \dots$$

$$\dots \frac{u^{\theta^{m-1}}_{n-m+1}(i_{n-m+1})}{u^{\theta^m}_{n-m}(j)} \tag{2.23}$$

and apply (2.17) repeatedly; here $i_1, i_2, \ldots i_{n-m+1}$ are arbitrary members of X .

Q.E.D.

From (0.3)

$$u_n = \max_i [\sum_k P_{0,n}(i,k)]$$

and we deduce

$$u_{n+m} \leq u_n \cdot u_m^{\theta^n} \tag{2.24}$$

Corollary 2.4. For (P_n) weakly ergodic

$$\lim_{n\to\infty} \frac{u_n}{u_{n-1}^{\theta}} = L$$

Proof. Pass to the limit in (2.17), take the maximum over i , and note that (since X is finite) this can be passed inside the limit; finally take the minimum over j , again passing inside the limit.

Q.E.D.

Remark 2.4. Define $t_n(i,j)$ by

$$\frac{u_n(i)}{u_n(j)} = \frac{h(i)}{h(j)} [1 + t_n(i,j)] ,$$

and note that (2.15) implies

$$\frac{1}{K^2} \leq [1 + t_n(i,j)] \leq K^2 , \quad n = 0,1,\ldots, \quad (i,j)\in(X\times X) .$$

and according to (2.12) $t_n(i,j) = \bar{O}(e_n)$.

3. Conditioning. We define

$$^n\hat{P}_{k,m}(i,j) = P_{k,m}(i,j)u^{\theta^m}_{n-m}(j)/u^{\theta^k}_{n-k}(i) ,$$

$$0 \leq k \leq m \leq n , \quad (i,j) \in X\times X . \tag{3.1}$$

An interpretation was given after (0.4).

We assume now that (P_n) is weakly ergodic. From the first lines in the proof of Corollary 2.2 and the definition of $t_n(i,j)$ in Remark 2.4

$$\frac{u_n(i)}{u^{\theta}_{n-1}(j)} = \frac{\sum_k P_0(i,k)h^{\theta}(k)[1+t^{\theta}_n(k,j)]}{h^{\theta}(j)}$$

Let $t_n = \max_{i,j} |t_n(i,j)|$ and use (2.18) to obtain

$$\frac{Lh(i)}{h^{\theta}(j)} [1-t^{\theta}_n] \leq \frac{u_n(i)}{u^{\theta}_{n-1}(j)} \leq \frac{Lh(i)}{h^{\theta}(j)} [1+t^{\theta}_n]$$

Now define $s_n(i,j)$ by

$$\frac{u_n(i)}{u^{\theta}_{n-1}(j)} = \frac{Lh(i)}{h^{\theta}(j)}\,[1+s_n(i,j)]\ . \tag{3.2}$$

The preceding relations show $|s_n(i,j)| \le t^{\theta}_n$ and so by Remark 2.4

$$s_n(i,j) = \overline{O}(e^{\theta}_{n-1})\ ,\ \left(\frac{1}{K^{\theta}}\right)^2 \le [1+s_n(i,j)] \le (K^{\theta})^2\ ,$$
$$n = 1,2,\ldots;\ (i,j)\in(X\times X)\ . \tag{3.3}$$

Proceeding as in (2.23) we obtain for arbitrary $i_1,i_2,\ldots,i_{n-1}$

$$\begin{aligned} {}^n\hat{P}_{k,m}(i,j) &= \frac{P_{k,m}(i,j)h^{\theta^m}(j)}{h^{\theta^k}(i)L^{\theta^k}\cdot L^{\theta^{k+1}}\cdots L^{\theta^{m-1}}}\{[1+s^{\theta^k}_{n-k}(i,j)]\cdot \\ &\quad\cdot[1+s^{\theta^{k+1}}_{n-k-1}(i_1,i_2)]\ldots[1+s^{\theta^{m-1}}_{n-m+1}(i_{n-1},j)]\} \\ &\longrightarrow \frac{P_{k,m}(i,j)h^{\theta^m}(j)}{h^{\theta^k}(i)L^{\theta^k}\cdot L^{\theta^{k+1}}\cdots L^{\theta^{m-1}}} =: \hat{P}_{k,m}(i,j) \end{aligned} \tag{3.4}$$

as $n\to\infty$.

Note ${}^n\hat{P}_{k,m}$, hence also $\hat{P}_{k,m}$ is a stochastic matrix and

$$\hat{P}_{k,m+t} = \hat{P}_{k,m}\,\hat{P}_{m,m+t} \quad \text{for} \quad k \le m \le m+t\ .$$

Let $\hat{P}_k = \hat{P}_{k,k+1}$. Then $(\hat{P}_k)$ is a sequence of stochastic matrices. When the P_k are sub-stochastic, one may interpret $\hat{P}_k$ as the transition probability at time k for the Markov chain obtained from the sub-Markov chain (P_n) by conditioning it to survive forever.

The sequence $(\hat{P}_k)$ will also be weakly ergodic. Indeed as $m \to \infty$

$$\hat{P}_{0,m}(i,j) = \frac{P_{0,m}(i,j)h^{\theta^m}(j)}{h(i)L\cdot L^{\theta}\ldots L^{\theta^{m-1}}} \sim \frac{u_m(i)\alpha_m(j)h^{\theta^m}(j)}{h(i)L\cdot L^{\theta}\ldots L^{\theta^{m-1}}} \sim$$

$$\frac{u_m(0)\alpha_m(j)h^{\theta^m}(j)}{h(0)L\cdot L^{\theta}\ldots L^{\theta^{m-1}}} := \hat{\alpha}(j) \ . \tag{3.5}$$

Consider (3.4) with $k = 0$ and define

$$s_{m,n} := \max_{1\leq k\leq m} \ \max_{(i,j)\in X\times X} \ |\, s^{\theta^k}_{n-k}(i,j)| \ , \quad 0 \leq m \leq n \ .$$

Writing $\{\ldots\}$ for the expression which appears in brackets in (3.4) with $k = 0$ we obtain

$$\epsilon_{m,n} := \|\{\ldots\}-1\| \leq m(1+s_{m,n})^{m-1}s_{m,n} \ , \ 0 \leq m \leq n \ . \tag{3.6}$$

Since $\hat{P}_{0,m}$ is stochastic, (3.4) implies

$$|{}^{n}\hat{P}_{0,m}(i,j) - \hat{P}_{0,m}(i,j)| \leq \epsilon_{m,n} \, . \qquad (3.7)$$

Now let $0 < t < 1$, and suppose there exist $M > 0$ and $\rho \in (0,1)$ such that

$$s_{m,n} \leq M\rho^{n-m} \, , \; 0 \leq m \leq tn \, , \; n = 0,1,\ldots \, . \qquad (3.8)$$

then it follows from (3.6) that as $n \to \infty$

$$\sum_{0 \leq m \leq nt} \epsilon_{m,n} \longrightarrow 0 \, . \qquad (3.9)$$

<u>Theorem 3.1</u>. Let (P_n) be weakly ergodic and assume that for each $t \in (0,1)$ there exists an $M > 0$ and $\rho \in (0,1)$ such that (3.8) holds. Then as $n \to \infty$

$$\sum_{0 \leq m \leq nt} |{}^{n}\hat{P}_{0,m}(i,j) - \hat{P}_{0,m}(i,j)| \longrightarrow 0 \, ,$$
$$(i,j) \in X \times X \, , \; 0 < t < 1 \qquad (3.10)$$

and

$$\frac{1}{n+1} \sum_{m=0}^{n} |{}^{n}\hat{P}_{0,m}(i,j) - \hat{P}_{0,m}(i,j)| \longrightarrow 0 \, ,$$
$$(i,j) \in X \times X \, . \qquad (3.11)$$

<u>Proof</u>. By (3.7) and the fact that (3.8) implies (3.9) we obtain (3.10). For $0 < t < 1$, write

$$\frac{1}{n+1}\sum_{m=0}^{n} |{}^{n}\hat{P}_{0,m}(i,j) - \hat{P}_{0,m}(i,j)| = \frac{1}{n+1}\sum_{0\le m\le nt} + \frac{1}{n+1}\sum_{nt\le m\le n}$$

As $n \to \infty$ the first term on the right goes to zero, by (3.10) and since ${}^{n}\hat{P}_{0,m}$ and $\hat{P}_{0,m}$ are stochastic matrices the second term on the right is bounded by $(1-t)$. Since t can be chosen as near one as desired, (3.11) follows.

Q.E.D.

4. Random Sequences. Now we consider the case where (P_n) is a stationary random sequence of non-negative matrices. We may take as our probability space Y the space of all sequences $y = (P_n, n=0,1,\ldots)$ of non-negative $N\times N$ matrices. Each P_n can be considered as a point in $\mathbb{R}^{N^2}$, and on this space we have the class of Borel sets $\mathcal{B}$. Then Y is a subset of $\mathbb{R}^{N^2}\times\mathbb{R}^{N^2}\times\ldots$, and on this space we introduce the product σ-field $\mathcal{B}\times\mathcal{B}\times\ldots$; the restriction of this σ-field to Y will be denoted by $\mathcal{Y}$. We then consider a probability space $(Y,\mathcal{Y},\nu)$ with ν a probability measure invariant under the shift θ (see the introduction). Now P_n is a random variable with value P_n^y for $y\in Y$. Other quantities introduced in the earlier sections, such as $u_n(i)$, $h(i)$, etc. are also random variables; of course this implies that the functions $y \longrightarrow u_n^y(i)$, $y \longrightarrow h^y(i)$, etc. are measurable. We shall say a property holds a.s. (almostly

surely) if it holds for υ-a.e.y. We will say (P_n) is weakly ergodic if (P^y_n) is weakly ergodic a.s.

We assume for the remainder of this section that (P_n) is weakly ergodic. Then $(\hat{P}_n)$ is also weakly ergodic, and stationary, $\hat{P}_n = \hat{P}_0^{\hat{\theta}^n}$, and of course the matrices $\hat{P}_n$ are stochastic. Now results about stationary random Markov chains can be applied. We will use a result of R. Cogburn [2], which extends earlier work of K. Nowrotzki [6]. From [2, Theorem 5.1 and Corollary 3.2], where weak ergodicity is not assumed, one obtains a probability measure $\bar{\upsilon}$ on the product space $(X\times Y)$ with $\bar{\upsilon}(X\times\cdot) = \upsilon(\cdot)$ such that

$$\lim_{n\to\infty} \frac{1}{n+1} \sum_{m=0}^{n} \hat{P}^y_{0,m}(i,j) =: \hat{\pi}^y_i(j) \qquad (4.1)$$

exists for $\bar{\upsilon}$ - a.e. (i,y). Since we are assuming weak ergodicity the limit in (4.1) can not depend on i and we obtain

$$\lim_{u\to\infty} \frac{1}{n+1} \sum_{m=0}^{n} \hat{P}^y_{0,m}(i,j) = \hat{\pi}^y(j) \text{ a.s.} \qquad (4.2)$$

Lemma 4.1. Let (P_n) be weakly ergodic. For each $t\in(0,1)$ there exist positive finite valued random variables A, M, ρ with $0 < \rho < 1$ such that

$$b(P_{m,n}) \leq A\rho^{n-m}, \quad 0 \leq m \leq nt, \quad n = 0,1,\ldots \quad \text{a.s.} \qquad (4.7)$$

and such that (3.8) holds a.s.

<u>Proof</u>. Let $0 < t < 1$. By (3.3) and (2.10)

$$s^{\theta^k}_{n-k}(i,j) = \bar{O}(e^{\theta^k}_{n-k+1}) = \bar{O}(b(P_{k+1,n})) \,,\quad 0 \leq k \leq n \,.$$

Since $b(P_{k,n})$ is an increasing function of k, we obtain

$$s_{m,n} = \max_{k\leq m} \max_{i,j} |s^{\theta^k}_{n-k}| = \max_{k\leq m} \bar{O}(b(P_{k+1,n})) = \bar{O}(b(P_{m+1,n}))$$

Hence the assertion in the Lemma concerning (3.8) will follow from (4.7), and this in turn will be a consequence of the following claim:

$$\overline{\lim_{n\to\infty}} \max_{0\leq m\leq nt} \left[\frac{\log b(P_{m+1,n})}{n-m+1}\right] < 0 \,,\quad \text{a.s.} \qquad (4.8)$$

We turn to the proof of (4.8). It will suffice to prove (4.8) under the additional assumption that υ is ergodic. For in any case υ has a representation as an integral average of ergodic measures (see [5]), and since (P_n) is a.s. weakly ergodic with respect to υ it must have this property also for the ergodic measures entering into the integral average. So we may assume now that υ is ergodic.

By the assumption of weak ergodicity $b(P_{0,n}) \longrightarrow 0$ as $n \to \infty$ a.s. So there exists a positive integer k^* (non-random) such that

$$\nu\{y : b(P^y_{0,k^*}) < 1\} > 0 .$$

Assume first that $k^* = 1$. By (2.7)

$$\log b(P_{m+1,n}) \leq \sum_{j=m+1}^{n-1} \log b(P_j) \leq \sum_{j=m+1}^{n-1} B_j \quad (4.9)$$

with $B_j := \max(\log b(P_j), -1)$. Now it suffices to show that (4.8) holds with the expression in brackets replaced by

$$\frac{1}{n-m-1} \sum_{j=m+1}^{n-1} B_j$$

Note $-1 \leq B_j \leq 0$ and B_j has negative expectation because $k^* = 1$. Since ν is ergodic the desired inequality can now be easily deduced from Birkhoff's ergodic Theorem.

If $k^* > 1$ the argument is reduced to the case already treated by grouping: in (4.9) consider $b(Q_j)$ in place of $b(P_j)$ with $Q_j = P_{jk,(j+1)k^*}$.

<u>Theorem 4.2</u>. Let (P_n) be weakly ergodic and $\hat{\pi}$ as defined in (4.2). Then as $n \to \infty$

$$\frac{1}{n+1} \sum_{m=0}^{n} {}^n\hat{P}_{0,m}(i,j) \longrightarrow \hat{\pi}(j) \quad \text{a.s.} , \quad (i,j) \in X \times X .$$

<u>Proof</u>. By Lemma 4.1, the hypothesis (3.8) of Theorem 3.1 holds a.s. Applying the conclusion of Theorem 3.1 and (4.1) we obtain (4.9).

Q.E.D.

<u>5. Behavior of L^{θ^n}</u>. We continue to assume that (P_n) is a stationary random sequence, with ν the underlying probability measure, and that (P_n) is weakly ergodic a.s. We will also assume

$$\int \log^+ u_1 d\nu < \infty . \tag{5.1}$$

Note that (u_n) is a sub-multiplicative process according to (2.24). Together with (5.1), the subadditive ergodic theorem implies that as $n \to \infty$

$$\begin{aligned} \frac{1}{n} \log u_n &\to \gamma \in [-\infty,\infty) \text{ a.s., and} \\ \frac{1}{n} \int \log u_n d\nu &\to \int \gamma d\nu , \end{aligned} \tag{5.2}$$

and if ν is ergodic γ is constant a.s. For a proof of the subadditive ergodic theorem see [4]. Now recall (2.25) and define s_n by

$$\frac{u_n}{u_{n-1}^{\theta}} = L(1+s_n) . \tag{5.3}$$

Comparing s_n with the quantities $s_n(i,j)$ introduced in (3.2) one verifies that

$$|s_n| \le \max_{i,j} |s_n(i,j)| . \tag{5.4}$$

Theorem 5.1. Assume (P_n) is weakly ergodic and that (5.1) holds. Then

$$\int \log L d\upsilon = \int \gamma d\upsilon \tag{5.5}$$

and as $n \to \infty$

$$\frac{1}{n} \sum_{k=0}^{n-1} \log L^{\theta} \longrightarrow \gamma \quad \text{a.s.} \tag{5.6}$$

Proof: Since (u_n) is a sub-multiplicative process

$$\frac{u_n}{u_{n-m}^{\theta^m}} = \frac{u_n}{u_{n-1}^{\theta}} \cdot \frac{u_{n-1}^{\theta}}{u_{n-2}^{\theta^2}} \cdots \frac{u_{n-m+1}^{\theta^{m-1}}}{u_{n-m}^{\theta^m}} \le u_m , \quad 0 \le m \le n . \tag{5.7}$$

Fixing m and letting $n \to \infty$ we obtain from Corollary 2.4 that

$$L \cdot L^{\theta} \ldots L^{\theta^{m-1}} \le u_m ,$$

hence

$$\frac{1}{m}\sum_{k=0}^{m-1} \log L^{\theta^k} \leq \frac{1}{m} \log u_m \,. \tag{5.8}$$

The last relation with $m = 1$ and assumption (5.1) show

$$\int \log^+ L d\nu \leq \int \log^+ u_1 \, d\nu < \infty \tag{5.9}$$

Again using (5.8) and the stationarity of ν we obtain

$$\int \log L d\nu \leq \frac{1}{m} \int \log u_m \, d\nu \,, \tag{5.10}$$

and passing to the limit, remembering (5.2), we obtain

$$\int \log L d\nu \leq \int \gamma d\nu \,. \tag{5.11}$$

This proves (5.5) in case $\int \gamma d\nu = -\infty$. In any case, since (5.9) holds it follows from the ergodic theorem that as $n \to \infty$,

$$\frac{1}{n}\sum_{k=0}^{n-1} \log L^{\theta^k} \longrightarrow \lambda \in [-\infty, \infty) \,, \quad \lambda = \lambda^\theta \text{ a.s.} \tag{5.12}$$

and

$$\int \log L \, d\nu = \int \lambda d\nu \,. \tag{5.13}$$

To complete the proof of the theorem it must be shown that $\lambda = \gamma$ a.s. Appealing again to the ergodic decomposition of ν it suffices to treat the case where ν is ergodic.

So assume now that ν is ergodic. Then γ and λ are a.s. constant and $\lambda \leq \gamma$ by (5.11) and (5.13). So we need only treat the case $\gamma > -\infty$. Fix $t \in (0,1)$ and let m be the largest integer less than or equal to tn. Since $u_0 = 1$ we may write

$$u_n = \prod_{k=0}^{n-1} \frac{u_{n-k}^{\theta^k}}{u_{n-k-1}^{\theta^{k+1}}} = \prod_{k=0}^{m} \cdot \prod_{k=m+1}^{n-1}$$

Take logarithms, use (5.3) in the first product on the right, and divide by n to obtain

$$\frac{1}{n} \log u_n = \frac{1}{n} \sum_{k=0}^{m} \log L^{\theta^k} +$$

$$\frac{1}{n} \sum_{k=0}^{m} \log (1 + s_{n-k}^{\theta^k}) + \frac{1}{n} \log u_{n-m-1}^{\theta^{m+1}} . \qquad (5.14)$$

As $n \to \infty$, the left member in (5.14) approaches γ, assumed finite. On the right hand side the first term converges to $t\lambda$. The second term on the right goes to 0; this follows from (5.4) and the fact that (3.8) holds by Lemma 4.1. The last term in (5.14), as the only remaining term, must now also converge a.s. to a non-random limit; the eventuality $\lambda = -\infty$ and the limit of the last term equals $+\infty$ remains to be ruled out. Putting $z_n = (n-m)^{-1} \log u_{n-m}$, we know $z_n \longrightarrow \gamma$ a.s. as $n \to \infty$. The last term in (5.14) equals $[(n-m)/n] z^{\theta^m}_{n-m}$. Since $z^{\theta^m}_{n-m}$ has the same distribution as z_{n-m} the last term must converge in distribution to $(1-t)\gamma$; so in fact it must converge to $(1-t)\gamma$ a.s.. So the second term in (5.14) must converge to $t\gamma$, that is $\gamma = \lambda$.

Q.E.D.

References

[1] Birkhoff, G., Lattice Theory (3rd edition). Amer. Math. Soc. Colloq. Publicns. Vol. 25, 1967, Providence, RI.

[2] Cogburn, R., The ergodic theory of Markov chains in random environments. Z.f. Wahrsh. verw. Geb. 66, 109-128, 1984.

[3] Daroch, J.N. and Seneta, E. On quasi-stationary distributions in absorbing discrete-time finite Markov chains, J. Appl. Prob. L., 88-100 (1965).

[4] Liggett, T.M., Interacting particle systems, Springer-Verlag, 1985, New York.

[5] Maitra, A., Integral representations of invariant meaasures. Trans. Amer. Math. Soc. 229, 209-225, (1977).

[6] Nowrotzki, K., Ein zufälliger Ergodensatz für eine Familie stochastischer Matritzen ohnes gemeinsames invariantes Verteilungsgesetz. Math. Nachrichten 70, 17-28, (1976).

[7] Seneta, E., Non-negative matrices and Markov Chains, Second edition, Springer-Verlag, 1980, New York.

[8] Seneta, E. and Verre-Jones, D., On quasi stationary distributions in discrete-time Markov chains with a denumerable infinity of states. J. Appl. Prob. 3, 403-434 (1966).

The r-quick Version of the Strong Law for Stationary ϕ-mixing Sequences

Magda Peligrad*

University of Cincinnati, Department of Mathematical Sciences

Cincinnati, Ohio, U.S.A.

Abstract

In this note we estimate the rate of convergence in the Marcinkiewicz-Zygmund strong law for partial sums S_n of ϕ-mixing sequences of random variables. The results are similar to the independent case and do not require an estimate of the ϕ-mixing coefficients.

1. Introduction and Notations.

In (1964) Baum and Katz studied the speed of convergence for Marcinkiewicz-Zygmund strong law for partial sums of an i.i.d. sequence. Motivated by applications to sequential analysis of time series, to the renewal theory or for the study of randomly selected sums, the Baum and Katz's theorem was the subject of extensions to dependent sequences of random variables. Among them we mention the

*Partially supported by a NSF grant

ISBN 0-12-231050-0

papers by Lai (1977), Hipp (1982), Peligrad (1985*), Berbee (1987). All these papers deal with weakly dependent (mixing) sequences of random variables and all of them require a knowledge of the speed of convergence to 0 of the mixing coefficients. In this paper we extend the Baum and Katz's result from i.i.d. to ϕ-mixing sequence without any additional assumption imposed on the ϕ-mixing coefficients.

Let (Ω, K, P) be a probability space and let $\{X_n\}_{n\geq 1}$ be a sequence of random variables. Denote by $F_n^m = \sigma(X_k, n \leq k \leq m)$ and define the ϕ-mixing coefficients by

$$\phi_m = \sup_{n\geq 1} \sup_{A\varepsilon F_1^n,\ B\varepsilon F_{n+m}^\infty,\ P(A)\neq 0} |P(B|A)-P(B)|$$

and

$$\rho_m = \sup_{n\geq 1} \sup_{f\varepsilon L_2(F_1^n),\ g\varepsilon L_2(F_{n+m}^\infty)} |corr(f,g)|$$

We say that the sequence is ϕ-mixing if $\phi_m \to 0$ as $m \to \infty$ and we say that the sequence is ρ-mixing if $\rho_m \to 0$ as $m \to \infty$. According to Iosifescu Theodorescu (1969), Lemma 1.1.7,

$$\rho_m \leq 2\, \phi_m^{1/2} \quad \text{for every } m \geq 1,$$

so every ϕ-mixing sequence is ρ-mixing.

In the sequel we shall also use the following notations: $S_n = \sum_{i=1}^{n} X_i$, $M_n = \max_{1\leq i\leq n} |S_i|$, $[x]$ denotes the greatest integer smaller than x, << is used instead of Vinogradov symbol O. I(A) denotes the indicator function of the set A.

The aim of this paper is to establish the following:

Theorem 1.1. Let $\{X_n\}_n$ be a strictly stationary ϕ-mixing sequence of random variables, $\alpha p > 1$, $\alpha > 1/2$ and assume that $EX_1 = 0$ for $\alpha \le 1$.

Then the following statements are equivalent:

(a) $E|X_1|^p < \infty$

(b) $\sum_n n^{p\alpha-2} P(\max_{1\le i\le n} |S_i| > \varepsilon n^\alpha) < \infty$ for all $\varepsilon > 0$

(c) $\sum_n n^{p\alpha-2} P(\max_{1\le i\le n} |X_i| > \varepsilon n^\alpha) < \infty$ for all $\varepsilon > 0$

(d) $\sum_n n^{p\alpha-2} P(\max_{i\ge n} |S_i|/i^\alpha > \varepsilon) < \infty$ for all $\varepsilon > 0$

Remark 1.2. We can formulate (d) in the terms of r-quick convergence (see Strassen (1967), and Lai (1976))

(d) $\Leftrightarrow$ $S_n/n^\alpha \to 0$ $(p\alpha-1)$ quickly

Remark 1.3. Under the independence assumption one can add to the four equivalent statements another one

(e) $\sum_n n^{p\alpha-2} P(|S_n| > \varepsilon n^\alpha) < \infty$ for all $\varepsilon > 0$.

Because $P(|S_n| > \varepsilon n^\alpha) \le P(\max_{1\le i\le n} |S_i| > \varepsilon n^\alpha)$, (e) is implied by (b) and therefore is a consequence of (a). In the opposite direction we can prove that (e) together with an assumption imposed to the size of the first mixing coefficient implies (a). More precisely (a) is implied by either one of the following relations:

(e′) $\sum_n n^{p\alpha-2} P(|S_n| > \varepsilon n^{\alpha}) < \infty$ for all $\varepsilon > 0$

and in addition $\phi_1 < 1/4$.

(e″) $\sum_n n^{p\alpha-2} \max_{1\le i\le n} P(|S_i| > \varepsilon n^{\alpha}) < \infty$ for all $\varepsilon > 0$

and in addition $\phi_1 < 1$.

2. Preliminary results.

The proof of the Theorem 1.1 requires some preliminary results on the behaviour of the partial sums of a ϕ-mixing sequence and of the maximum summand, as well as the estimation of the moments of the partial sums. The results we present here are useful for dealing with the truncated random variables.

The following lemma is a precisation of Lemma (3.4) in Peligrad (1982). For related results, see also Bradley (1981).

Lemma 2.1. Assume $\{X_n\}_n$ is a sequence of centered random variables such that $EX_i^2 < \infty$ for every $1 \le i \le n$. Then

$$ES_n^2 \le d_n \cdot n \cdot \max_{1\le i\le n} EX_i^2$$

where $d_n = 8000 \exp\left(3 \sum_{i=1}^{[\log_2 n]} \rho_n(2^i)\right)$ □

Next lemma lists two maximal inequalities for partial sums. They are variants of Lemma 3.1 and relation (3.7) in Peligrad (1985).

Lemma 2.2. Let $\{X_n\}_n$ be a sequence of random variables. Suppose for some $a > 0$ and $q \in N$

$$\phi_q + \max_{1\le i\le n} P(|S_n - S_i| > a/2) \le \eta < 1$$

Then for every $x > a$ and $n > q$ the following relations hold:

$$P(M_n > 3x) \le (1-\eta)^{-1} P(|S_n| > 2x) + (1-\eta)^{-1} P(\max_{1\le i\le n} |X_i| > x/2(q-1)) \tag{2.1}$$

and

$$P(|S_n| > 2x) \le \eta P(M_n > x) + P(\max_{1\le i\le n} |X_i| > x/2q) \tag{2.2}$$

The following lemma shows that (2.1) simplifies under certain conditions.

Lemma 2.3. Let $\{X_k\}_k$ be a sequence of random variables such that $\phi_1 < 1/4$. Let $\{X_k^1\}_k$ be an independent copy of $\{X_k\}_k$ and consider the symmetrized sequence $\{\tilde{X}_k\}_k$ such that for each k, $\tilde{X}_k = X_k - X_k^1$. Denote $\tilde{S}_n = \sum_{i=1}^{n} \tilde{X}_i$ and $\tilde{M}_n = \max_{1\le i\le n} |\tilde{S}_i|$. Then, for every $x > 0$

$$P(\tilde{M}_n > x) \le 2(1/2 - 2\phi_1)^{-1} P(|\tilde{S}_n| > x).$$

Proof. According To Theorem 3.2 in Bradley (1986) the ϕ-mixing coefficients for $\{\tilde{X}_k\}_k$, which we denote by $\{\tilde{\phi}_k\}_k$, cannot exceed twice the size of the ϕ-mixing coefficients for $\{X_k\}_k$. So $\tilde{\phi}_1 \le 2\phi_1$. Now the proof follows step by step the proof of Levy's inequality (see Loève, C. P. Lèvy Inequalities, p. 259). The only modification is that instead of independence is used the definition of $\tilde{\phi}_1$.

A combination of the relations (2.1) and (2.2) gives by Tchebyshev's inequality:

<u>Lemma 2.4.</u> Consider the conditions of Lemma 2.1 fulfilled and denote by $B_n^2 = d_n \cdot n \cdot \max_{1\le i\le n} EX_i^2$. Consider there is an integer q and a positive A such that $\eta = \phi_q + 4A^{-2} < 1$. Then for every $x > AB_m$ and every $m \ge q$,

$$P(M_m > 3x) \le \eta(1-\eta)^{-1} P(M_m > x) + 2(1-\eta)^{-1} P(\max_{1\le i\le m} |X_i| > x/2q) \tag{2.3}$$

□

In the following we estimate the moments of the partial sums. The notations are the same as in the Lemmas 2.1 and 2.4.

<u>Lemma 2.5.</u> Let $\{X_n\}_n$ be a ϕ-mixing sequence of random variables, centered, and suppose for k real, $k \ge 2$, $E|X_i|^k < \infty$ for each $i \ge 1$. Then, for every $n \ge q$,

$$EM_n^k \le C_1(n\cdot d_n \cdot \max_{1\le i\le n} EX_i^2)^{k/2} + C_2 E \max_{1\le i\le n} |X_i|^k \tag{2.4}$$

where C_1 and C_2 can be taken to be

$$C_1 = (1-3^k\eta(1-\eta)^{-1})^{-1} 3^k A^k$$

$$C_2 = 2((1-\eta)-3^k\eta)^{-1} (6q)^k$$

Here A and q are chosen such that $1 - \eta > 0$ and $1 - 3^k\eta(1-\eta)^{-1} > 0$. (Recall $\eta = \phi_q + 4A^{-2}$.)

<u>Proof</u>. It is easy to see we have:

$$EM_n^k = \int_0^\infty kx^{k-1} P(M_n > x)dx \le$$

$$3^k(A^kB_n^k + \int_{AB_n}^\infty kx^{k-1} P(M_n > 3x)dx)$$

Using now the estimate for $P(M_n > 3x)$ provided by Lemma 2.2 we get

$$EM_n^k \le 3^k(A^kB_n^k + \eta(1-\eta)^{-1} EM_n^k +$$

$$2(1-\eta)^{-1}(2q)^k E \max_{1\le i\le n} |X_i|^k)$$

This relation gives the result of this lemma with the above mentioned constants. □

For the ϕ-mixing sequences, the maximum up to n, has a similar behavior to an independent sequence.

Lemma 2.6. Let $\{X_n\}_n$ be a sequence of random variables. Then for every n:

$$(1-\phi_1)\, P(\max_{1\le i\le n} X_i^* \ge x) \le P(\max_{1\le i\le n} X_i \ge x) \le (1+\phi_1)\, P(\max_{1\le i\le n} X_i^* \ge x)$$

where $\{X_i^*\}$ is a sequence of independent random variables such that for each i, X_i^* is distributed as X_i.

Proof. The proof of this lemma for the stationary case can be found in Peligrad (1988). The extension to the nonstationary case is immediate. Alternatively, one can use the proof of Lemma 3 of O'Brien (1974) by taking there $A_i = P(X_i \le x)$ □

For the situation when we know only that $\lim \phi_n < 1$, the quantity $1 - \phi_1$ can be 0. So for such situations we need the following precisation: For every $p \ge 1$ and $n \ge p$

$$(1-\phi_p)\, P(\max_{1\le k\le [n/p]} X_{kp}^* > x) \le P(\max_{1\le i\le n} X_i > x)$$
$$\le (1+\phi_p) \sum_{i=1}^{p} P(\max_{0\le k\le [n/p]} X_{kp+i}^* > x) \tag{2.5}$$

The proof of (2.5) follows by Lemma 2.6 and the relation

$$P(\max_{1\le k\le [n/p]} X_{kp} > x) \le P(\max_{1\le i\le n} X_i > x) \le \sum_{i=1}^{p} P(\max_{0\le k\le [n/p]} X_{kp+i} > x)$$

Proof of Theorem 1.1.

Some implications are immediate or are similar to facts proved elsewhere. In order to prove (a) $\Leftrightarrow$ (c) we choose and fix $p \geq 1$ such that $\phi_p < 1$. According to (2.5), because the sequence is stationary,

$$(1-\phi_p)\, P(\max_{1\le k\le [n/p]} |X_k^*| > \varepsilon n^{\alpha}) \le P(\max_{1\le i\le n} |X_i| \ge \varepsilon n^{\alpha})$$

$$\le p(1+\phi_p)\, P(\max_{1\le k\le [n/p]+1} |X_k^*| > \varepsilon n^{\alpha})$$

where $(X_k^*)_k$ is an i.i.d. sequence, X_1^* being distributed as X_1. Now (a) $\Leftrightarrow$ (c) is reduced to the i.i.d. case and follows from Theorem 3, Baum and Katz (1965). (b) $\Rightarrow$ (d) by Lemma 4 of Lai (1977) and (d) $\Rightarrow$(a) by Lemmas 4 and 5 of Lai (1977).

Obviously (b) $\Rightarrow$ (c) because

$$P(\max_{1\le i\le n} |X_i| > \varepsilon n^{\alpha}) \le P(\max_{1\le i\le n} |S_i| > \varepsilon n^{\alpha}/2).$$

We prove now that (a) $\Rightarrow$ (b).

We consider only the case $p \geq 1$ because the case $p < 1$ was established in Peligrad (1985*) without mixing assumptions. Without loss of generality we assume that the random variables are centered at expectations for all $\alpha > 1/2$ because for $\alpha > 1$ and n large enough we have:

$$P(M_n > \varepsilon n^{\alpha}) \le P(\max_{1\le i\le n} |\sum_{j=1}^{i} (X_j - EX_j)| > \varepsilon n^{\alpha} - nE|X_1|)$$

$$\leq P\left(\max_{1\leq i\leq n} \left|\sum_{j=1}^{i} (X_j - EX_j)\right| > (\varepsilon/2)n^{\alpha}\right)$$

Define

$$X_i^n(1) = X_i I(|X_i| > n^{\alpha}) - EX_i I(|X_i| > n^{\alpha})$$

and

$$X_i^n(2) = X_i I(|X_i| \leq n^{\alpha}) - EX_i I(|X_i| \leq n^{\alpha})$$

Put $S_n(j) = \sum_{i=1}^{n} X_i^n(j)$ and $M_n(j) = \max_{1\leq k\leq n} |S_k(j)|$ for $j = 1, 2$.

We note that

$$\sum_n n^{p\alpha-2} P(M_n > \varepsilon n^{\alpha}) \leq \sum_{j=1}^{2} \sum_n n^{p\alpha-2} P(M_n(j) > \varepsilon n^{\alpha}/2).$$

Let us denote by $b_k = P(k \leq |X_1| < k + 1)$ and we note that

$$E|X_1|^p < \infty \iff \sum_k k^p b_k < \infty. \tag{2.6}$$

We have successively, after changing the order of summation,

$$\sum_n n^{p\alpha-2} P(M_n(1) > \varepsilon n^{\alpha}) \leq$$

$$\varepsilon^{-1} \sum_n n^{p\alpha-\alpha-1} E|X_1^n(1)| \leq 2\varepsilon^{-1} \sum_n n^{p\alpha-\alpha-1} E|X_1| I(|X_1| > n^{\alpha}) \leq$$

$$2\varepsilon^{-1} \sum_n n^{p\alpha-\alpha-1} \sum_{k\geq n^{\alpha}-1} (k+1) b_k \ll \sum_k (k+1)^p b_k,$$

which is finite by (2.6).

Let now k be an integer such that k = 3 if $p < 2$ and $k > (p\alpha-1)/(\alpha-1/2)$ if $p \geq 2$.

By taking into account that $X_j^{(n)}(2)$ are centered, by Tchebyshev's inequality and then by Lemma 2.5, we get:

$$\sum_n n^{p\alpha-2} P(M_n(2) > \varepsilon n^{\alpha}) \ll \sum_n n^{p\alpha-2-\alpha k} E((M_n(2))^k$$

$$\ll \sum_n n^{p\alpha-2-\alpha k} (n\, d_n\, EX_1^2(2))^{k/2} +$$

$$\sum_n n^{p\alpha-\alpha k-1} E|X_1(2)|^k = I + II$$

Now, it is easy to see that, after changing the order of summation, we have:

$$II \ll \sum_n n^{p\alpha-k\alpha-1} \sum_{j \leq n^{\alpha}} j^k\, b_j \ll \sum_j j^p b_j$$

$$\ll E|X_1|^p$$

Now we have

$$I \ll \sum_n n^{p\alpha-2-\alpha k+k/2} \quad d_n^{k/2} (E\, X_1^2(2))^{k/2}$$

Because d_n is a function which is slowly varying as $n \to \infty$, and because when $p \geq 2$, $E\, X_1^2(2) < \infty$, the series converges for our choice of $k > (p\alpha-1)/(\alpha-1/2)$. When $p < 2$, then $E\, X_1^2(2) \ll n^{\alpha(2-p)} E|X_1|^p$ and the series is convergent for k = 3.

Proof of the Remark 1.3. First we prove that (e') implies (a). As in the Lemma 2.3 we consider $\{\tilde{X}_k\}_k$ a symmetrization of $\{X_k\}_k$. By the weak symmetrization inequality (see Loève, p. 257). (e') implies

$$\sum_n n^{p\alpha-2} P(|\tilde{S}_n| > \varepsilon\ n^{\alpha}) < \infty \quad \text{for all } \varepsilon > 0$$

whence by Lemma 2.3

$$\sum_n n^{p\alpha-2} P(\tilde{M}_n > \varepsilon\ n^{\alpha}) < \infty \text{ for all } \varepsilon > 0.$$

Now, by the implication (b) ⇒ (a) of Theorem 1.1, we get $E|\tilde{X}_1|^p < \infty$ and therefore we have $E|X_1|^p < \infty$.

Now we prove that (e") implies (a). We note first that (e") implies

$\max_{1\le i\le n} P(|S_i| > \varepsilon n^{\alpha}) \to 0$ as $n \to \infty$ for every $\varepsilon \ge 0$, because $p\alpha$ is strictly larger than 1 and

$$n^{p\alpha-1} \max_{1\le j\le n} P(|S_j| > \varepsilon n^{\alpha}) << \sum_{k=n}^{2n} k^{p\alpha-2} \max_{1\le j\le k} P(|S_j| > \varepsilon (k/2)^{\alpha}) \to 0$$

as $n \to \infty$ for every $\varepsilon > 0$.

By Lemma 1.1.6 in Iosifescu Theodorescu (1969)

$$P(M_n > \varepsilon n^{\alpha}) \le (1-\phi_1 - \max_{1\le k\le n} P(|S_k| > \varepsilon n^{\alpha}/2))^{-1} P(|S_n| > \varepsilon n^{\alpha}/2)$$

and by (e") and by the above consideration we get $\sum_n n^{p\alpha-2} P(M_n > \varepsilon n^{\alpha}) < \infty$ for every $\varepsilon > 0$. By Theorem 1.1, (a) follows.

Acknowledgement. I would like to express my gratitude to the referee for pointing out some gaps in a previous form of this paper.

References

Baum, L.E., Katz, M., Convergence rates in the law of large numbers. Trans. Amer. Math. Soc. 120, 108-123 (1965).

Bradley, R., A sufficient condition for linear growth of variances in a stationary random sequence, Proc. Amer. Math. Soc., 83, 586-589 (1981).

Bradley, R., Basic properties of strong mixing conditions, in Dependence in Prob. and Stat.; Progress in Prob. and Stat., Birkhäuser, 165-193, (1986).

Berbee, H.C.P., Convergence rates in the strong law for bounded mixing sequences, Prob. Theory and Rel. Fields 74, 255-270 (1987).

Chow, Y.S. and Lai, T.L., Some one-sided theorems on the tail distribution of sample sums with applications to the last time and largest excess of boundary crossings, Trans. Amer. Math. Soc. 208, 51-72 (1975).

Hipp, C., Convergence rates of the strong law for stationary mixing sequences, Z. Wahr. Verw. Geb. 49, 49-62 (1979).

Iosifescu, M. and Theodorescu, R., Random processes and learning, New York, Springer (1969).

Lai, T.L., On r-quick convergence and a conjecture of Strassen, Ann. Probab. 4, 612-627 (1976).

Lai, T.L., Convergence rates and r-quick version of the strong law for stationary mixing sequences, Ann. Probab. 5, 693-706 (1977).

Loève, M., Probability Theory 1, 4th Edition, Springer-Verlag (1977).

O'Brien, G.L., The maximum term of uniformly mixing stationary processes, Z. Wahr. Verw. Gebiete 30, 57-63 (1974).

Peligrad, M., Invariance principles for mixing sequences of random variables, Ann. Probab. 10, 968-981 (1982).

Peligrad, M., An invariance principle for ϕ-mixing sequences, Ann. Probab. 13, 1304-1313 (1985).

Peligrad, M., Convergence rates of the strong law for stationary mixing sequences, Z. Wahr. Verw. Gebiete 70, 307-314 (1985*).

Peligrad, M., On Ibragimov-Iosifescu conjecture for ϕ-mixing sequences (preprint, 1988).

Strassen, V., Almost sure behaviour of sums of independent random variables and martingales, Proc. Fifth Berkeley Symp. Math. Stat. Probab. 2, 315-343 (1967).

ALMOST EVERYWHERE CONVERGENCE OF SOME NONHOMOGENEOUS AVERAGES

Karl Petersen

IRMAR, Laboratoire de Probabilités
Université de Rennes I
35042 Rennes, France

Department of Mathematics
CB#3250, Phillips Hall
University of North Carolina
Chapel Hill, NC 27599 USA

1. **Introduction.** The most frequently encountered averages in ergodic theory are

$$\frac{1}{n}\sum_{k=0}^{n-1} f(T^k x) \qquad \text{and} \qquad \frac{1}{t}\int_0^t f(T_s x)\, dx,$$

for a measure-preserving transformation (m.p.t.) T or measure-preserving flow $\{T_s: -\infty<s<\infty\}$ on a measure space $(X,\mathfrak{B},\mu)$ and measurable function f on X. In the case of the Ergodic Theorem, one looks for limits as n or t tends to infinity, while the Local Ergodic Theorem deals with the limit as $t\to 0^+$. Here we wish to discuss two unusual kinds of averages:

$$H_n f(x) = \sum_{k=-n}^{n}{}' \frac{f(T^k x)}{k},$$

or, more generally,

$$H_{n,\theta} f(x) = \sum_{k=-n}^{n}{}' \frac{e^{ik\theta} f(T^k x)}{k},$$

ISBN 0-12-231050-0

where $-\pi \leq \theta < \pi$ and the primes indicate that the term $k=0$ is omitted; and

$$S_n f(x) = \frac{1}{n} \sum_{k=0}^{n-1} f(T_{G(k)} x),$$

with G a function that possibly takes non-integral values. We will concentrate on the example $G(k) = \sqrt{k}$, although most of what we have to say will apply equally as well to a wide class of Young's functions, including all p'th roots $p>1$. Combinations of the first kind are nonhomogeneous in that the weights assigned to the values of f at different points along the orbit of x are not all the same; here also cancellations caused by looking both backwards and forwards in time play an important role. Combinations of the second kind are nonhomogeneous in that the spacings between the sampling times $G(k)$ are not all the same; here growth and convexity properties of G are significant. Almost everywhere existence of the ergodic Hilbert transform $Hf(x) = \lim_{n\to\infty} H_n f(x)$ and rotated ergodic Hilbert transform $H_\theta f(x) = \lim_{n\to\infty} H_{n,\theta} f(x)$ are well known [6, 14]; we will discuss a sort of continuity property of the limit as $\theta \to 0$ and note the relationship of this property to the Carleson-Hunt Theorem on the a.e. convergence of the Fourier series of L^2 functions. A.e. convergence of combinations of the second kind has been proved only recently, and the connections of this result with the standard ergodic theorems are still not clear. We will find a condition necessary and sufficient for a flow of unitary operators on L^2 (perhaps not generated by a measure-preserving flow) to satisfy a pointwise ergodic theorem along the sequence of square roots of integers, and we will give examples of unitary flows which satisfy the pointwise ergodic theorem for all $f \in L^2$ and the local ergodic theorem for a fixed $f_0 \in L^2$ but which do not satisfy the pointwise ergodic theorem along square roots for this f_0. In the setting of unitary transformations or flows on L^2, one may regard both the continuity theorem for $H_\theta f$ and the ergodic theorem along square roots as strict strengthenings of the usual Pointwise Ergodic Theorem: unitary groups on L^2 that are generated by measure-preserving actions have some extra properties that are not shared by all unitary actions which satisfy the ergodic theorems.

Acknowledgements: I would like to thank Michael Taylor for help with the proof of Lemma 2.4, James Campbell for contributing to the formulation of Proposition 2.6, Vitaly Bergelson for suggesting Theorem 3.1, the University of Rennes I for hospitality while part of this work was underway, and CNRS and NSF (through grants DMS-8400730 and DMS-8620132) for financial support. Some of these ideas have been presented in more detail elsewhere, while the rest are in the process of development.

2. The Ergodic Theorem, spectral measures, Hilbert transforms, and Fourier series. Conditions necessary and sufficient for a unitary action (a single operator U or a flow $\{U_t\}$) to satisfy the Ergodic Theorem or Local Ergodic Theorem have been given by Gaposhkin, in terms of the spectral measure E of the action. We will use the same letter to denote the spectral measure either of an operator or of a flow. In the case of a single operator U, E is a projection-valued measure on the Borel sets of $[-\pi,\pi)$ such that

$$U = \int_{-\pi}^{\pi} e^{i\lambda} dE(\lambda);$$

for a flow $\{U_t\}$ the spectral measure E is defined for Borel subsets of $\mathbb{R}$, and

$$U_t = \int_{-\infty}^{\infty} e^{i\lambda t} dE(\lambda), \quad -\infty<t<\infty.$$

2.1. Theorem (Gaposhkin, [7-10]). (1) A unitary action on $L^2[-\pi,\pi)$ and function $f \in L^2[-\pi,\pi)$ satisfy the Pointwise Ergodic Theorem if and only if

$$E\{\lambda: \ 0<|\lambda|<2^{-r}\}f(x) \to 0 \quad \text{a.e.}$$

(2) A unitary flow on $L^2[-\pi,\pi)$ and function $f \in L^2[-\pi,\pi)$ satisfy the Local Ergodic Theorem if and only if

$$E\{\lambda: |\lambda|>2^r\}f(x) \to 0 \quad \text{a.e..}$$

Similar statements hold for normal operators and even functions of normal operators (see [11]), but we will focus on just the unitary case.

Since a unitary operator U that is generated by a measure-preserving (invertible) transformation T according to the formula $Uf = f \circ T$ is known to satisfy the Ergodic Theorem, part (1) of this theorem tells us something about the spectral measures of measure-preserving transformations. The questions naturally arise whether one can see directly that for the spectral measure E of a m.p.t. T we have $E\{\lambda: \ 0<|\lambda|<2^{-r}\}f(x) \to 0$ a.e. for all $f \in L^2$, and whether or not this property holds for all sequences $\epsilon_r \to 0$ and not just geometrically decreasing sequences such as 2^{-r}. These questions can be answered affirmatively by studying a continuity property of the rotated ergodic Hilbert transform. This property is proved by means of a lemma which is equivalent to the Carleson-Hunt strong 2,2 estimate for the supremum of the partial sums of the Fourier series of an L^2 function.

2.2. Proposition [4]. Let T: $X \rightarrow X$ be a m.p.t. on a measure space $(X, \mathfrak{B}, \mu)$, $f \in L^2(X, \mathfrak{B}, \mu)$, and $\epsilon_k \rightarrow 0^+$. Then $E\{\lambda: \ 0 < |\lambda| < \epsilon_k\} f(x) \rightarrow 0$ a.e. if and only if $H_{\epsilon_k} f(x) \rightarrow Hf(x) + i\pi E\{0\} f(x)$ a.e. .

Proof. We have

$$\sum_{k=-n}^{n}{}' \frac{f(T^k x)}{k} = \int_{-\pi}^{\pi} \sum_{k=-n}^{n}{}' \frac{e^{ik\lambda}}{k} \, dE(\lambda) \, f(x) = i \int_{-\pi}^{\pi} q_n(\lambda) \, dE(\lambda) \, f(x),$$

where

$$q_n(\lambda) = \frac{1}{i} \sum_{k=-n}^{n}{}' \frac{e^{ik\lambda}}{k}$$

is uniformly bounded on $[-\pi, \pi)$ and converges pointwise as $n \rightarrow \infty$ to

$$\eta(\lambda) = \begin{cases} \pi - \lambda & \text{if } 0 < \lambda \leq \pi \\ 0 & \text{if } \lambda = 0 \\ -(\pi + \lambda) & \text{if } -\pi \leq \lambda < 0 \, . \end{cases}$$

It follows that

$$Hf = \int_{-\pi}^{\pi} i\eta(\lambda) \, dE(\lambda) \, f \quad \text{and} \quad H_\epsilon f = \int_{-\pi}^{\pi} i\eta(\lambda + \epsilon) \, dE(\lambda) \, f,$$

so that

$$\frac{1}{i}[H_\epsilon f - Hf] = \int_{-\pi}^{\pi} [\eta(\lambda + \epsilon) - \eta(\lambda)] \, dE(\lambda) \, f$$

$$= -\epsilon f + \pi[E\{-\epsilon\} f + E\{0\} f] + 2\pi E(-\epsilon, 0) f,$$

from which the result is apparent.

2.3. Continuity Theorem [4]. Let T: $X \to X$ be a m.p.t. on a measure space $(X,\mathfrak{B},\mu)$, $f \in L^2(X,\mathfrak{B},\mu)$, and $\epsilon_k \to 0^+$. Then the following two conditions (equivalent by the preceding proposition) hold:

(1) $E\{\lambda:\ 0<|\lambda|<\epsilon_k\}f(x) \to 0$ a.e.;

(2) $H_{\epsilon_k}f(x) \to Hf(x) + i\pi E\{0\}f(x)$ a.e. .

A dense set in L^2 of functions for which the conclusion of this theorem holds is easily manufactured by forming the ranges of the projections $E\{|\lambda|>\epsilon\}$. The needed maximal inequality asserts weak 2,2 for the double maximal function

$$H^{**}f(x) = \sup_{n,\theta} \left| \sum_{k=-n}^{n}{}' \frac{e^{ik\theta} f(T^k x)}{k} \right| .$$

Using transference, it is enough to consider an analogous maximal operator on l^2, and when we consider a sequence in l^2 as the sequence of Fourier coefficients of a function $h \in L^2[-\pi,\pi)$ and subtract off some terms that are already known to be strong 2,2, we are left with proving the following estimate.

2.4. Harmonic Analysis Maximal Lemma [4]. For $h \in L^2[-\pi,\pi)$ and $j \in \mathbb{Z}$, let

$$I^* h(j) = \sup_{\epsilon>0} \left| \int_{-\epsilon}^{\epsilon} h(x)\, e^{-ijx}\, dx \right| .$$

Then there is a constant C such that

$$\|I^* h\|_{l^2(\mathbb{Z})} \le C\|h\|_{L^2[-\pi,\pi)} \quad \text{for all } h \in L^2[-\pi,\pi).$$

This can be recognized to be the analogue, on the other side of the Fourier transform, of the following result basic to the proof of a.e. convergence of Fourier series of L^2 functions.

2.5. Carleson-Hunt Maximal Lemma [5, 12]. For $\hat{h} \in l^2(\mathbb{Z})$ and $x \in [-\pi, \pi)$, let

$$S^* \hat{h}(x) = \sup_{n>0} \left| \sum_{j=-n}^{n} \hat{h}(j)\, e^{ijx} \right| .$$

Then there is a constant C such that

$$\|S^* \hat{h}\|_{L^2[-\pi,\pi)} \leq C \|\hat{h}\|_{l^2(\mathbb{Z})} \text{ for all } h \in L^2[-\pi,\pi).$$

In fact, each of these two lemmas can be used to prove the other fairly readily. Thus the proof of the Continuity Theorem is completed by invoking the Carleson-Hunt Lemma. Moreover, keeping in mind that it is enough to prove the Harmonic Analysis Lemma just for step functions constant on intervals of length $1/N$, one can formulate an elementary-looking equivalent that may be susceptible to proof by direct geometric and combinatorial arguments. (See [2] for another discrete equivalent of the Carleson-Hunt Lemma.)

2.6. Discrete Equivalent of Carleson-Hunt Lemma. For $N=1, 2, \ldots$ and $r=1, \ldots, N$ let $\omega_r = e^{2\pi ir/N}$. Then there is a constant C such that for any N, any $v_1, \ldots, v_N \geq 0$ and any choice of $N_r = 1, \ldots, N$ for $r=1, \ldots, N$,

$$\sum_{r=1}^{N} \frac{1}{r^2} \left| (1 - \omega_r) \sum_{m=1}^{N_r} v_m \omega_r^{\ m} \right|^2 \leq C \frac{1}{N} \sum_{m=1}^{N} v_m^{\ 2} .$$

3. **The Ergodic Theorem with sampling.** If along the orbits of a flow one uses a sequence of sampling times like $G(k)=\sqrt{k}$, then an a.e. convergence result for Cesaro averages will involve elements of both the Pointwise Ergodic Theorem and Local Ergodic Theorem, since $G(k)\to\infty$ but $G(k+1)-G(k)\to 0$ as $k\to\infty$; in this respect, theorems of this kind differ from subsequence ergodic theorems such as the statement for the sequence of squares recently proved by Bourgain [3]. (Nevertheless, it is possible that a.e. convergence might hold for all sampling functions G which have the right convexity properties and grow no slower nor faster than a power of k.) We note that for $G(k)=\lceil\sqrt{k}\rceil$ the result is trivial by partial summation, while for $G(k)=\log k$ a.e. convergence of the averages does not always hold, since the sequence $\log k$ is not correctly distributed mod 1 in any subinterval of [0, 1].

3.1. <u>Ergodic Theorem along Square Roots</u> [15]. Let $\{T_s\colon -\infty<s<\infty\}$ be a measure-preserving flow on a probability space $(X,\mathcal{B},\mu)$ and let $f\in L^1(X,\mathcal{B},\mu)$. Then

$$\lim_{n\to\infty}\frac{1}{n}\sum_{k=0}^{n-1} f(T_{\sqrt{k}}x) = \lim_{t\to\infty}\frac{1}{t}\int_0^t f(T_s x)\,ds \quad a.e..$$

In order to prove this theorem, we begin by writing

$$\frac{1}{n}\sum_{k=0}^{n-1} f(T_{\sqrt{k}}x) = \frac{f(x)}{n} + \frac{1}{n}\int_0^{n-1} f(T_{\sqrt{t}}x)\,dt$$

$$+\frac{1}{n}\int_0^{n-1} T_{\sqrt{t}}(T_{\sqrt{\lceil t\rceil}-\sqrt{t}}f - f)(x)\,dt\,.$$

The first term tends to 0, and change of variables and integration by parts show that the second tends to the usual limit in the Ergodic Theorem, so the question is whether the third term, which involves averaging along the orbit of x (in the manner of the Ergodic Theorem) an increasingly smaller quantity (of the kind encountered in the Local Ergodic Theorem), tends to 0 as $n \to \infty$. This clearly holds for functions f which for fixed x have $f(T_t x)$ uniformly continuous in t, a dense set in L^1, so again it is enough to prove a maximal estimate, such as the following one.

3.2. Maximal Lemma along Square Roots [15]. Let $\{T_s: -\infty<s<\infty\}$ be a measure-preserving flow on a probability space $(X,\mathcal{B},\mu)$. Let $\mathcal{D}$ be the subset of $L^1(X,\mathcal{B},\mu)$ consisting of all functions f of the form

$$f(x) = \int_{-\infty}^{\infty} g(T_s x)\, \phi(s)\, ds ,$$

where g is bounded and measurable on $(X,\mathcal{B},\mu)$ and ϕ is a continuous function of bounded variation on $\mathbb{R}$ with compact support. Then there is a constant C such that for all $f \in \mathcal{D}$ and all $\lambda>0$,

$$\mu\{x \in X: \sup_{m \geq 1} |\tfrac{1}{m} \sum_{k=0}^{m-1} f(T_{\sqrt{k}}x)| > \lambda\} \leq C \sqrt{\frac{\|f\|_1}{\lambda}} .$$

The maximal lemma and ergodic theorem along square roots both extend to p'th roots and indeed to a fairly wide class of Young's functions $G(k)$; see [15] for the details. Because of the use of Jensen's Inequality in the proof, we do not know whether this theorem will extend to flows on infinite measure spaces or flows of positive contractions on L^1. Also, we have not been able to prove the stronger maximal *theorem* (as opposed to maximal *lemma*) version, in which the integral on the right side is taken not over all of X but only over the set where the maximal function is greater than λ. If serious applications to the sampling of stationary processes, recurrence in measure-preserving flows, number theory, or properties of dynamical systems should emerge, further investigations such as the sampling variants of the ergodic Hilbert transform, the Wiener-Wintner Theorem, or the Chacon-Ornstein Theorem might be justified.

4. **Sampling for unitary flows.** It is natural to ask whether the ergodic theorem along square roots is anything new or whether it perhaps follows, say by changes of variables and integration by parts, from the Pointwise Ergodic Theorem and Local Ergodic Theorem. In this section we will describe several examples which show that this is not the case. We first construct (Example 4.1) a flow of unitary operators $\{U_t\}$ on $L^2[0, 1]$ which satisfies the Pointwise Ergodic Theorem for all $f \in L^2$ but such that the ergodic theorem along square roots fails for some $f_0 \in L^2$. With a bit more effort, this example can be modified in such a way that the Pointwise Ergodic Theorem holds for all $f \in L^2$ and the Local Ergodic Theorem also holds for a particular f_0, yet the ergodic theorem along square roots fails for that same f_0. We give a condition on the spectral measure of a flow, analogous to Gaposhkin's conditions, necessary and sufficient for the flow and a given $f \in L^2$ to satisfy the ergodic theorem along square roots. This condition is only partially satisfactory because it still involves a kernel, but it does help us to clarify further the relationships among these three theorems. In particular, we can use the condition to construct a flow

(Example 4.6) which satisfies both the Pointwise Ergodic Theorem and the ergodic theorem along square roots for all $f \in L^2$, but for which the Local Ergodic Theorem fails for some $f_0 \in L^2$.

4.1. Example. All our examples will be discrete-spectrum unitary flows on $L^2[0, 1]$ defined by choosing an orthonormal system $\{\phi_n\}$ and set of eigenvalues $\{\lambda_n\}$ and letting, for any square-summable sequence $\{c_n\}$,

$$U_t \sum_{n=1}^{\infty} c_n \phi_n = \sum_{n=1}^{\infty} c_n e^{i\lambda_n t} \phi_n , \quad -\infty < t < \infty .$$

(If $\{\phi_n\}$ is not complete, the flow is defined only on a subspace.) Typically, $\{\phi_n\}$ will be a system of divergence as in the Menchov-Rademacher theorem, in that an expansion $\sum c_n \phi_n$ with square-summable coeffcients need not converge a.e. if $\sum c_n{}^2 \log^2 n = \infty$ (see [1]).

For this first example, let $\{\phi_n\}$ be such a system of divergence and $\{c_n\}$ a square-summable sequence of coefficients for which the associated othogonal series does not converge a.e.. Increasing sequences $\{\lambda_j\}$ of real numbers and $\{n_j\}$ of positive integers are chosen inductively as follows. If n_s and λ_s have been chosen for $s<j$, choose $\lambda_j > j$ so that

$$\left| e^{i\lambda_j \sqrt{s}} - 1 \right| < \frac{1}{j} \quad \text{for all } s=1, \ldots , n_{j-1}.$$

That this is possible follows from known facts about simultaneous Diophantine approximation; it is seen very easily by noting that the orbit of the vector 0 in the n_{j-1}-torus is recurrent under the almost periodic flow $T_t(x) = x + t\,(\sqrt{1}, \sqrt{2}, \ldots , \sqrt{n_{j-1}})$. Then choose n_j so that

$$\left|\frac{1}{n_j}\sum_{s=1}^{n_j} e^{i\lambda_k\sqrt{s}}\right| < \frac{1}{j} \quad \text{for all } k=1, \ldots, j.$$

This is possible because for each fixed k the sequence $\{\lambda_k\sqrt{s}\}$ is equidistributed mod 1 and so satisfies the Weyl criterion; in fact, the left side can be bounded by by a constant times $|\lambda_k|/\sqrt{n_j}$ (see [13], p. 14).

If $f=\sum_{k=1}^{\infty} c_k\phi_k$, then

$$S_{n_j}f = \frac{1}{n_j}\sum_{s=1}^{n_j} U_{\sqrt{s}}f = \sum_{k=1}^{j} c_k\left[\frac{1}{n_j}\sum_{s=1}^{n_j} e^{i\lambda_k\sqrt{s}}\right]\phi_k +$$

$$\sum_{k=j+1}^{\infty} c_k\left[\frac{1}{n_j}\sum_{s=1}^{n_j} e^{i\lambda_k\sqrt{s}}\right]\phi_k,$$

and in the first term, α_j, the expression inside the brackets is always within $1/j$ of 0, while in the second, β_j, it is always within $1/j$ of 1. Thus

$$\sum_j\|\alpha_j\|^2 \leq \sum_j\sum_{k=1}^{j} |c_k|^2\frac{1}{j^2} \leq C\|f\|_2^2 < \infty,$$

so that $\alpha_j \rightarrow 0$ a.e.. If

$$\gamma_j = \sum_{k=j+1}^{\infty} c_k\phi_k \quad ,$$

then $\sum\|\beta_j-\gamma_j\|_2^2 < \infty$, so that $\beta_j-\gamma_j\rightarrow 0$ a.e.. Since $\{\gamma_j\}$ diverges a.e., so does $\{\beta_j\}$ and hence also $\{S_{n_j}f\}$.

This flow satisfies the Pointwise Ergodic Theorem because clearly it fulfills condition (1) of the theorem of Gaposhkin. One may also compute directly that

$$\frac{1}{t}\int_0^t U_s \sum_{k=1}^{\infty} c_k \phi_k \, ds = \sum_{k=1}^{\infty} c_k \frac{1}{t}\int_0^t e^{i\lambda_k s}\, ds\, \phi_k$$

$$= \frac{1}{t}\sum_{k=1}^{\infty} c_k \frac{e^{i\lambda_k t} - 1}{i\lambda_k} \phi_k ,$$

so that if

$$b_k = c_k \frac{e^{i\lambda_k t} - 1}{i\lambda_k} ,$$

then $\sum |b_k|^2 \log^2 k < \infty$ and hence, by the Menchov-Rademacher Theorem, $\sum b_k \phi_k$ converges a.e. and direct estimates show that $\frac{1}{t} \sum b_k \phi_k \to 0$ a.e. as $t \to \infty$.

4.2. Example. By modifying Example 4.1 somewhat, one can arrange to have the Pointwise Ergodic Theorem hold for all f, the Local Ergodic Theorem hold for a particular f_0, but the square root theorem fail for that same f_0. The idea is to add to the spectral measure of $\{U_t\}$ point masses at $\nu_j \in \mathbf{R}$ which are projections onto orthonormal functions ψ_j in such a way that divergence of the root averages is not affected but Gaposhkin's condition $E\{\lambda: |\lambda| > 2^r\} f(x) \to 0$ a.e. for the Local Ergodic Theorem becomes satisfied.

First rename the ϕ_k of Example 4.1 as u_k. Extend them to [-1, 1] as even functions and normalize. Let $v_k(x) = (\sin \pi k x)/\sqrt{\pi}$, so that the v_k are orthonormal on [-1, 1] and $v_k \perp u_k$ for all k. Let $\phi_k = (u_k + v_k)/\sqrt{2}$ and $\psi_k = (u_k - v_k)/\sqrt{2}$. Define

$$U_t \sum_{n=1}^{\infty} (a_n \phi_n + b_n \psi_n) = \sum_{n=1}^{\infty} (a_n e^{i\lambda_n t} \phi_n + b_n e^{i\nu_n t} \psi_n),$$

where the ν_k, λ_k, and n_j are chosen subject to the following conditions:

$\lambda_k < \nu_k < \lambda_{k+1}$;

$2\lambda_k < \nu_k$, $2\nu_k < \lambda_{k+1}$;

$n_j > j^5$, $n_{j+1} - n_j$ is increasing, and $\left| \frac{1}{n_j} \sum_{s=1}^{n_j} e^{i\nu_k \sqrt{s}} \right| < \frac{1}{j}$ for all k;

n_j and λ_j still satisfy the conditions of Example 4.1.

To see that this is possible, we need a quantitative statement about recurrence of flows on the torus more precise than the one used in Example 1.

4.3. Block Recurrence Lemma. Let $l_1, \ldots, l_k$ be positive integers and $n = l_1 + \ldots + l_k$. Let $\alpha_1, \ldots, \alpha_n$ be distinct elements of $[0, 1)$, let $\alpha = (\alpha_1, \ldots, \alpha_n)$, and let μ be the unique invariant measure on the orbit closure of 0 in the flow $T_t x = x + t\alpha$ (mod 1 coordinatewise) on the n-torus. Let $l_0 = 0$, and for each $r = 1, \ldots, k$ let

$$f_r(x) = \sum_{j=l_0+\ldots+l_{r-1}+1}^{l_1+\ldots+l_r} e^{ix_j}$$

and, for $u \geq 0$, $A_r(u) = \{x \in X: |f_r(x)| < \sqrt{l_r + u}\}$. Then there is an absolute constant c such that $\mu(A_1(c) \cap \ldots \cap A_k(c)) > 0$ (and therefore the orbit of 0 visits $\bigcap_r A_r(c)$ infinitely many times).

This lemma is proved by averaging $|f_r(x)|^2$ along the orbit of 0, applying the Ergodic Theorem, and using induction on r. We use it to find arbirarily large ν_k satisfying

$$\left|\frac{1}{n_j}\sum_{s=1}^{n_j} e^{i\nu_k\sqrt{s}}\right| < \frac{1}{j} \text{ for all } k\geq j\ .$$

Now if the c_n are as before and $f=\sum(c_n\phi_n - c_n\psi_n)$, then f still fails to satisfy the ergodic theorem along square roots, while all L^2 functions continue to satisfy the Pointwise Ergodic Theorem. Also,

$$E(2^n, \infty)f = \sum_{k>k(n)} c_k v_k \ \pm \text{ possibly one term,}$$

which tends to 0 a.e. because $\{v_k\}$ is a system of convergence, so that in this situation f also satisfies the Local Ergodic Theorem.

The relationships among the Pointwise Ergodic Theorem, Local Ergodic Theorem, and ergodic theorem along square roots for unitary flows on L^2 can be further clarified by establishing a condition on the spectral measure of the flow, analogous to those given by Gaposhkin, necessary and sufficient for it to satisfy the ergodic theorem along square roots. This task is complicated by the fact that the kernel involved,

$$K_n(\lambda) = \frac{1}{n}\sum_{s=1}^{n} e^{i\lambda\sqrt{s}} \ ,$$

is an almost periodic function of $\lambda\in\mathbf{R}$, unlike the 2π-periodic function of λ that results from erasing the square roots which one encounters when considering the Ergodic Theorem. Still, this kernel does satisfy useful estimates such as (for $2^r \leq n_1 < n_2 < 2^{r+1}$)

$$|K_n(\lambda)| \leq \frac{C}{\sqrt{n}}\left[|\lambda| + \frac{1}{|\lambda|}\right]$$

and

$$\left|K_{n_1}(\lambda) - K_{n_2}(\lambda)\right| \leq C\max\left\{|\lambda|\sqrt{n_1}\left[\frac{1}{n_1} - \frac{1}{n_2}\right], \frac{1}{|\lambda|\sqrt{n_2}}\right\},$$

$$\left|K_{n_1}(\lambda) - K_{n_2}(\lambda)\right| \leq 2\frac{n_2 - n_1}{n_2},$$

which allow one to employ piecewise estimates and dyadic decompositions in the manner of Gaposhkin to establish the following result.

4.4. Spectral Condition for Ergodic Theorem along Square Roots. The unitary flow $\{U_t\}$ and function $f \in L^2$ satisfy the pointwise ergodic theorem along square roots if and only if

$$E\{0<|\lambda|<2^{-r/2}\}f + \int_{\{|\lambda|>2^{r+1},\ |K_m(\lambda)|\geq C2^{-r/4}\}} K_m(\lambda)\,dE(\lambda)f \to 0$$

a.e. as $m \to \infty$, $2^r \leq m < 2^{r+1}$.

4.5. Corollary. If for the unitary flow $\{U_t\}$ the ergodic theorem along square roots holds for all $f \in L^2$, then the Pointwise Ergodic Theorem holds for all $f \in L^2$.

4.6. Example. By using the above condition on the spectral measure, we can construct an example of a unitary flow for which the pointwise ergodic theorem along square roots as well as the usual Pointwise Ergodic Theorem for the flow are both satisfied for all $f \in L^2$, yet the Local Ergodic Theorem fails for some $f_0 \in L^2$. If we do not introduce any spectral mass in the interval (-1/2, 1/2), then the Pointwise Ergodic Theorem will always be satisfied, and the first term in the preceding theorem will always be 0. We can also keep the set over which the integral in the second term is taken always empty by placing point masses only at points $\lambda_n \in (2^{n+1}, 2^{n+2}]$ such that $|K_m(\lambda_n)| < C \cdot 2^{-r/4}$ for $1 \leq r \leq n$ and $2^r < m \leq 2^{r+1}$. (The existence of such points is proved again by a mean-value argument, this time combined with the piecewise estimate-dyadic decomposition technique). If the spectral measure E has projection onto ϕ_k at λ_k, then

$$E(2^{n+1}, \infty) f = \sum_{k=n}^{\infty} c_k \phi_k \quad ,$$

which need not converge to 0 a.e. if $\{\phi_n\}$ is a system of divergence, so the Local Ergodic Theorem will not hold for all $f \in L^2$.

These results taken together tend to show that the spectral measures of measure-preserving transformations and flows have some special smoothness properties. It may be possible to prove many almost everywhere convergence theorems for measure-preserving actions by a two-step method, in which one first establishes a condition on the spectral measure of a unitary action necessary and sufficient for the convergence result to hold for all $f \in L^2$ and then shows that all measure-preserving actions satisfy this condition. This second step might even follow in a variety of situations from a single strengthening of Theorem 2.3.

REFERENCES

1. G. Alexits, *Convergence Problems of Orthogonal Series*, Pergamon Press, Oxford and New York, 1961.

2. B. M. Baishanski, On Carleson's convergence theorem for L^2 functions, *Harmonic Analysis in Euclidean Spaces*, Proc. Symp. Pure Math. 35, Part 1, A.M.S., Providence, R.I., 1979, 167-170.

3. J. Bourgain, On the Maximal Ergodic Theorem for certain subsets of the integers, preprint.

4. J. Campbell and K. Petersen, The spectral measure and Hilbert transform of a measure-preserving transformation, to appear in Trans. Amer. Math. Soc.

5. L. Carleson, On convergence and growth of partial sums of Fourier series, Acta Math. 116 (1966), 135-157.

6. M. Cotlar, A unified theory of Hilbert transforms and ergodic theorems, Rev. Mat. Cuyana 1 (1955), 105-167.

7. V. F. Gaposhkin, Criteria for the strong law of large numbers for some classes of second-order stationary processes and homogeneous random fields, Theory Prob. Appls. 22 (1977), 286-310.

8. V. F. Gaposhkin, A theorem on the convergence almost everywhere of a sequence of measurable functions, and its applications to sequences of stochastic integrals, Math. USSR Sbornik 33 (1977), 1-17.

9. V. F. Gaposhkin, The Local Ergodic Theorem for groups of unitary operators and second order stationary processes, Math. USSR Sbornik 39 (1981), 227-242.

10. V. F. Gaposhkin, Individual Ergodic Theorem for normal operators in L_2, Funct. Anal. Appls. 15 (1981), 14-18.

11. V. F. Gaposhkin, An ergodic theorem for functions of normal operators, Funct. Anal. Appls. 18 (1984), 1-5.

12. R. Hunt, On the convergence of Fourier series, *Orthogonal Expansions and their Continuous Analogues*, D.T. Haimo, ed., S. Ill. Univ. Press, Carbondale, 1968, 235-255.

13. L. Kuipers and H. Niederreiter, *Uniform Distribution of Sequences*, John Wiley & Sons, New York, 1974.

14. K. Petersen, Another proof of the existence of the ergodic Hilbert transform, Proc. Amer. Math. Soc. 88 (1983), 39-43.

15. K. Petersen, The Ergodic Theorem with time compression, to appear in Journal d'Analyse Math.

SIMPLE SYMMETRIC RANDOM WALK IN Z^d

P. Révész

Technical University of Vienna, Institute of Statistics

Vienna, Austria

1 Introduction

Let $X_1, X_2, \ldots$ be a sequence of independent, indentically distributed random vectors taking values from Z^2 with distribution

$$\mathbf{P}\{X_1 = (0,1)\} = \mathbf{P}\{X_1 = (0,-1)\} = \mathbf{P}\{X_1 = (1,0)\} = \mathbf{P}\{X_1 = (-1,0)\} = 1/4$$

and let

$$S_0 = 0 = (0,0) \text{ and } S(n) = S_n = X_1 + X_2 + \cdots + X_n \quad (n = 1, 2, \ldots)$$

i.e. $\{S_n\}$ is the simple symmetric random walk on the plane. Further let

$$\xi(x,n) = \#\{k : 0 \leq k \leq n, S_k = x\}$$

$(n = 1, 2, \ldots; x = (i,j); i,j = 0, \pm 1, \pm 2, \ldots)$ be the local time of the random walk. We say that the disc

$$Q(N) = \{x = (i,j) : \|x\| = (i^2 + j^2)^{1/2} \leq N\}$$

is covered by the random walk in time n if

$$\xi(x,n) > 0 \text{ for every } x \in Q(N).$$

ISBN 0-12-231050-0

Let $R(n)$ be the largest integer for which $Q(R(n))$ is covered in n.

We have proved:

THEOREM A (Erdős - Révész, 1988) *For any* $\varepsilon > 0$ *we have*

$$R(n) \geq \exp\left(\frac{(\log n)^{1/2}}{(\log\log n)^{3/4+\varepsilon}}\right) \ a.s.$$

for all but finitely many n.

In the present paper we prove:

THEOREM 1

$$R(n) \leq \exp(2(\log n)^{1/2} \log\log\log n) \ a.s.$$

for all but finitely many n.

In some sense the following theorem also shows that Theorem A is not far from the best possible one.

THEOREM 2 *For any* $C > 0$ *we have*

$$\mathbf{P}\{R(n) \leq \exp(C(\log n)^{1/2})\} > 1 - \exp(-\frac{C^2}{4}).$$

Consequently for any $C > 0$

$$R(n) \leq \exp(C(\log n)^{1/2}) \ i.o. \ a.s.$$

Theorem 2 suggests the following

CONJECTURE. There exists a distribution function $H(x)$ with $H(0) = 0$ for which

$$\lim_{n\to\infty} \mathbf{P}\{\frac{\log R(n)}{(\log n)^{1/2}} < x\} = H(x).$$

We also ask about the largest integer $\tilde{R}(n)$ for which the disc $Q(\tilde{R}(n))$ is "homogeneously" covered in n. We prove

THEOREM 3 *Let* $\tilde{R}(n) = (\log n)^K$. *Then for any* $K > 0$ *we have*

$$\lim_{n\to\infty} \sup_{x \in Q(\tilde{R}(n))} \left| \frac{\xi(x,n)}{\xi(0,n)} - 1 \right| = 0 \ a.s.$$

The limit behaviour of $\xi(0,n)$ was investigated by Erdős - Taylor (1960). They proved:

THEOREM B

$$\lim_{n\to\infty} \mathbf{P}\{\xi(0,n) < u\log n\} = 1 - e^{-\pi u} \quad (u > 0)$$

and for any $\varepsilon > 0$

$$\frac{\log n}{(\log\log n)^{1+\varepsilon}} \le \xi(0,n) \le (1+\varepsilon)\log n \log\log\log n \ a.s.$$

for all but finitely many n.

Theorems 3 and B imply that every point of the disc $Q(\tilde{R}(n))(\tilde{R}(n) = (\log n)^K)$ will be covered at least $\log n(\log\log n)^{-1-\varepsilon}$ times.

Very likely this latter statement is true for a much bigger $\tilde{R}(n)$ as well.

It is worth while to note that for fixed x a much stronger statement than that of Theorem 3 can be obtained. In fact we have

THEOREM 4

$$\limsup_{n\to\infty} \frac{|\xi(0,n) - \xi(x,n)|}{(2\xi(0,n)\log\log\xi(0,n))^{1/2}} = \sigma(x) \ a.s.$$

where $\sigma(x)$ *is a positive constant depending only on* x. *(Its exact value is given in the proof.)*

REMARK 1. In order to see that for fixed x Theorem 4 is stronger than Theorem 3 apply Theorem B. In fact Theorems B and 4 combined, imply: for any fixed x

$$\lim_{n\to\infty} \left(\frac{\log n}{(\log\log n)^{1+\varepsilon}}\right)^{1/2} \left| \frac{\xi(x,n)}{\xi(0,n)} - 1 \right| = 0 \text{ a.s.}$$

Let

$$I(x,n) = \begin{cases} 1 & \text{if } \xi(x,n) > 0, \\ 0 & \text{if } \xi(x,n) = 0, \end{cases} \tag{1}$$

$$K(N,n) = (N^2\pi)^{-1} \sum_{x\in Q(N)} I(x,n) \tag{2}$$

i.e. $K(N,n)$ is the density of the points of $Q(N)$ covered by the random walk $\{S_k, 0 \leq k \leq n\}$.

We (Erdős - Révész, 1988) also formulated the following trivial result: for any $0 < \alpha < 1/2$

$$\limsup_{n \to \infty} K(n^\alpha, n) \geq 1 - 2\alpha \text{ a.s.} \tag{3}$$

Here we prove that $K(n^\alpha, n)$ has a limit distribution. In fact we prove:

THEOREM 5 *For any $0 < \alpha < 1/2$ there exists a distribution function $G_\alpha(x)$ with $G_\alpha(0) = 0, G_\alpha(1) = 1$ and*

$$\lim_{n \to \infty} \mathbf{P}\{K(n^\alpha, n) < x\} = G_\alpha(x) \quad (-\infty < x < \infty).$$

Theorems A and 1 describe the area of the largest disc around the origin covered by the random walk $\{S_k, k \leq n\}$.

In $Z^d (d \geq 3)$ the analogous problem is clearly meaningless since the largest covered ball around the origin is finite with probability one.

However one can ask in any dimension about the radius of the largest ball (not surely around the origin) covered by the simple symmetric random walk in time n. Formally speaking let

$$Q(N, u) = \{x : \|x - u\| \leq N\}$$

and $R^*(n) = R^*(n,d)$ be the largest integer for which there exists a r.v. $u = u(n) \in Z^d$ such that $Q(R^*(n), u)$ is covered by the random walk in time n i.e.

$$\xi(x, n) \geq 1 \text{ for any } x \in Q(R^*(n), u).$$

Then we formulate our

THEOREM 6 *For any n big enough, $d \geq 3$ and $\varepsilon > 0$ we have*

$$C(d)(\log n)^{1/d} \leq R^*(n) \leq (\log n)^{\frac{2d-3}{(d-1)(d-2)}+\varepsilon} \quad a.s.$$

with some $C(d) > 0$.

REMARK 2. I conjecture that the lower (trivial) estimate in Theorem 6 is exact enough and the upper estimate is poor. In fact my CONJECTURE is the following: for any n big enough and $\varepsilon > 0$ we have

$$(\log n)^{(d-1)^{-1}-\varepsilon} \leq R^*(n) \leq (\log n)^{(d-1)^{-1}+\varepsilon} \text{ a.s.}$$

Theorems on the rate of escape claim that $\|S_n\|$ is about $n^{1/2}$ if n is big enough and $d \geq 3$. In fact we have

THEOREM C (Dvoretzky - Erdős (1950), Erdős - Taylor (1960)). *Let $d \geq 3$. Then for any $\varepsilon > 0$ and n big enough we have*

$$n^{1/2}(\log n)^{-\frac{1+\varepsilon}{d+2}} \leq \|S_n\| \leq (1+\varepsilon)(\frac{2}{d} n \log\log n)^{1/2} \ a.s.$$

This rate of escape suggests that the sphere - surface $\{x : \|x\| = R\}$ is visited about R times by the random walk if R is big enough. In order to formulate such a result introduce the following notations:

$$Z(R) = \{x : | \, \|x\| - R \, | \leq 1\},$$

$$J(x) = \begin{cases} 0 & \text{if } \xi(x,n) = 0 \text{ for every } n = 1, 2, \ldots, \\ 1 & \text{otherwise} \end{cases}$$

and

$$\check{\theta}(R) = \sum_{x \in Z(R)} J(x)$$

i.e. $J(x) = 1$ if $x \in Z^d$ is visited by the random walk at all and $\check{\theta}(R)$ is the number of visited points in $Z(R)$. Then we have

CONJECTURE. *There exists a distribution function $H(x)$ for which $H(0) = 0$ and*

$$\lim_{R \to \infty} \mathbf{P}\{\frac{\check{\theta}(R)}{R} < x\} = H(x) \quad (-\infty < x < \infty).$$

Unfortunately I cannot settle this conjecture but I present Lemma 3 as an indication that this conjecture should be true. The analogue question for a Wiener process will be treated in Section 6.

2 Lemmas

LEMMA 1 *For any $d \geq 3$ there exists a positive constant C_d such that*

$$\mathbf{P}\{S_n = x \text{ for some } n\} = \mathbf{P}\{J(x) = 1\} = \frac{C_d + o(1)}{R^{d-2}} \quad (R \to \infty)$$

where $R = \|x\|$.

REMARK 3. The analogous lemma for Wiener process is known (c.f. Lemma C).

PROOF. Clearly

$$\mathbf{P}\{S_n = x\} = \sum_{k=0}^{n} \mathbf{P}\{S_k = x, S_j \neq x, j = 0, 1, 2, \ldots, k-1\}\mathbf{P}\{S_{n-k} = 0\}$$

and

$$\sum_{n=0}^{\infty} \mathbf{P}\{S_n = x\} =$$
$$\sum_{n=0}^{\infty}\sum_{k=0}^{n} \mathbf{P}\{S_k = x, S_j \neq x, j = 0, 1, 2, \ldots, k-1\}\mathbf{P}\{S_{n-k} = 0\} =$$
$$\sum_{n=0}^{\infty} \mathbf{P}\{S_n = 0\}\sum_{k=0}^{\infty} \mathbf{P}\{S_k = x, S_j \neq x, j = 0, 1, 2, \ldots, k-1\}.$$

Since

$$\mathbf{P}\{S_n = x\} = n^{-d/2}(a_d \exp(-\frac{b_d R^2}{n}) + o(1)) \text{ if } \|x\| = R \leq O(n^{1/2})$$

with some $a_d > 0, b_d > 0$ a simple calculation gives that

$$\sum_{n=0}^{\infty} \mathbf{P}\{S_n = x\} = (K_d + o(1))R^{2-d} \quad (R \to \infty).$$

with some $K_d > 0$. Taking into account that

$$\sum_{n=0}^{\infty} \mathbf{P}\{S_n = 0\} < \infty$$

we obtain

$$\mathbf{P}\{J(x) = 1\} = \sum_{k=0}^{\infty} \mathbf{P}\{S_k = x, S_j \neq x, j = 0, 1, 2, \ldots, k-1\} = (C_d + o(1))R^{2-d}.$$

Hence we have the Lemma.

Let $d = 2$ and introduce the following notations

$$\begin{aligned} m_k = m_k(x_1, x_2, \ldots, x_k; n) &= \mathbf{E}(I(x_1, n) \ldots I(x_k, n)), \\ \nu(z) = \nu_z &= \min\{k : k > 0, S_k = z\}. \end{aligned}$$

i.e. m_k is the probability that the points $x_1, x_2, \ldots, x_k$ are visited in the first n steps. Then we have

LEMMA 2 *For any* $0 < q < 1, k = 2, 3, \ldots$ *we have*

$$m_1(u, n - qn)[\mathcal{M}(qn) + (1 - k)m_k(x_1, x_2, \ldots, x_k; qn)] \leq$$
$$m_k(x_1, x_2, \ldots, x_k; n) \leq \frac{m_1(v; n)\mathcal{M}(n)}{1 + m_1(v, n)(k - 1)}$$

where u *and* v *are arbitrary lattice points of* R^2 *with*

$$\|u\| = \max_{1 \leq i < j \leq k} \|x_i - x_j\| \text{ and } \|v\| = \min_{1 \leq i < j \leq k} \|x_i - x_j\|$$

and

$$\mathcal{M}(k, n) = \mathcal{M}(n) = \sum_{i=1}^{n} m_{k-1}(x_1, x_2, \ldots, x_{i-1}, x_{i+1}, \ldots, x_k; n).$$

REMARK 4. $m_1(v, n)$ resp. $m_1(u, n - qn)$ depend on the exact location of v resp. u not only on $\|v\|$ resp. $\|u\|$. Hence the exact meaning of the above inequality is : consider all v resp. u with $\|v\| = \min_{1 \leq i < j \leq k} \|x_i - x_j\|$ resp. $\|u\| = \max_{1 \leq i < j \leq k} \|x_i - x_j\|$ and take that v for which the right-hand side of the above inequality is the largest resp. that u for which the left-hand side is the smallest.

In order to present the proof in an intelligible form we prove Lemma 2 at first in case $k = 2$. That is we prove

LEMMA 2* *For any* $0 < q < 1$ *we have*

$$m_1(x - y; n - qn)[m_1(x; qn) + m_1(y; qn) - m_2(x, y; qn)] \leq$$
$$m_2(x, y; n) \leq \frac{m_1(x - y; n)[m_1(x; n) + m_1(y; n)]}{1 + m_1(x - y; n)}. \tag{4}$$

PROOF.

$$m_2(x,y;n) = \mathbf{P}(I(x,n)=1, I(y,n)=1) =$$
$$\sum_{k=0}^{n} \mathbf{P}(I(x,n)=1, I(y,n)=1 \mid \nu_x = k < \nu_y)\mathbf{P}(\nu_x = k < \nu_y) +$$
$$\sum_{k=0}^{n} \mathbf{P}(I(x,n)=1, I(y,n)=1 \mid \nu_y = k < \nu_x)\mathbf{P}(\nu_y = k < \nu_x) =$$
$$\sum_{k=0}^{n} \mathbf{P}(I(y,n)=1 \mid \nu_x = k < \nu_y)\mathbf{P}(\nu_x = k < \nu_y) +$$
$$\sum_{k=0}^{n} \mathbf{P}(I(x,n)=1 \mid \nu_y = k < \nu_x)\mathbf{P}(\nu_y = k < \nu_x) =$$
$$\sum_{k=0}^{n} \mathbf{P}(I(y-x,n-k)=1)\mathbf{P}(\nu_x = k < \nu_y) +$$
$$\sum_{k=0}^{n} \mathbf{P}(I(x-y,n-k)=1)\mathbf{P}(\nu_y = k < \nu_x).$$

Consequently we have

$$m_2(x,y;n) \leq$$
$$\mathbf{P}(I(x-y,n)=1)\mathbf{P}(\cup_{k=0}^{n}\{\{\nu_x = k < \nu_y\} + \{\nu_y = k < \nu_x\}\}) =$$
$$\mathbf{P}(I(x-y,n)=1)\mathbf{P}(I(x,n)=1 \text{ or } I(y,n)=1) =$$
$$m_1(x-y;n)[m_1(x;n) + m_1(y,n) - m_2(x,y;n)]$$

what implies the upper part of (4).

We also have

$$m_2(x,y;n) \geq$$
$$\sum_{k=0}^{qn} \mathbf{P}(I(y-x,n-k)=1)\mathbf{P}(\nu_x = k < \nu_y) +$$
$$\sum_{k=0}^{qn} \mathbf{P}(I(x-y,n-k)=1)\mathbf{P}(\nu_y = k < \nu_x) \geq$$
$$\mathbf{P}(I(x-y,n-qn)=1)\mathbf{P}(I(x,qn)=1 \text{ or } I(y,qn)=1) =$$
$$m_1(x-y;n-qn)[m_1(x;qn) + m_1(y;qn) - m_2(x,y;qn)].$$

Hence we have Lemma 2*.

PROOF OF LEMMA 2. Let P_k resp. $P_k(r)$ be the set of the permutations of the integers $1,2,\ldots,k$ resp. $1,2,\ldots,r-1,r+1,\ldots,k$. Further let

$$A = A(i_1, i_2, \ldots, i_k; j) =$$

$$\{\nu(x_{i_1}) < \nu(x_{i_2}) < \ldots < \nu(x_{i_{k-1}}) = j < \nu(x_{i_k})\}.$$

Then we have

$$\begin{aligned}
&m_k(x_1, x_2, \ldots, x_k; n) = \\
&\quad \sum_{(i_1,\ldots,i_k)\in P_k} \mathbf{P}(I(x_1, n) = \ldots = I(x_k, n) = 1 \mid A)\mathbf{P}(A) = \\
&\quad \sum \mathbf{P}(I(x_{i_k}, n) = 1 \mid A)\mathbf{P}(A) = \\
&\quad \sum \mathbf{P}(I(x_{i_k} - x_{i_{k-1}}, n - j) = 1)\mathbf{P}(A).
\end{aligned}$$

Consequently

$$\begin{aligned}
&m_k \leq \mathbf{P}(I(v, n) = 1) \sum \mathbf{P}(A) = \\
&\quad \mathbf{P}(I(v, n) = 1) \sum_{r=1}^{k} \sum_{(i_1,\ldots,i_{k-1})\in P_k(r)} \mathbf{P}(A) = \mathbf{P}(I(v, n) = 1) \times \\
&\quad \mathbf{P}(\sum_{r=1}^{k} \{I(x_1, n) = \ldots = I(x_{r-1}, n) = I(x_{r+1}, n) = \ldots = I(x_k, n) = 1\}) = \\
&\quad m_1(v, n)[\mathcal{M}(n) + m_k(-\binom{k}{2} + \binom{k}{3} - \cdots (-1)^{k+1}\binom{k}{k})] = \\
&\quad m_1(v, n)[\sum_{r=1}^{k} m_{k-1}(x_1, \ldots, x_{r-1}, x_{r+1}, \ldots, x_k; n) + (1 - k)m_k]
\end{aligned}$$

what implies the upper part of the inequality of Lemma 2.

We also have

$$\begin{aligned}
&m_k \geq \sum_{\substack{(i_1,\ldots,i_k)\in P_k \\ 1\leq j\leq qn}} \mathbf{P}(I(x_{i_k} - x_{i_{k-1}}, n - j) = 1)\mathbf{P}(A) \geq \\
&\quad \mathbf{P}(I(u, n - qn) = 1)\mathbf{P}(\sum_{r=1}^{k} \sum_{\substack{(i_1,\ldots,i_{k-1})\in P_k(r) \\ 1\leq j\leq qn}} A) = \\
&\quad \mathbf{P}(I(u, n - qn) = 1) \times \\
&\quad \mathbf{P}(\sum_{r=1}^{k} \{I(x_1, qn) = \ldots = I(x_{r-1}, qn) = I(x_{r+1}, qn) = \ldots = I(x_k, qn) = 1\}) = \\
&\quad m_1(u, qn)[\mathcal{M}(qn) + (1 - k)m_k].
\end{aligned}$$

Hence we have Lemma 2.

Let $d \geq 3$ and

$$\mu_k = \mu_k(x_1, x_2, \ldots, x_k) = \mathbf{E}(J(x_1)J(x_2)\ldots J(x_k)).$$

Then we have

LEMMA 3

$$\mu_2(x,y) = \frac{\mu_1(x-y)(\mu_1(x)+\mu_1(y))}{1+\mu_1(x-y)}$$

and in general

$$\frac{\mu_1(u)\sum_{i=1}^k \mu_{k-1}(x_1,x_2,\ldots,x_{i-1},x_{i+1},\ldots,x_k)}{1+\mu_1(u)(k-1)} \leq$$
$$\mu_k(x_1,x_2,\ldots,x_k) \leq$$
$$\frac{\mu_1(v)\sum_{i=1}^k \mu_{k-1}(x_1,x_2,\ldots,x_{i-1},x_{i+1},\ldots,x_k)}{1+\mu_1(v)(k-1)}$$

where u and v are arbitrary lattice points of $R^d (d \geq 3)$ with

$$\|u\| = \max_{1\leq i<j\leq k} \|x_i - x_j\| \text{ and } \|v\| = \min_{1\leq i<j\leq k} \|x_i - x_j\|.$$

Proof is essentially the same as that of Lemma 2. Hence it will be omitted.

3 Proofs of Theorems 1 - 2

In the proofs there will be frequently used the following

LEMMA A (Erdős - Taylor, 1960).

$$\mathbf{P}\{I(x,n)=0\} =$$
$$\begin{cases} \frac{2\log\|x\|}{\log n}\left(1+O\left(\frac{\log_3\|x\|}{\log\|x\|}\right)\right) & if\ 20 < \|x\| < n^{1/3}, \\ 1-2\frac{\log(n^{1/2}/\|x\|)}{\log n}\left(1+O\left(\frac{\log_2 n^{1/2}/\|x\|}{\log_3 n^{1/2}/\|x\|}\right)\right) & if\ n^{1/6} < \|x\| < \frac{n^{1/2}}{20}. \end{cases}$$

Let

$$N = N(n) = \exp((\log n)^{1/2}\log_3 n)$$

and

$$k = k(n) = \exp((\log n)^{1/2}).$$

Then for any $n = 1,2,\ldots$ and for any $0 < \varepsilon < 1$ there exists a sequence $x_1 = x_1(n), x_2 = x_2(n), \ldots, x_k = x_k(n) (k = k(n))$ such that

$$N-1 \leq \|x_i\| \leq N \quad (i = 1,2,\ldots,k),$$
$$N^{1-\varepsilon} \leq \|x_i - x_j\| \leq N \quad (1 \leq i < j \leq k).$$

Now we formulate our

LEMMA 4

$$m_k(x_1,x_2,\ldots,x_k;n)\le\exp(-(2-4\varepsilon)\log_3 n)(x_i=x_i(n),i=1,2,\ldots,k=k(n)).$$

PROOF. By Lemmas A and 2 we have

$$m_k\le\frac{(1-\frac{2\log N^{1-\varepsilon}}{\log n}(1+O(\frac{\log_3 N^{1-\varepsilon}}{\log N^{1-\varepsilon}})))M(n)}{1+(k-1)(1-\frac{2\log N^{1-\varepsilon}}{\log n}(1+O(\frac{\log_3 N^{1-\varepsilon}}{\log N^{1-\varepsilon}})))}\le$$
$$\frac{1}{k}(1-\frac{2}{k}\frac{\log N^{1-\varepsilon}}{\log n}(1+O(\frac{\log_3 N^{1-\varepsilon}}{\log N^{1-\varepsilon}})))M(n)\le$$
$$\exp(-\frac{2}{k}\frac{\log N^{1-\varepsilon}}{\log n}(1+O(\frac{\log_3 N^{1-\varepsilon}}{\log n^{1-\varepsilon}})))\frac{1}{k}M(n)\le\ldots\le$$
$$\exp(-2\frac{\log N^{1-\varepsilon}}{\log n}(1+O(\frac{\log_3 N^{1-\varepsilon}}{\log N^{1-\varepsilon}}))\log k)\le$$
$$\exp(-(2-4\varepsilon)\frac{(\log n)^{1/2}\log_3 n}{\log n}(\log n)^{1/2})\le$$
$$\exp(-(2-4\varepsilon)\log_3 n).$$

Hence we have Lemma 4.

PROOF OF THEOREM 1. Let

$$n_j=[\exp(e^j)],\tilde N(n)=\exp(2(\log n)^{1/2}\log_3 n),k_{j+1}=\exp((\log n_{j+1})^{1/2}).$$

Then

$$\mathbf{P}\{R(n_{j+1})\ge\tilde N(n_j)\}\le\mathbf{P}\{R(n_{j+1})\ge N(n_{j+1})\}\le$$
$$m_{k_{j+1}}(x_1,x_2,\ldots,x_{k_{j+1}};n_{j+1})\le(j+1)^{-(2-4\varepsilon)}$$

where $x_i=x_i(n_{j+1})(i=1,2,\ldots,k_{j+1})$. Hence

$$R(n_{j+1})\le\tilde N(n_j)\quad\text{a.s.}$$

for all but finitely many j. Let $n_j\le n<n_{j+1}$. Then

$$R(n)\le R(n_{j+1})\le\tilde N(n_j)\le\tilde N(n)$$

which proves Theorem 1.

Let

$$M = M(n) = \exp(C(\log n)^{1/2}) \quad (C > 0)$$

and

$$K = K(n) = C^* \exp(\frac{C}{4}(\log n)^{1/2}) \quad (C^* > 0).$$

Then for any $n = 1, 2, \ldots$ there exists a sequence $y_1 = y_2(n), y_2 = y_2(n), \ldots, y_K = y_K(n)$ such that

$$\begin{aligned} M - 1 &\leq \|y_i\| \leq M \quad (i = 1, 2, \ldots, K), \\ M^{3/4} &\leq \|y_i - y_j\| \leq M \quad (1 \leq i < j \leq K). \end{aligned}$$

if C^* is small enough. Now we formulate our

LEMMA 4*

$$m_K(y_1, y_2, \ldots, y_K; n) \leq \exp(-\frac{C^2}{4}).$$

PROOF. In the same way as we proved Lemma 4 we obtain

$$m_K \leq \frac{1}{K}(1 - \frac{2}{K}\frac{\log M^{3/4}}{\log n}(1 + O(\frac{\log_3 M^{3/4}}{\log M^{3/4}})))M(K, n) \leq \ldots \leq$$
$$\exp(-2\frac{\log M^{3/4}}{\log n}(1 + O(\frac{\log_3 M^{3/4}}{\log M^{3/4}}))\log K) \leq \exp(-\frac{C^2}{4}).$$

Hence we have Lemma 4*.

PROOF OF THEOREM 2. Clearly

$$\mathbf{P}\{R(n) \geq \exp(C(\log n)^{1/2})\} \leq m_K \leq \exp(-\frac{C^2}{4})$$

which proves Theorem 2.

4 Proofs of Theorems 3 - 4

Introduce the following notations:

$$p(0 \rightsquigarrow x) = \mathbf{P}\{\inf\{n : n \geq 1, S_n = 0\} > \inf\{n : n \geq 1, S_n = x\}\} =$$
$$\mathbf{P}\{\{S_n\} \text{ reaches } x \text{ before returning to } 0\}.$$

Let $\rho_1(0 \rightsquigarrow x), \rho_2(0 \rightsquigarrow x), \ldots$ resp. $\rho_1(x \rightsquigarrow 0), \rho_2(x \rightsquigarrow 0), \ldots$ be the first, second,... waiting times to reach x from 0, resp. to reach 0 from x, i.e.

$$\begin{aligned}
\rho_1(0 \rightsquigarrow x) &= \inf\{n : n \geq 1, S_n = x\}, \\
\rho_1(x \rightsquigarrow 0) &= \inf\{n : n \geq \rho_1(0 \rightsquigarrow x), S_n = 0\} - \rho_1(0 \rightsquigarrow x), \\
\rho_2(0 \rightsquigarrow x) &= \inf\{n : n \geq \rho_1(0 \rightsquigarrow x) + \rho_1(x \rightsquigarrow 0), S_n = x\} - \\
&\quad (\rho_1(0 \rightsquigarrow x) + \rho_1(x \rightsquigarrow 0)), \\
\rho_2(x \rightsquigarrow 0) &= \inf\{n : n \geq \rho_1(0 \rightsquigarrow x) + \rho_1(x \rightsquigarrow 0) + \rho_2(0 \rightsquigarrow x), S_n = 0\} - \\
&\quad (\rho_1(0 \rightsquigarrow x) + \rho_1(x \rightsquigarrow 0) + \rho_2(0 \rightsquigarrow x)), \ldots
\end{aligned}$$

Let $\tau(0 \rightsquigarrow x, n)$ be the number of $0 \rightsquigarrow x$ excursions completed before n i.e.

$$\tau(0 \rightsquigarrow x, n) = \max\{i : \sum_{j=1}^{i-1} (\rho_j(0 \rightsquigarrow x) + \rho_j(x \rightsquigarrow 0)) + \rho_i(0 \rightsquigarrow x) \leq n\}.$$

Finally let $\Xi_1, \Xi_2, \ldots$ be the local time of 0 during the first,second,...$0 \rightsquigarrow x$ excursions i.e.

$$\begin{aligned}
\Xi_1 &= \xi(0, \rho_1(0 \rightsquigarrow x)), \\
\Xi_2 &= \xi(0, \rho_1(0 \rightsquigarrow x) + \rho_1(x \rightsquigarrow 0) + \rho_2(0 \rightsquigarrow x)) - \Xi_1, \\
\Xi_3 &= \xi(0, \rho_1(0 \rightsquigarrow x) + \rho_1(x \rightsquigarrow 0) + \rho_2(0 \rightsquigarrow x) + \rho_2(x \rightsquigarrow 0) + \rho_3(0 \rightsquigarrow x)) - \\
&\quad (\Xi_1 + \Xi_2), \ldots
\end{aligned}$$

Observe that $\Xi_1, \Xi_2, \ldots$ are i.i.d.r.v.'s with distribution

$$\mathbf{P}(\Xi_1 = k) = \mathbf{P}(\xi(0, \rho_1(0 \rightsquigarrow x)) = k) = (1 - p(0 \rightsquigarrow x))^{k-1} p(0 \rightsquigarrow x) (k = 1, 2, \ldots).$$

Consequently

$$\begin{aligned}
\mathbf{E}\Xi_1 &= (p(0 \rightsquigarrow x))^{-1}, \\
\mathbf{E}(\Xi_1 - (p(0 \rightsquigarrow x))^{-1})^2 &= (p(0 \rightsquigarrow x))^{-2}(1 - p(0 \rightsquigarrow x))
\end{aligned}$$

and

$$\limsup_{n \to \infty} \frac{\Xi_1 + \Xi_2 + \cdots + \Xi_n - (H_1 + H_2 + \cdots + H_n)}{(2n \log\log n)^{1/2}} = 2^{1/2} \frac{(1 - p(0 \rightsquigarrow x))^{1/2}}{p(0 \rightsquigarrow x)}$$

where

$$H_1 = \xi(x, \rho_1(0 \rightsquigarrow x) + \rho_1(x \rightsquigarrow 0)),$$

$$H_2 = \xi(x, \rho_1(0 \rightsquigarrow x) + \rho_1(x \rightsquigarrow 0) + \rho_2(0 \rightsquigarrow x) + \rho_2(x \rightsquigarrow 0)) - H_1, \dots$$

Since

$$\sum_{k=0}^{\tau(0 \rightsquigarrow x, n)} \Xi_k \leq \xi(0, n) \leq \sum_{k=0}^{\tau(0 \rightsquigarrow x, n)+1} \Xi_k,$$

$$\lim_{k \to \infty} k^{-1/2} \Xi_k = 0$$

and the sequence $\tau(0 \rightsquigarrow x, n)$ takes every positive integer we have

$$\limsup_{n \to \infty} \frac{\xi(0, n) - \xi(x, n)}{(2\tau(0 \rightsquigarrow x, n) \log \log \tau(0 \rightsquigarrow x, n))^{1/2}} = 2^{1/2} \frac{(1 - p(0 \rightsquigarrow x))^{1/2}}{p(0 \rightsquigarrow x)} \quad \text{a.s.}$$

By the law of large numbers

$$\lim_{n \to \infty} \frac{\xi(0, n) - \tau(0 \rightsquigarrow x, n)(p(0 \rightsquigarrow x))^{-1}}{\tau(0 \rightsquigarrow x, n)} = 0 \text{ a.s.}$$

Hence

$$\limsup_{n \to \infty} \frac{\xi(0, n) - \xi(x, n)}{(2\xi(0, n) \log \log \xi(0, n))^{1/2}} = \left(\frac{2(1 - p(0 \rightsquigarrow x))}{p(0 \rightsquigarrow x)}\right)^{1/2} = \sigma(x)$$

and we have Theorem 4.

In the proof of Theorem 3 the following three lemmas will be used.

LEMMA B (Erdős - Révész, 1988) *Let*

$$\alpha(r) = \mathbf{P}\{\inf\{n : \|S_n\| \geq r\} < \inf\{n : n \geq 1, S_n = 0\}\}.$$

Then

$$\lim_{r \to \infty} \alpha(r) \log r = \frac{\pi}{2}.$$

LEMMA 5 *There exists a positive constant C such that*

$$p(0 \rightsquigarrow x) \geq \frac{C}{\log \|x\|}$$

for any $x \in Z^2$ with $\|x\| \geq 2$. Further

$$\liminf_{\|x\| \to \infty} p(0 \rightsquigarrow x) \log \|x\| \geq \frac{\pi}{12}.$$

PROOF. Let $x = \|x\| e^{i\varphi}$. Then by Lemma B the probability that the particle crosses the arc $\|x\| e^{i\psi}(\varphi - \pi/3 < \psi < \varphi + \pi/3)$ before returning to 0 is larger than $(1-\varepsilon)\pi(6 \log \|x\|)^{-1}$. Since starting from any point of the arc $\|x\| e^{i\psi}(\varphi - \pi/3 < \psi < \varphi + \pi/3)$ the probability that the particle hits x before 0 is larger than 1/2 we obtain the lemma.

LEMMA 6 *For any $K > 0$ we have*

$$\lim_{n\to\infty} \sup_{\|x\| \le (\log n)^K} \left| \frac{\xi(0,n) - \tau(0 \rightsquigarrow x, n)(p(0 \rightsquigarrow x))^{-1}}{\tau(0 \rightsquigarrow x, n)} \right| = 0 \ a.s.$$

PROOF. By central limit theorem one can obtain that uniformly for any x with $\|x\| \le (\log n)^K$

$$\mathbf{P}\left\{\frac{\tau(0 \rightsquigarrow x) - \xi(0,n)p(0 \rightsquigarrow x)}{(\xi(0,n)p(0 \rightsquigarrow x)(1 - p(0 \rightsquigarrow x)))^{1/2}} < z\right\} \to \Phi(z)$$

which implies Lemma 6.

PROOF OF THEOREM 3. Since Ξ_1 and H_1 are independent with distribution

$$\mathbf{P}(\Xi_1 = k) = \mathbf{P}(H_1 = k) = (1 - p(0 \rightsquigarrow x))^{k-1} p(0 \rightsquigarrow x) \ (k = 1, 2, \ldots)$$

we obtain that

$$\mathbf{E}\exp(t(\Xi_1 - H_1)) = p^2(1 - (1-p)(e^t + e^{-t}) + (1-p)^2)^{-1}.$$

Hence a simple calculation implies that for any $x \in Z^2$ we have

$$\mathbf{E}\exp \frac{\Xi_1 + \Xi_2 + \cdots + \Xi_n - (H_1 + H_2 + \cdots + H_n)}{(2n(p(0 \rightsquigarrow x))^{-2}(1 - p(0 \rightsquigarrow x)))^{1/2}} \le C$$

where C is an absolute positive constant. The above inequality together with the Chebishev inequality and the Borel - Cantelli lemma imply that for any $\varepsilon > 0$ and $K > 0$

$$\lim_{n\to\infty} \sup_{\|x\| \le n^K} \frac{|\Xi_1 + \Xi_2 + \cdots + \Xi_n - (H_1 + H_2 + \cdots + H_n)|}{(2n(p(0 \rightsquigarrow x))^{-2}(1 - p(0 \rightsquigarrow x)))^{1/2}(\log n)^{1+\varepsilon}} = 0 \text{ a.s.}$$

Since by Theorem B and Lemma 5 $\tau(0 \rightsquigarrow x, n) \ge (\log n)^{1/2}\|x\|^{-\varepsilon}$ a.s. for all but finitely many n we have

$$\lim_{n\to\infty} \sup_{\|x\| \le (\log n)^{K/2}\|x\|^{-\varepsilon K}} \frac{|\Xi_1 + \cdots + \Xi_{\tau(0 \rightsquigarrow x, n)} - (H_1 + \cdots + H_{\tau(0 \rightsquigarrow x, n)})|}{2\tau(0 \rightsquigarrow x, n)(p(0 \rightsquigarrow x))^{-2})^{1/2}(\log \tau(0 \rightsquigarrow x, n))^{1+\varepsilon}} =$$

$$\lim_{n\to\infty} \sup_{\|x\| \le (\log n)^{K/2}\|x\|^{-\varepsilon K}} \left| \frac{\xi(x,n)}{\xi(0,n)} - 1 \right| \frac{(\xi(0,n)p(0 \rightsquigarrow x))^{1/2}}{2^{1/2}(\log(\xi(0,n)p(0 \rightsquigarrow x)))^{1+\varepsilon}} = 0$$

which implies Theorem 3.

5 Proof of Theorem 5

Lemma A implies

LEMMA 7 *Let* $n^\alpha(\log n)^{-\beta} \le \|x\| \le Cn^\alpha (0 < \alpha < 1/2, \beta \ge 0, C \ge 1)$. *Then*

$$\mathbf{P}\{I(x,n)=1\} = 1 - 2\alpha + O\left(\frac{\log\log n}{\log n}\right).$$

LEMMA 8 *Let* x *and* y *be two lattice-points of* R^2 *such that*

$$n^\alpha(\log n)^{-\beta} \le \|x\|, \|y\|, \|x-y\| \le Cn^\alpha \quad (0 < \alpha < 1/2, \beta \ge 0, C \ge 1).$$

Then

$$\lim_{n\to\infty} m_2(x,y;n) = \frac{(1-2\alpha)^2}{1-\alpha}. \tag{5}$$

PROOF. Lemmas 2* and 7 imply

$$m_2(x,y;n) \le \frac{m_1(x-y;n)[m_1(x;n)+m_1(y;n)]}{1+m_1(x-y;n)} =$$
$$\frac{2(1-2\alpha+O(\frac{\log\log n}{\log n}))^2}{2-2\alpha+O(\frac{\log\log n}{\log n})} = \frac{(1-2\alpha)^2}{1-\alpha} + O\left(\frac{\log\log n}{\log n}\right). \tag{6}$$

Similarly

$$m_2(x,y;n) \ge m_1(x-y;n-qn)[m_1(x;qn)+m_1(y;qn)-m_2(x,y;qn)] \ge$$
$$(1-2\alpha+O(\frac{\log\log n}{\log n}))[2(1-2\alpha)+O(\frac{\log\log n}{\log n}) - m_2(x,y;n)].$$

Consequently

$$m_2(x,y;n) \ge \frac{(1-2\alpha)^2}{1-\alpha} - O\left(\frac{\log\log n}{\log n}\right). \tag{7}$$

(6) and (7) together imply (5).

The next lemma is an extension of Lemma 8.

LEMMA 9 *Let $x_1, x_2, \ldots, x_k$ $(k = 1, 2, \ldots)$ be a sequence of lattice points of R^2 such that*

$$n^\alpha(\log n)^{-\beta} \le \|x_i - x_j\|, \|x_i\| \le Cn^\alpha$$

$(0 < \alpha < 1/2, \beta \ge 0, C \ge 1, 1 \le i < j \le k)$. *Then*

$$\lim_{n\to\infty} m_k(x_1, x_2, \ldots, x_k; n) = (1-2\alpha)^k \Pi_{j=2}^k (1 - (1 - \frac{1}{j})2\alpha)^{-1}.$$

PROOF. Lemmas 2 and 7 imply

$$m_k \le \frac{1 - 2\alpha + O(\frac{\log\log n}{\log n})}{1 + (k-1)(1 - 2\alpha + O(\frac{\log\log n}{\log n}))} M(n) =$$
$$\frac{1 - 2\alpha + O(\frac{\log\log n}{\log n})}{1 - (1 - \frac{1}{k})(2\alpha - O(\frac{\log\log n}{\log n}))} \frac{1}{k} M(n).$$

By induction we obtain

$$m_k \le (1 - 2\alpha + O(\frac{\log\log n}{n}))^k \Pi_{j=2}^k (1 - (1 - \frac{1}{j})(2\alpha - O(\frac{\log\log n}{\log n})))^{-1}. \qquad (8)$$

Similarly

$$m_k \ge (1 - 2\alpha + O(\frac{\log\log n}{\log n}))[M(qn) + (1-k)m_k].$$

Consequently

$$m_k \ge \frac{1 - 2\alpha + O(\frac{\log\log n}{\log n})}{1 - (1 - \frac{1}{k})(2\alpha - O(\frac{\log\log n}{\log n}))} \frac{1}{k} M(qn)$$

and by induction we obtain Lemma 9.

PROOF OF THEOREM 5. Let $A(s, n)$ be the set of all possible s-tuples of $Q(n^\alpha)$ with the property

$$\|x\|, \|x_i - x_j\| \ge n^\alpha(\log n)^{-\beta} \quad (i \ne j).$$

Then by Lemma 8

$$\mathbf{E}(K(n^\alpha, n))^s =$$
$$(\frac{1}{n^{2\alpha}\pi})^s \mathbf{E}(\sum_{x\in Q(n^\alpha)} I(x, n))^s =$$
$$((\frac{1}{n^{2\alpha}\pi})^s + o(1)) \times$$
$$\sum_{(x_1, x_2, \ldots, x_s)\in A(s,n)} \mathbf{E}\Pi_{j=1}^s I(x_j, n) \to (1-2\alpha)^s \Pi_{j=1}^s (1 - (1 - \frac{1}{j})2\alpha)^{-1}.$$

Consequently (since $0 \le K(n^\alpha, n) \le 1$) we have Theorem 5 with

$$\int_0^1 x^s dG_\alpha(x) = (1-2\alpha)^s \Pi_{j=1}^s (1-(1-\frac{1}{j})2\alpha)^{-1} =$$
$$\Pi_{j=1}^s (1+\frac{2\alpha}{1-2\alpha}\frac{1}{j})^{-1} = O(s^{-\frac{2\alpha}{1-2\alpha}})$$

for any fixed $0 < \alpha < 1/2$ as $s \to \infty$.

REMARK 5. The above equality easily implies that $G_\alpha(1-\varepsilon) < 1$ for any $0 < \alpha < 1/2$ and $\varepsilon > 0$ and in turn $\limsup_{n\to\infty} K(n^\alpha, n) = 1$ a.s. for any $0 < \alpha < 1/2$.

It is natural to ask the following

QUESTION. Is it true for any $0 < \alpha < 1/2$ and $\varepsilon > 0$ that

$$G_\alpha(\varepsilon) > 0.$$

Similarly it is natural to have the following

CONJECTURE. For any $0 \le \beta < 1$

$$\lim_{n\to\infty} K(\exp(\log n)^\beta, n) = 1 \text{ a.s.}$$

6 Proof of Theorem 6

It is well-known that in a coin tossing sequence of size N the length of the longest head-run is more than $(1-\varepsilon)(\log N)/\log 2$ with probability one for all but finitely many N (see e.g. Erdős - Révész, 1976). Similarly, given an arbitrary sequence $x_1, x_2, \ldots, x_n$ of the vectors $\pm e_1, \pm e_2, \ldots, \pm e_d$ (where $e_1, e_2, \ldots, e_d$ are the orthogonal unit vectors of R^d) the sequence $X_1, X_2, \ldots, X_N; N = \exp((1+\varepsilon)(\log 2d)n)$ will contain the block $x_1, x_2, \ldots, x_n$ with probability one if n is big enough i.e. there exists a $1 \le j \le N-n$ such that $X_{j+l} = x_l (l = 1, 2, \ldots, n)$. Since a suitable sequence $x_1, x_2, \ldots, x_n$ covers a ball of radius $C_d n^{1/d}$ we obtain the lower inequality of Theorem 6.

In order to get the upper inequality we present two lemmas.

LEMMA 10 *For any $0 < \alpha < 1$ and $L > 0$ there exists a sequence $x_1, x_2, \ldots, x_T$ of the points of Z^d such that*

$$L \leq \|x_i\| < L + 1 \quad (i = 1, 2, \ldots, T),$$

$$\|x_i - x_j\| \geq L^\alpha \quad (i, j = 1, 2, \ldots, T; i \neq j),$$

$$T = CL^{(1-\alpha)(d-1)}$$

where $C = C_d$ is a positive constant depending on d Only.

PROOF is trivial.

LEMMA 11

$$\mathbf{P}\{Q(L,0) \subset R^d \textit{ is covered eventually}\} \leq$$
$$\mathbf{P}\{x_1, x_2, \ldots, x_T \textit{ are covered eventually}\} \leq$$
$$T!L^{-\alpha(d-2)T}(C_d + o(1))^T$$

where $x_1, x_2, \ldots, x_T$ and T resp. C_d are defined in Lemma 10 resp. Lemma 1.

Consequently for any $n > 0$ and $0 < \alpha < 1$ we have

$$\mathbf{P}\{\exists u = u(n) \in Z^d \textit{ such that } S_n \in Q(L, u), Q(L, u) \textit{ is covered eventually}\} \leq$$
$$C^*(C_d + o(1))^T L^d T! L^{-\alpha(d-2)T} \tag{9}$$

PROOF is trivial utilizing Lemma 1.

PROOF OF THEOREM 6. Choosing

$$\alpha = \frac{d-1}{2d-3} + \delta \ (0 < \delta < \frac{\varepsilon(d-1)(d-2)^2}{(2d-3)^2 + \varepsilon(d-1)(d-2)(2d-3)})$$

(where ε in the statement of Theorem 6) we have

$$C^*(C_d + o(1))^T L^d T! L^{-\alpha(d-2)T} \leq$$
$$C^* L^d((C_d + o(1))TL^{-\alpha(d-2)})^T \leq$$
$$C^* L^d((C_d + o(1))CL^{(1-\alpha)(d-1)-\alpha(d-2)})^{CL^{(1-\alpha)(d-1)}} =$$
$$C^* L^d(C_d + o(1))CL^{-\delta(2d-3)}CL^D$$

where

$$D = (\frac{d-2}{2d-3} - \delta)(d-1).$$

Replacing L by $(\log n)^{\frac{2d-3}{(d-1)(d-2)}+\varepsilon}$ we obtain that the probability in (9) is less than or equal to

$$C^*(\log n)^{d(\frac{2d-3}{(d-1)(d-2)}+\varepsilon)}(C_d + o(1))C(\log n)^{-\delta(2d-3)}(\frac{2d-3}{(d-1)(d-2)} + \varepsilon)^{C(\log n)^Q}$$

where

$$Q = D(\frac{2d-3}{(d-1)(d-2)} + \varepsilon) > 1.$$

Hence we have Theorem 6 by Borel - Cantelli lemma.

Theorem 6 says that the path of the random walk in its first n steps covers relatively big balls. It is natural to ask where these big covered balls are located in R^d.

Let $\rho(n)$ be the largest integer for which there exists a r.v. $u = u(n) \in Z^d$ such that $\|u\| \leq n$ and $Q(\rho(n), u)$ is covered by the random walk eventually i.e.

$$J(x) = 1 \text{ for any } x \in Q(\rho(n), u).$$

As a trivial consequence of Theorems C and 6 we obtain

THEOREM 6* *For any n big enough, $d \geq 3$ and $\varepsilon > 0$ we have*

$$C(d)(\log n)^{1/d} \leq \rho(n) \leq (\log n)^{\frac{2d-3}{(d-1)(d-2)}+\varepsilon} \quad a.s.$$

with some $C(d) > 0$.

7 Speed of escape (Wiener process)

Let $W(t) = \{W_1(t), W_2(t), \ldots, W_d(t)\} (d \geq 3)$ be a Wiener process.

DEFINITION. $W(t)$ is crossing the sphere - surface $\{x : \|x\| = R\}$ $\theta = \theta(R)$ times if $\theta(R)$ is the largest integer for which there exists a random sequence $0 < \alpha_1 = \alpha_1(R) < \beta_1 = \beta_1(R) < \alpha_2 = \alpha_2(R) < \beta_2 = \beta_2(R) \ldots < \alpha_\theta = \alpha_\theta(R) < \beta_\theta =$

$\beta_\theta(R) < \infty$ such that

$$
\begin{aligned}
\|W(t)\| &< R \quad \text{if} \quad t < \alpha_1, \\
\|W(\alpha_1)\| = R, R-1 < \|W(t)\| &< R+1 \quad \text{if} \quad \alpha_1 \le t < \beta_1, \\
\|W(\beta_1)\| = R-1 \text{ or } R+1, \|W(t)\| &\ne R \quad \text{if} \quad \beta_1 \le t < \alpha_2, \\
\|W(\alpha_2)\| = R, R-1 < \|W(t)\| &< R+1 \quad \text{if} \quad \alpha_2 \le t < \beta_2, \\
\|W(\beta_2)\| = R-1 \text{ or } R+1, \|W(t)\| &\ne R \quad \text{if} \quad \beta_2 \le t < \alpha_3, \ldots, \\
\|W(\alpha_\theta)\| = R, R-1 < \|W(t)\| &< R+1 \quad \text{if} \quad \alpha_\theta \le t < \beta_\theta, \\
\|W(\beta_\theta)\| = R+1, \ \|W(t)\| &> R \quad \text{if} \quad t \ge \beta_\theta.
\end{aligned}
$$

$(\theta(R))^{-1}$ will be called the SPEED OF ESCAPE in R.

THEOREM 7

$$\lim_{R\to\infty} \mathbf{P}\{\frac{(d-2)\theta(R)}{2R} < t\} = 1 - e^{-t} \quad (t \ge 0).$$

The proof is based on the following two lemmas.

LEMMA C (F.B. Knight, 1981 p. 103). *For any $d \ge 3$*

$$\mathbf{P}(W(t) \in Q(r,u) \text{ eventually}) = (\frac{r}{R})^{d-2}$$

provided $\|u\| = R \ge r$. Further for any $0 < a < x < b < \infty$

$$p(a,x,b) = \mathbf{P}\{B(a,b,t) \mid \|W(t)\| = x\} = \frac{x^{2-d} - b^{2-d}}{a^{2-d} - b^{2-d}}$$

where

$$B(a,b,t) = \{\inf\{s : s > t, \|W(s)\| = a\} < \inf\{s : s > t, \|W(s)\| = b\}\}.$$

LEMMA 12

$$\mathbf{P}\{\theta(R) = k\} = \lambda(1-\lambda)^{k-1} \quad (k = 1, 2, \ldots)$$

where

$$\lambda = p(R)(1 - (\frac{R}{R+1})^{d-2})$$

and

$$p(R) = p(R-1, R, R+1).$$

PROOF. Clearly we have

$$\begin{aligned}
\mathbf{P}\{\theta(R)=1\} &= \lambda \\
\mathbf{P}\{\theta(R)=2\} &= q(R)p(R)(1-(\frac{R}{R+1})^{d-2})+ \\
& p(R)(\frac{R}{R+1})^{d-2}p(R)(1-\frac{R}{R+1})^{d-2})= \\
& p(R)(1-(\frac{R}{R+1})^{d-2})[q(R)+p(R)(\frac{R}{R+1})^{d-2}]=\lambda(1-\lambda)
\end{aligned}$$

where $q(R)=1-p(R)$. Similarly

$$\mathbf{P}(\theta(R)=k)=$$
$$\sum_{j=0}^{k-1}\binom{k-1}{j}(q(R))^j(p(R)(\frac{R}{R+1})^{d-2})^{k-1-j}p(R)(1-(\frac{R}{R+1})^{d-2})=$$
$$\lambda(q(R)+p(R)(\frac{R}{R+1})^{d-2})^{k-1}=\lambda(1-\lambda)^{k-1}.$$

Hence we have Lemma 12.

Observe that

$$\mathbf{E}\theta(R)=\frac{1}{\lambda}=\frac{1}{p(R)}\frac{1}{1-(10\frac{1}{R+1})^{d-2}}=\frac{R+1+o(1)}{p(R)(d-2)}.$$

Since $p(R)\to 1/2(R\to\infty)$ we have

$$\lim_{R\to\infty}\frac{\mathbf{E}\theta(R)}{R}=\frac{2}{d-2}. \tag{10}$$

Lemma 12 together with (10) easily implies Theorem 7. Studying the properties of the process $\{\theta(R),R>0\}$ a natural question is the following: does a sequence $0<R_1<R_2\ldots$ exist for which

$$\lim_{n\to\infty}R_n=\infty \text{ and } \theta(R_i)=1\ i=1,2,\ldots$$

The answer of this question is affirmative. In fact we prove a much stronger theorem. In order to formulate this theorem we introduce the following

DEFINITION. Let $\psi(R)$ be the largest integer for which there exists a positive integer $u=u(R)\le R$ such that

$$\theta(k)=1 \text{ for any } u\le k\le u+\psi(R).$$

It is natural to say that the speed of escape in the interval $(u,u+\psi(R))$ is maximal.

THEOREM 8

$$\psi(r) \geq \frac{\log\log R}{\log 2} \quad i.o.\ a.s.$$

REMARK 6. I do not know the analogue of this Lemma for simple symmetric random walk. Lemmas B and 5 can be considered as analogous forms in special cases.

PROOF OF THEOREM 8. Let

$$A(R) = \{\theta(k) = 1 \text{ for every } R \leq k \leq R + [\frac{\log\log R}{\log 2}]\}.$$

Then

$$\mathbf{P}(A(R)) = \frac{1+o(1)}{\log R}\frac{d-2}{2R}$$

and

$$\mathbf{P}(A(R)A(R+S)) = \frac{1+o(1)}{\log R}\frac{1}{S - \frac{\log\log R}{\log 2}}\frac{1}{\log R}\frac{d-2}{2R}$$

if $\frac{\log\log R}{\log 2} < S = o(R)$. In case $S \geq O(R)$ the events $A(R)$ and $A(R+S)$ are asymtotically independent. Hence

$$\sum_{R=1}^{n} \mathbf{P}(A(R)) = (\frac{d-2}{2} + o(1))\log\log n$$

and for any $\varepsilon > 0$ if n is big enough we have

$$\sum_{R=1}^{n} \sum_{S=[\frac{\log\log R}{\log 2}]} \mathbf{P}(A(R)A(R+S)) = (\frac{d-2}{2})^2(\log\log n)^2(1+\varepsilon)$$

which implies Theorem 8 by Borel - Cantelli lemma.

CONJECTURE.

$$\lim_{R\to\infty} \frac{\psi(R)}{\log\log R} = \frac{1}{\log 2} \text{ a.s.}$$

Theorem 8 clearly implies that $\theta(R) = 1$ i.o. a.s.

It is natural to ask: how big $\theta(R)$ can be. An answer of this question is

THEOREM 9 *For any $\varepsilon > 0$ we have*

$$\theta(R) \leq 2(d-2)^{-1}(1+\varepsilon)R\log R \ a.s.$$

if R is big enough and

$$\theta(R) \geq (1-\varepsilon)R\log\log\log R \ i.o.\ a.s.$$

Since this result is far from the best possible one and the proof is trivial we omit it.

ACKNOWLEDGEMENT. I am indebted to the referee who pointed out a number of mistakes in the original manuscript and even gave ideas how to correct them. My best thanks to Prof. E. Bolthausen for his valuable suggestions.

REFERENCES

Dvoretzky, A. - Erdős, P. (1950) Some problems on random walk in space. *Proc. Second Berkeley Symposium*, 353-368.

Erdős, P. - Révész, P. (1976) On the length of the longest head-run. *Topics in Information Theory. Coll. Math. Soc. J. Bolyai* **16**, 219-228. Ed. Csiszár, I. - Elias, P.

Erdős, P. - Révész, P. (1988) On the area of the circles covered by a random walk. *J. Mult. Anal.* **27**, 169-180.

Erdős, P. - Taylor, S.J. (1960) Some problems concerning the structure of random walk paths. *Acta Math. Acad. Sci. Hung.* **11**, 137-162.

Knight, F.B. (1981) *Essentials of Brownian motion and diffusion.* Math. Surveys No. 18 Am. Math. Soc. Providence R.I.

ALMOST SUPERADDITIVE ERGODIC THEOREMS IN THE MULTIPARAMETER CASE

K. Schürger
University of Bonn, Department of Economics
Bonn, FR Germany

1. Introduction

Derriennic [4] obtained a pointwise ergodic theorem for processes satisfying an almost subadditivity condition (cf. (2.2) below) which is weaker than Kingman's [7] subadditivity condition (cf. also [10]). There are several possibilities to extend Kingman's notion of subadditivity to the multiparameter case; cf. e.g. Smythe [12], Akcoglu and Krengel [1] and Schürger [11].

The main purpose of the present paper is to show that in the multiparameter case a pointwise ergodic theorem (Theorem 3.2 in Section 3) can be obtained for processes satisfying a *strong almost superadditivity condition* (cf. Definition 2.1 in Section 2) which is weaker than the superadditivity condition introduced in [1]. We also show (cf. Theorem 3.4) that some of the hypotheses in our pointwise ergodic theorem can be considerably relaxed if the process under consideration has monotone sample paths. In Section 4 we show that our pointwise ergodic theorem (Theorem 3.2) applies e.g. to the travelling salesman problem, starting with any stationary integrable point process Π on $\mathbb{R}^d$. The almost sure convergence theorem obtained was previously known only when Π is a Poisson point process with intensity one (cf. [2],[13]).

ISBN 0-12-231050-0

2. Notations and Basic Definitions

If $u = (u_1, \ldots, u_d)$ and $v = (v_1, \ldots, v_d)$ are vectors of $\mathbb{R}^d$, $u \leq v$ $(u < v)$ means $u_i \leq v_i$ $(u_i < v_i)$, $1 \leq i \leq d$ (instead of $u \leq v$ and $u < v$ we also write $v \geq u$ and $v > u$, respectively). Let $e = (1, \ldots, 1) \in \mathbb{Z}_+^d$; 0 will also denote the null vector $(0, \ldots, 0) \in \mathbb{Z}_+^d$. Put $[u, v[= \{w \in \mathbb{R}^d : u \leq w < v\}$, $u, v \in \mathbb{R}^d$. The index set of the multiparameter processes under consideration will be $\mathcal{I} = \mathcal{I}(d) = \{[u, v[: u < v,\ u, v \in \mathbb{Z}_+^d\}$, $d \geq 1$. Let $I + u = \{v + u : v \in I\}$, $u \in \mathbb{R}^d$, $I \in \mathcal{I}(d)$, and $\alpha I = \{\alpha u : u \in I\}$, $\alpha \in \mathbb{R}_+$, $I \in \mathcal{I}(d)$. In the sequel let $X = (X_I) \subset L^1$, and $Y = (Y_I) \subset L_+^1$, $I \in \mathcal{I}(d)$, be any two families of real random variables defined on a common probability space $(\Omega, \mathcal{A}, P)$. Put $\overline{X}_I = \frac{1}{\mu(I)} X_I$, $\overline{Y}_I = \frac{1}{\mu(I)} Y_I$, $I \in \mathcal{I}(d)$ (μ denoting the d-dimensional Lebesgue measure).

Definition 2.1. *Let $d \geq 1$. We say that $X = (X_I)$ satisfies the strong almost superadditivity condition (SASC) w.r.t. $Y = (Y_I)$ if the following condition is satisfied:*

For any disjoint sets $I_1, \ldots, I_k \in \mathcal{I}(d)$ such that $I_1 \cup \ldots \cup I_k \in \mathcal{I}(d)$, we have

$$X_{I_1 \cup \ldots \cup I_k} \geq \sum_{j=1}^{k} (X_{I_j} - Y_{I_j}). \tag{2.1}$$

If (2.1) holds in the case when $Y_I \equiv 0$, $I \in \mathcal{I}(d)$, we arrive at a notion of subadditivity introduced in [1] (cf. also [8, p. 201]).

Remark 2.2. In the case $d = 1$, Derriennic [4] (see also [9],[10]) has introduced the following kind of almost subadditivity of a process $X = (X_I) \subset L^1$ w.r.t. a process $Y = (Y_I) \subset L_+^1$ (thereby weakening Kingman's [7] subadditivity condition):

$$X_{[s,u[} \leq X_{[s,t[} + X_{[t,u[} + Y_{[t,u[} \text{ for all integers } 0 \leq s < t < u. \tag{2.2}$$

Repeated application of (2.2) shows that then $(-X_I)$ is strongly almost superadditive in our sense. A different extension of (2.2) to the multiparameter case has been introduced in [11].

Let $U = (U_I)$ and $V = (V_I)$, $I \in \mathcal{I}(d)$, be families of real random variables defined on a common probability space. U will be called *stationary* if the finite-dimensional distributions of the random variables U_{I+u}, $I \in \mathcal{I}(d)$, do not depend on $u \in \mathbb{Z}_+^d$. U and V are called *jointly stationary* if the R^2-valued process $((U_I, V_I))$ is stationary.

Let $(I(t)) \subset \mathcal{I}$ be a sequence such that

$$I(t) = [0, n(t)[,\ t \geq 1, \text{ and } n(t) \to \infty \text{ as } t \to \infty \tag{2.3}$$

(here, $n(t) \to \infty$ means that $n_i(t) \to \infty, 1 \le i \le d$). The sequence $(I(t))$ is called *regular* (or, more precisely, *K-regular*) if there exists an increasing sequence $(I'(t)) \subset \mathcal{I}$ and a constant $K > 0$ such that for $t = 1, 2, \ldots$

$$I(t) \subset I'(t), \ \mu(I'(t)) \le K\mu(I(t)). \tag{2.4}$$

Let $(n(t)) \subset \mathbb{Z}_+^d$ be a sequence such that $n(t) \ge e, \ t \ge 1$, and $n(t) \to \infty$ as $t \to \infty$. We will say that $(n(t))$ (or the corresponding sequence $([0, n(t)[) \subset \mathcal{I}(d))$ *remains in a sector of* $\mathbb{Z}_+^d$, provided we have $(n(t)) \subset S_c(d)$ for some constant $c \ge 1$. Here, the *c-sector* $S_c(d)$ $(c \ge 1)$ of $\mathbb{Z}_+^d$ is given by

$$S_c = S_c(d) = \left\{ u \in \mathbb{Z}_+^d \ : \ u \ge e, \ \frac{u_i}{u_j} \le c, \ 1 \le i, j \le d \right\}.$$

3. Almost superadditive ergodic theorems

The proofs of our convergence results will be based on a rather elementary maximal ergodic inequality (see Lemma 3.1 below) generalizing a corresponding result of Akcoglu and Krengel [1]; see also [8, Theorem 6.2.6].

Lemma 3.1. *Suppose X satisfies the SASC w.r.t. Y. Let X and Y be jointly stationary and suppose that $X_{[w,w+e[} \ge Y_{[w,w+e[}$, $w \in \mathbb{Z}_+^d$. Let $I(t)$, $1 \le t \le t_0$, and $I'(1) \subset I'(2) \subset \ldots \subset I'(t_0)$ be sets in $\mathcal{I}(d)$ such that (2.4) holds for $1 \le t \le t_0$. Finally, let $A, B \in \mathcal{I}(d)$ be any sets such that $I(t) + u \subset B$, $u \in A$, $1 \le t \le t_0$. Then we have for $\alpha > 0$*

$$P\left\{ \max_{1 \le t \le t_0} (\overline{X}_{I(t)} - \overline{Y}_{I(t)}) \ge \alpha \right\} \le \frac{2^d K}{\alpha \mu(A)} E[X_B] \tag{3.1}$$

(the constant K figuring in (2.4)).

The proof of Lemma 3.1 (which will be omitted) parallels that given in [1], [8], and uses a modification of a certain covering lemma of Wiener [15].

We will say that a process $Y = (Y_I) \subset L_+^1$ satisfies *Condition* REG (SEC) if for each sequence $(I(t)) \subset \mathcal{I}$ which is regular (remains in a sector of $\mathbb{Z}_+^d$) we have that

$$\lim_{t \to \infty} \overline{Y}_{I(t)} = 0 \text{ a.s.}$$

The following result which is our main convergence theorem, generalizes Theorem (2.7) of [1] (see also [8, Theorem 6.2.9]); its proof will be based on Lemma 3.1.

Theorem 3.2. *Suppose X satisfies the SASC w.r.t. Y, and let X and Y be jointly stationary. Suppose the limit*

$$\lim_{t\to\infty} E\left[\overline{X}_{[0,te[}\right] = \gamma(X) \tag{3.2}$$

exists and is finite. Finally assume that

$$\liminf_{t\to\infty} E\left[\overline{Y}_{[0,te[}\right] = 0. \tag{3.3}$$

(i) If, for the sequence $I(t) = [0, te[,\ t = 1, 2, \ldots,$ we have

$$\lim_{t\to\infty} \overline{Y}_{I(t)} = 0 \text{ a.s.}, \tag{3.4}$$

then

$$\lim_{t\to\infty} \overline{X}_{I(t)} = X_\infty \text{ exists a.s.} \tag{3.5}$$

(ii) If Y satisfies REG (SEC), the almost sure limit X_∞ in (3.5) exists for each sequence $(I(t))$ which is regular (remains in a sector of $\mathbb{Z}_+^d$), where X_∞ does not depend on $(I(t))$.

Proof. Suppose Y satisfies REG. Let $\varepsilon > 0$ be given. For the rest of the proof fix any integer $s \geq 1$ satisfying

$$\left|E\left[\overline{X}_{[0,te[}\right] - \gamma(X)\right| < \varepsilon,\ t = s, s+1, \ldots \tag{3.6}$$

and

$$E\left[\overline{Y}_{[0,se[}\right] < \varepsilon \tag{3.7}$$

(the existence of s follows from (3.2) and (3.3)). Let $(I(t))$ be a fixed K-regular sequence. Let the family $Z = (Z_I),\ I \in \mathcal{I}$, be given by

$$Z_I = X_I - \sum_{w \in I \cap \mathbb{Z}_+^d} \left(X_{[w,w+e[} - Y_{[w,w+e[}\right),\ I \in \mathcal{I}. \tag{3.8}$$

An application of Tempel'man's [14] ergodic theorem (see also [8, Theorem 6.2.8]) yields that

$$\lim_{t\to\infty} \left(\overline{X}_{I(t)} - \overline{Z}_{I(t)}\right) \text{ exists a.s.}$$

For the proof of (3.5) it therefore suffices to show that

$$\lim_{t\to\infty} \overline{Z}_{I(t)} \text{ exists a.s.} \tag{3.9}$$

(note that the SASC implies $Z_I \geq 0,\ I \in \mathcal{I}$). Hence, if we put

$$\overline{Z} = \limsup_{t\to\infty} \overline{Z}_{I(t)},\ \underline{Z} = \liminf_{t\to\infty} \overline{Z}_{I(t)},$$

we have to show

$$\overline{Z} < \infty \text{ a.s.} \tag{3.10}$$

and

$$\underline{Z} = \overline{Z} \text{ a.s.} \tag{3.11}$$

Put

$$\mathcal{I}_s = \mathcal{I}_s(d) = \left\{[su, sv[\in \mathcal{I}(d) :\ u < v,\ u, v \in \mathbb{Z}_+^d\right\} \tag{3.12}$$

(s figuring in (3.6) and (3.7)). Let $I_*(t),\ t \geq 1$, denote the largest set in $\mathcal{I}_s$ such that $I_*(t) \subset I(t)$ (we may assume $[0, se[\ \subset I(t),\ t \geq 1$). Similarly, let $I^*(t),\ t \geq 1$, denote the smallest set in $\mathcal{I}_s$ such that $I(t) \subset I^*(t)$. Clearly

$$\lim_{t\to\infty} \frac{\mu\big(I_*(t)\big)}{\mu\big(I^*(t)\big)} = 1. \tag{3.13}$$

The sequences $\big(J(t)\big) \subset \mathcal{I}_s$ and $\big(K(t)\big) \subset \mathcal{I}$, given by $J(2t-1) = sK(2t-1) = I_*(t),\ J(2t) = sK(2t) = I^*(t),\ t \geq 1$, are both $4^d K$-regular. Let us put

$$\overline{Z}_s = \limsup_{t\to\infty} \overline{Z}_{J(t)},\ \underline{Z}_s = \liminf_{t\to\infty} \overline{Z}_{J(t)}.$$

Let us show

$$\overline{Z} \leq \overline{Z}_s \text{ a.s.} \tag{3.14}$$

and

$$\underline{Z} \geq \underline{Z}_s \text{ a.s.} \tag{3.15}$$

Clearly, (3.14) is a consequence of

$$\overline{Z} \leq \limsup_{t\to\infty} \overline{Z}_{I^*(t)} \text{ a.s.} \tag{3.16}$$

In order to verify (3.16), first note that Z satisfies the SASC w.r.t. Y, which yields

$$Z_{I^*(t)} \geq Z_{I(t)} - Y_{I(t)} + \sum_{w\in(I^*(t)\setminus I(t))\cap\mathbb{Z}_+^d} \big(Z_{[w,w+e[} - Y_{[w,w+e[}\big)\,.$$

Applying Tempel'man's ergodic theorem and taking into account (3.13) we arrive at (3.16) (recall that Y is supposed to satisfy REG). Relation (3.15) can be obtained in a similar way. It can be seen from the rest of the proof (read backwards) that $\overline{Z}_s < \infty$ a.s. which, by (3.14), implies (3.10). If V_I, $I \in \mathcal{I}$, are any random variables, let $V^{(s)} = V_{sI}$, $I \in \mathcal{I}$, and put

$$_sU_I = Z_I^{(s)} - \sum_{w \in I \cap \mathbb{Z}_+^d} \left(Z_{[w,w+e[}^{(s)} - 2Y_{[w,w+e[}^{(s)} \right), \; I \in \mathcal{I}. \tag{3.17}$$

Using Tempel'man's ergodic theorem once more and noting that Y satisfies REG, gives that

$$\overline{Z}_s - \underline{Z}_s \leq \limsup_{t \to \infty} \frac{1}{\mu(J(t))} \left({}_sU_{K(t)} - Y_{K(t)}^{(s)} \right) \text{ a.s.}$$

By (3.14) and (3.15), this implies

$$P\{\overline{Z} - \underline{Z} \geq \alpha\} \leq P\left\{ \sup_{t \geq 1} ({}_s\overline{U}_{K(t)} - \overline{Y}_{K(t)}^{(s)}) \geq \alpha s^d \right\}, \; \alpha > 0 \tag{3.18}$$

(note that $\mu(J(t)) = s^d \mu(K(t))$, $t \geq 1$). Fix any integer $t_0 \geq 1$, and let $1 \leq t_1 < t_2$ be any integers such that

$$K(t) + u \subset B, \; u \in A, \; 1 \leq t \leq t_0, \tag{3.19}$$

where we put $A = [0, t_1 e[$, $B = [0, t_2 e[$. Since $_sU$ satisfies the SASC w.r.t. $Y^{(s)}$, and

$$_sU_I \geq \sum_{w \in I \cap \mathbb{Z}_+^d} Y_{[w,w+e[}^{(s)}, \; I \in \mathcal{I},$$

we can apply Lemma 3.1 which yields (taking into account (3.19)) that, for all $\alpha > 0$,

$$P\left\{ \max_{1 \leq t \leq t_0} ({}_s\overline{U}_{K(t)} - \overline{Y}_{K(t)}^{(s)}) \geq \alpha s^d \right\} \leq \left(\frac{8t_2}{t_1} \right)^d \frac{K}{\alpha \mu(sB)} E[{}_sU_B]. \tag{3.20}$$

An easy calculation yields

$$E[{}_sU_B] = E[X_{sB}] - \mu(B) E\left[X_{[0,se[}\right] + 2\mu(B) E\left[Y_{[0,se[}\right]$$

which, by (3.6) and (3.7), implies

$$\frac{1}{\mu(sB)} E[{}_sU_B] \leq 4\varepsilon. \tag{3.21}$$

Since the integers $1 \leq t_1 < t_2$ satisfying (3.19) can be chosen in such a way that t_2/t_1 is arbitrarily close to one, we deduce from (3.18), (3.20) and (3.21)

$$P\{\overline{Z} - \underline{Z} \geq \alpha\} \leq \frac{4 \cdot 8^d \varepsilon K}{\alpha}, \ \varepsilon > 0, \alpha > 0.$$

This proves (3.11) (clearly, (3.10) is an easy consequence of (3.20) and (3.14)). The above reasoning also proves the first part of Theorem (3.2) since in the case $I(t) = [0, te[$, $t \geq 1$, each of the sets $I_*(t)$ and $I^*(t)$ is of the form $[0, t'e[$ for some integer $t' \geq 0$. Finally let us suppose that Y satisfies SEC. In this case, the proof given above shows that the limit in (3.5) exists almost surely for each regular sequence $(I(t))$ which remains in a sector of $\mathbb{Z}_+^d$. Now consider any sequence $(I(t))$, $I(t) = [0, n(t)[$, $t \geq 1$, which remains in a c-sector of $\mathbb{Z}_+^d$. Arguing similarly as in [8, p. 203], let $t_1, t_2, \ldots$ be a permutation of the numbers $1, 2, \ldots$ such that $\max\{n_j(t_i) : 1 \leq j \leq d\}$, $i = 1, 2, \ldots$, is increasing. Since $(I(t_i))$ is c^d-regular, $(\overline{X}_{I(t_i)})$ converges almost surely as $i \to \infty$, which, in turn, implies the almost sure convergence of $(\overline{X}_{I(t)})$. This completes the proof of Theorem 3.2.

Sometimes it may be difficult to verify Condition (3.4) of Theorem 3.2. It turns out, however, that (3.4) can be replaced by a rather weak moment condition on Y (see Theorem 3.4 below) provided $(X_{[0,te[})$ is *pathwise monotone*, i.e., each sample path of $(X_{[0,te[})$ is increasing or decreasing (the kind of monotonicity may depend on the sample path).

Definition 3.3. We will say that $X = (X_I)$ satisfies the *2-almost superadditivity condition* (2-ASC) w.r.t. $Y = (Y_I)$ provided X and Y have the following property:

For any disjoint sets $I_1, I_2 \in \mathcal{I}$ such that $I_1 \cup I_2 \in \mathcal{I}$ and $0 \in I_1$, we have

$$X_{I_1 \cup I_2} \geq X_{I_1} + X_{I_2} - Y_{I_2}. \tag{3.22}$$

Theorem 3.4. *Suppose X satisfies the SASC w.r.t. Y, and let X and Y be jointly stationary. Suppose the limit*

$$\lim_{t \to \infty} E\left[\overline{X}_{[0,te[}\right] = \gamma(X) \tag{3.23}$$

exists and is finite. Let there exist an integer $p \geq 2$ such that

$$\sum_{t=1}^{\infty} E\left[\overline{Y}_{[0,p^t re[}\right] < \infty, \quad r = 1, 2, \ldots \tag{3.24}$$

(i) We have that

$$\lim_{t \to \infty} \overline{X}_{[0,p^t re[} = X_\infty^{(r)} \textit{ exists a.s.}, r = 1, 2, \ldots \tag{3.25}$$

(ii) Additionally, let X satisfy the 2-ASC w.r.t. Y, and suppose that

$$\lim_{t \to \infty} E\left[\overline{Y}_{[0,te[}\right] = 0. \tag{3.26}$$

Furthermore assume that, for all $N \geq e$ and $1 \leq j \leq d-1$,

$$\limsup_{t \to \infty} E\left[\overline{Y}_{[0,N_1[\times \ldots \times [0,N_j[\times[0,t[\times\ldots\times[0,t[}\right] = \kappa_j(N_1, \ldots, N_j) < \infty, \tag{3.27}$$

where

$$\lim_{t \to \infty} \kappa_j(t, \ldots, t) = 0, \quad 1 \leq j \leq d-1. \tag{3.28}$$

Then we have that

$$\lim_{t \to \infty} \overline{X}_{[0,te[} = X_\infty \textit{ exists in } L^1, \tag{3.29}$$

so that, in particular,

$$X_\infty^{(r)} = X_\infty \textit{ a.s.}, \quad r = 1, 2, \ldots \tag{3.30}$$

($X_\infty^{(r)}$ figuring in (3.25)). Thus, if $(X_{[0,te[})$ is pathwise monotone,

$$\lim_{t \to \infty} \overline{X}_{[0,te[} = X_\infty \textit{ a.s.} \tag{3.31}$$

Proof. (i) Let $p \geq 2$ be as in (3.24). In order to prove (3.25), apply the reasoning in the proof of Theorem 3.2 to the sequence $(I(t))$ given by $I(t) = [0, p^t re[$, $t \geq 1$, for fixed $r \geq 1$. In fact, first note that, by (3.24),

$$\lim_{t \to \infty} \overline{Y}_{[0,p^t re[} = 0 \text{ a.s.}, \quad r = 1, 2, \ldots$$

It follows from (3.23) and (3.24) that the integer s (in (3.6) and (3.7)) can be chosen to be equal to $p^{t_0} r$ for a suitable integer $t_0 \geq 1$. Then, the sets $I_*(t)$ and $I^*(t)$ in the proof of Theorem 3.2 are both equal to $I(t)$, $t \geq t_0$. Proceeding further as in the proof of Theorem 3.2, we end up with (3.25).

(ii) The proof of (3.29) proceeds similarly as the proof of a more general result in [11]. If $(X_{[0,te[})$ is assumed to be pathwise monotone, (3.31) is a consequence of (3.25), (3.30) and [9, Theorem 3.2].

4. Applications

In this section we will show how Theorem 3.2 applies e.g. to the travelling salesman problem (cf. [2],[6],[13]). In the sequel, let Π be any stationary point process on $\mathbb{R}^d$ $(d \geq 2)$, and denote by $\Pi(A)$, $A \subset R^d$, the set of points of Π contained in A. Π will be called *integrable* if $E\big[\mid \Pi(A) \mid\big] < \infty$ for all bounded Borel sets $A \subset R^d$ ($\mid B \mid$ denoting the cardinal number of a set B). If $A \subset R^d$ is finite, let $L_1(A)$ denote the length of the shortest path through the points of A. We can now define a process $X = (X_I)$, $I \in \mathcal{I}(d)$, by putting

$$X_I = L_1\big(\Pi(I)\big), \quad I \in \mathcal{I}(d) \tag{4.1}$$

in the case when Π is integrable.

Lemma 4.1. *Let Π be any stationary integrable point process on $\mathbb{R}^d$ $(d \geq 2)$. Then the process $(-X_I)$ (where X_I, $I \in \mathcal{I}(d)$, is given by (4.1)) satisfies the SASC w.r.t. the process $Y = (Y_I) \subset L^1_+$, given by*

$$Y_{[u,v[} = (2^d + 2)\sum_{i=1}^{d}(v_i - u_i), \quad [u,v[\; \in \mathcal{I}(d). \tag{4.2}$$

Proof. Fix any disjoint sets $I_j = [u^{(j)}, v^{(j)}[\; \in \mathcal{I}(d)$, $1 \leq j \leq k$, such that $I = I_1 \cup \ldots \cup I_k \in \mathcal{I}(d)$. If $\Pi(I_j) \neq \emptyset$, let π_j be a shortest path through the points in $\Pi(I_j)$. In order to construct a suitable path π_0 through the points in $\Pi(I)$, we first introduce an undirected multigraph G without loops as follows. The vertex set of G, consisting of points $Q_1, \ldots, Q_t$ (say), is the set of corners of the d-dimensional rectangles I_j, $1 \leq j \leq k$. Connect Q_r and Q_s $(r \neq s)$ by a line segment (*taken twice*) provided Q_r and Q_s belong to the *same* edge of some set I_j, *and* there does not exist any further vertex in $\{Q_1, \ldots, Q_t\}$ located between Q_r and Q_s. Clearly, G is connected, and each vertex of G has an even degree. Hence (cf. [3, p. 14]) there exists an Euler trail $\widetilde{\pi}_0$ through the vertices of G, containing each line segment of G exactly once. The desired path π_0 can now be constructed as follows. Walking along $\widetilde{\pi}_0$ and arriving of $u^{(j)}$ *for the first time* (assuming $\Pi(I_j) \neq \emptyset$), we visit, starting at $u^{(j)}$, one of the endpoints of π_j, walk along π_j, return to $u^{(j)}$ and continue to walk along $\widetilde{\pi}_0$. This shows (diam (A) denoting the diameter of a set $A \subset \mathbb{R}^d$)

$$X_{I_1 \cup \ldots \cup I_k} \leq \sum_{j=1}^{k} X_{I_j} + 2\sum_{j=1}^{k} diam(I_j) + 2^d \sum_{j=1}^{k}\sum_{i=1}^{d}\left(v_i^{(j)} - u_i^{(j)}\right)$$

which implies the assertion of the lemma.

Theorem 4.2. *Let Π be any stationary integrable point process on $\mathbb{R}^d$ $(d \geq 2)$, and let $X = (X_I)$ be given by (4.1). Then*

$$\lim_{t \to \infty} \overline{X}_{I(t)} \text{ exists a.s.}$$

for each sequence $(I(t)) \subset \mathcal{I}(d)$ which is regular or remains in a sector of $\mathbb{Z}_+^d$.

In order to prove Theorem 4.2, we need the following result which can be obtained in a straightforward way by utilizing a construction of Few [5] (its proof will be omitted).

Lemma 4.3. *If $L(t)$ denotes the length of a shortest path through $t \geq 1$ points contained in a set $I = [u, v[\in \mathcal{I}(d)$ $(d \geq 2)$, we have, putting $w = v - u$,*

$$L(t) \leq 2(d+1)^{d-1} \frac{(w_1 + \ldots + w_d)^2}{\max_{1 \leq i \leq d} w_i} t^{\frac{(d-1)}{d}}$$

$$\leq 2d(d+1)^{d-1}(w_1 + \ldots + w_d) t^{\frac{(d-1)}{d}}. \tag{4.3}$$

Proof of Theorem 4.2. Let us show that $(-X_I)$ and (Y_I) (given by (4.2)) satisfy the hypotheses of Theorem 3.2. In view of Lemmas 4.1 and 4.3 only (3.2) remains to be verified. Put $a(t) = E[X_{[0,te[}]$, $t = 1, 2 \ldots$. The sequence $(a(t))$ satisfies the inequalities

$$a(km) \leq m^d a(k) + ckm^d, \quad k, m = 1, 2 \ldots, \tag{4.4}$$

where $c = (2^d + 2)d$. This can be easily seen by decomposing $[0, kme[$ into m^d congruent hypercubes and using the fact that $(-X_I)$ is stationary and satisfies the SASC w.r.t. (Y_I) given by (4.2). Putting $b = \liminf_{t \to \infty} t^{-d} a(t)$, we have (by (4.3)) $0 \leq b < \infty$. Now fix any $\varepsilon > 0$ and let $k \geq 1$ be such that

$$c < \varepsilon k, \quad a(k) < (b + \varepsilon) k^d. \tag{4.5}$$

For integers $t \geq 2k$ let $m = m(t) \geq 2$ be such that $(m-1)k < t \leq mk$. By (4.4),

$$\frac{1}{t^d} a(t) \leq \frac{1}{t^d} a(km) \leq \frac{1}{(m-1)^d k^d} \left(m^d a(k) + ckm^d \right), \quad t \geq 2k.$$

By (4.5), this entails

$$\limsup_{t \to \infty} \frac{1}{t^d} a(t) \leq b + 2\varepsilon.$$

This shows that Condition (3.2) is satisfied. Hence Theorem 4.2 follows from Theorem 3.2.

Remark 4.4 The almost sure convergence of $\overline{X}_{I(t)})$, $X_{I(t)} = L_1\left(\Pi\big(I(t)\big)\right)$, $t \geq 1$, was previously only known in the case when Π is a Poisson point process on $\mathbb{R}^d$ with intensity one, and $I(t) = [0, te[$, $t \geq 1$ (cf. [2],[6],[13]). We would like to point out that Theorem 3.2 can also be applied in order to obtain almost sure convergence results for the matching problem of Papadimitriou, Steiner trees [13] as well as minimal triangulations.

References

[1] **Akcoglu, M.A. and Krengel, U. (1981):** Ergodic theorems for superadditive processes. *J. Reine Angew. Math. 323, 53–67.*

[2] **Beardwood, J., Halton, J.H. and Hammersley, J.M. (1959):** The shortest path through many points. *Proc. Camb. Phil. Soc. 55, 299–327.*

[3] **Bollobás, B. (1979):** Graph Theory. *Springer-Verlag, Berlin.*

[4] **Derriennic, Y. (1983):** Un théorème ergodique presque sous-additif. *Ann. Prob. 11, 669–677.*

[5] **Few, L. (1955):** The shortest path and the shortest road through n points. *Mathematika 2, 141–144.*

[6] **Karp, R.M. and Steele, J.M. (1985):** Probabilistic analysis of heuristics. In *The Traveling Salesman Problem*, ed. E.L. Lawler, J.K. Lenstra, A.H.G. Rinnooy Kan and D.B. Shmoys. *Wiley, New York, 181–205.*

[7] **Kingman, J.F.C. (1968):** The ergodic theory of subadditive stochastic processes. *J. R. Statist. Soc. B 30, 499–510.*

[8] **Krengel, U. (1985):** Ergodic Theorems. *De Gruyter, Berlin.*

[9] **Schürger,K. (1986):** A limit theorem for almost monotone sequences of random variables. *Stoch. Proc. Appl. 21, 327–338.*

[10] **Schürger, K. (1988):** On Derriennic's almost subadditive ergodic theorem. *Acta Math. Hungar. 52. To appear.*

[11] **Schürger, K. (1988):** Almost subadditive multiparameter ergodic theorems. *Stoch. Proc. Appl. To appear.*

[12] **Smythe, R.T. (1976):** Multiparameter subadditive processes. *Ann. Prob. 4, 772–782.*

[13] **Steele, J.M. (1981):** Subadditive Euclidean functionals and nonlinear growth in geometric probability. *Ann. Prob. 9, 365–376.*

[14] **Tempel'man, A.A. (1972):** Ergodic theorems for general dynamical systems. *Trans. Moscow Math. Soc. 26, 94–132.*

[15] **Wiener, N. (1939):** The ergodic theorem. *Duke Math. J. 5, 1–18.*

On Convergence of Vector–Valued Mils Indexed by a Directed Set

Zhen–Peng Wang[1*] and Xing–Hong Xue[2]

[1] Department of Statistics, East China Normal University, Shanghai, China

[2] Department of Statistics, Columbia University, New York, NY 10027, USA

Abstract. Let E be a Banach space with the Radon–Nikodym property, and J a directed set. We show that if a stochastic basis $(\mathcal{F}_t, J)$ satisfies the Vitali condition V, then E–valued mils of class (d) converge essentially in the norm topology. In a real–valued case, the condition (d) is weakened to a one–sided condition.

1. Notations and Summary. Let E be a Banach space, $(\Omega, \mathcal{F}, P)$ a fixed probability space, J a directed set filtering to the right with the order $\leq$, and $(\mathcal{F}_t, J)$ a stochastic basis of $\mathcal{F}$, i.e., $(\mathcal{F}_t, J)$ is an increasing family of sub–σ–algebras of $\mathcal{F}$. Throughout this paper, (X_t) is an E–valued, Bochner integrable, $(\mathcal{F}_t, J)$–adaptive sequence. The sign $\bar{H}$ denotes the complement of an event H. As usual, the word a.e. is frequently omitted. A stopping time is a map $\tau : \Omega \to J$ such that $(\tau \leq t) \in \mathcal{F}_t,\ t \in J$. A stopping time τ is called simple if it takes finitely many values in J. Let T be the set of all simple stopping times. For $\tau \in T$, write $E^\tau = E^{\mathcal{F}_\tau}$, and let

$$J(\tau) = \{t : t \in J, t \geq \tau\},$$

$$T(\tau) = \{\sigma : \sigma \in T, \sigma \geq \tau\}.$$

AMS 1980 subject classifications: 60G48, 60B11.

Keywords and phrases: Mils, Banach space, Vitali condition, essential convergence, stochastic convergence.

*Research supported by the National Natural Science Foundation of China.

ISBN 0-12-231050-0

We denote $s.\lim(e.\lim)$ the stochastic limit (essential limit) in the norm topology of a family of E–valued r.v.'s.

Definition. (X_t) *is called*

(i) a pramart ([3]), if

$$\forall\, \epsilon > 0 \quad \lim_{\sigma \in T} \sup_{\tau \in T(\sigma)} P\{\|E^{\sigma} X_{\tau} - X_{\sigma}\| > \epsilon\} = 0;$$

(ii) a mil(1) (martingale in the limit ([5])), if

$$e.\lim_{s.\in J} e.\sup\nolimits_{t \in J(s)} \|E^{s} X_t - X_s\| = 0;$$

(iii) a mil(2), if

$$\forall\, \epsilon > 0 \quad \lim_{\sigma \in T} P\{e.\sup\nolimits_{t \in J(\sigma)} \|E^{\sigma} X_t - X_{\sigma}\| > \epsilon\} = 0;$$

(iv) a mil(3), if

$$\forall\, \epsilon > 0 \quad \lim_{\sigma \in T} \sup_{t \in J(\sigma)} P\{\|E^{\sigma} X_t - X_{\sigma}\| > \epsilon\} = 0.$$

Using the well–known localization procedure (see, e.g., [3]), we can easily show that (i) is equivalent to the following

$$\forall\, \epsilon > 0 \quad \lim_{\sigma \in T} \mathrm{P}\{e.\sup\nolimits_{\tau \in T(\sigma)} \|E^{\sigma} X_{\tau} - X_{\sigma}\| > \epsilon\} = 0.$$

Hence, pramart $\Longrightarrow$ mil(2) $\Longrightarrow$ mil(3). For $\sigma \in T$ define

$$g(\sigma) = e.\sup\nolimits_{t \in J(\sigma)} \|E^{\sigma} X_t - X_{\sigma}\|.$$

Then it is easy to see that for $\sigma \in T$ and $s \in J$

$$g(s) \geq g(\sigma) \quad \text{on} \quad (\sigma = s).$$

Thus, for any $s \in J$,

$$e.\sup_{\sigma \in T(s)} g(\sigma) = e.\sup_{t \in J(s)} g(t).$$

We write $s.\limsup(s.\liminf)$ for a stochastic upper (lower) limt (see [3], p 88). Then

$$\begin{aligned} s.\limsup_{\sigma \in T} g(\sigma) &\leq e.\limsup_{\sigma \in T} g(\sigma) = e.\lim_{\tau \in T}\{e.\sup_{\sigma \in T(\tau)} g(\sigma)\} \\ &= e.\lim_{s \in J}\{e.\sup_{\sigma \in T(s)} g(\sigma)\} = e.\lim_{s \in J}\{e.\sup_{t \in J(s)} g(t)\} = e.\limsup_{t \in J} g(t), \end{aligned}$$

and mil(1) $\Longrightarrow$ mil(2). Hence, we always have

$$mil(1) \Longrightarrow mil(2) \Longrightarrow mil(3).$$

If $(\mathcal{F}_t, J)$ satisfies the Vitali condition V, then ([3]) pramart $\Longrightarrow$ mil(1). It is also known that mil(1) $\not\Longrightarrow$ pramart ([3]), and mil(3) $\not\Longrightarrow$ mil(2) $\not\Longrightarrow$ mil(1) ([7]). But, if (X_t) is of class B, i.e., $\limsup_T \int \|X_\tau\| < \infty$, and if E has the Radon–Nikodym property, then mil(3) $\Longrightarrow$ pramart (Theorem 3 below). If $J = N \equiv \{1, 2, \cdots\}$, then ([7]) mil(3) is equivalent to the mil defined by Talagrand [6]:

$$\forall \epsilon > 0 \quad \lim_n \sup_{m \geq n} \mathrm{P}\{ \sup_{n \leq k \leq m} \|E^k X_m - X_k\| > \epsilon\} = 0.$$

This can be shown by the following fact: for any $\epsilon > 0$,

$$\sup_{m \geq n} \mathrm{P}\{\|E^{\sigma_n^m} X_m - X_{\sigma_n^m}\| > \epsilon\} = \sup_{m \geq n} \mathrm{P}\{ \sup_{n \leq k \leq m} \|E^k X_m - X_k\| > \epsilon\}$$
$$\geq \sup_{\sigma \in T(n)} \sup_{m \in T(\sigma)} \mathrm{P}\{\|E^\sigma X_m - X_\sigma\| > \epsilon\} \geq \sup_{m \geq n} \mathrm{P}\{\|E^{\sigma_n^m} X_m - X_{\sigma_n^m}\| > \epsilon\}.$$

where, $\sigma_n^m = \inf\{n \leq k \leq m : \|E^k X_m - X_k\| > \epsilon\}$ $(\inf\{\phi\} = m)$. Similarly, if we extend Talagrand's mil to processes indexed by a directed set J as the following:

$$\forall \epsilon > 0 \quad \lim_t \sup_{s \in J(t)} \mathrm{P}\{e. \sup_{s \geq u \in J(t)} \|E^u X_s - X_u\| > \epsilon\} = 0,$$

then it is easy to check that Talagrand's mil $\Longrightarrow$ mil(3).

Millet and Sucheston ([3], Theorem 7.2,) proved that if $(\mathcal{F}_t, J)$ satisfies the ordered Vitali condition V' (hence the Vitali condition V holds), then every L^1–bounded, real–valued mil(1) (X_t) converges essentially. Frangos ([2], Theorem 3.2) proved that if E is a separable dual space and J has a countable cofinal subset, and if the Vitali condition V holds, then every E–valued pramart (X_t) of class (d) (i.e., $\liminf_{t \in J} \int \|X_t\| < \infty$) converges essentially in the norm topology. When $J = N$, Talagrand [6] proved that if E has the Radon–Nikodym property, then every mil(3) (X_n) of class (d) converges almost everywhere in the norm topology. The papers mentioned above motivated the present work. We shall prove the following results:

Theorem 1. *Suppose that (X_t) is a real–valued mil(3) satisfying*

$$\min\{\liminf_{t \in J} \int X_t^+, \ \liminf_{t \in J} \int X_t^-\} < \infty.$$

Then $(X_\tau, \tau \in T)$ converges stochastically to a finite limit $s \lim_{\tau \in T} X_\tau$. Moreover, if $(\mathcal{F}_t, J)$ satisfies the Vitali condition V, then (X_t) converges essentially to a finite limit $e.\lim_J X_t$.

Theorem 2. *Let E be a Banach space with the Radon–Nikodym property. If (X_t) is an E–valued mil(3) of class (d), then $(X_\tau, \tau \in T)$ converges stochastically in the norm topology.*

Moreover, if $(\mathcal{F}_t, J)$ satisfies the Vitali condition V, then (X_t) converges essentially in the norm topology.

By Chatterji's Theorem ([1]) and Millet & Sucheston [3], Theorem 5.1, we get the following

Corollary. *Let E be a Banach space. We have $(i) \Longleftrightarrow (ii)$ and $(iii) \Longleftrightarrow (iv)$:*
(i) E has the Radon–Nikodym property;
(ii) for every E–valued mil(3) of class (d) $s.\lim_T X_\tau$ exists;
(iii) $(\mathcal{F}_t, J)$ satisfies the Vitali condition V;
(iv) every real–valued mil(3) of class (d) converges essentially.

Theorem 3. *Let E be a Banach space with the Radon–Nikodym property.*
(a) If (X_t) is of class B, then pramart $\Longleftrightarrow$ mil(3).
(b) If (X_t) is uniformly integrable, then (X_t) is a mil(3) if and only if $s.\lim_T X_\tau$ exists.

2. Proof of Theorem 1. Without loss of generality, we may and do assume that $\liminf_{t\in J} \int X_t^+ < \infty$. Assume that $(X_\tau, \tau \in T)$ does not converge stochastically. Then, there exist $a < b, c > 0$ such that $\mathrm{P}(A) > c$, where

$$A = \{s.\liminf_{\tau\in T} X_\tau < a < b < s.\limsup_{\tau\in T} X_\tau\}.$$

We may assume that $a = 0$ and $b = 1$. The above assumptions and Lemma 1 below imply that for any $\epsilon > 0, t \in J$, we can find $\sigma, \tau \in T(t)$ such that

$$\mathrm{P}\{(X_\sigma > 1) \cap A\} > \mathrm{P}(A) - \epsilon, \quad \mathrm{P}\{(X_\tau < 0) \cap A\} > \mathrm{P}(A) - \epsilon.$$

By the definition of the mil(3), there exists $(s_n) \subset J$ such that $s_n \le s_{n+1}$ and

$$\sup_{\sigma\in T(s_n)} \sup_{t\in J(\sigma)} \mathrm{P}\{|E^\sigma X_t - X_\sigma| > 1/3\} < \frac{c}{2^{n+3}}, \qquad n \ge 1.$$

Take $\sigma_1 \in T(s_1)$ and $A_1 \in \mathcal{F}_{\sigma_1}$ such that

$$\mathrm{P}\{(A \setminus A_1) \cup (A_1 \setminus A)\} < \frac{c}{2^6}, \quad \mathrm{P}(B_1) > \mathrm{P}(A) - \frac{c}{2^6},$$

where $B_1 = (X_{\sigma_1} > 1) \cap A_1$. Then pick $\tau_1 \in T(\sigma_1)$ such that $\mathrm{P}(C_1) > \mathrm{P}(A) - \frac{c}{2^5} > \mathrm{P}(B_1) - \frac{c}{2^4}$, where $C_1 = (X_{\tau_1} < 0) \cap B_1$. Assume that $(\sigma_i, \tau_i, B_i, C_i), 1 \leq i < k$, have been chosen, pick $\sigma_k \in T(s_k) \cap T(\tau_{k-1}), A_k \in \mathcal{F}_{\sigma_k}$, such that

$$\mathrm{P}\{(A \setminus A_k) \cup (A_k \setminus A)\} < \frac{c}{2^{k+5}}, \quad \mathrm{P}(B_k) > \mathrm{P}(A) - \frac{c}{2^{k+5}},$$

where $B_k = (X_{\sigma_k} > 1) \cap A_k$; then choose $\tau_k \in T(\sigma_k)$ such that

$$\mathrm{P}(C_k) > \mathrm{P}(A) - \frac{c}{2^{k+4}} > \mathrm{P}(B_k) - \frac{c}{2^{k+3}},$$

where $C_k = (X_{\tau_k} < 0) \cap B_k$. Thus we get $(\sigma_k, \tau_k, B_k, C_k), k \geq 1$. Since $\liminf_{t \in J} \int X_t^+ < \infty$, we can choose $t_n \in J(\tau_n)$ such that $\sup_n \int X_{t_n}^+ < \infty$. For any fixed $l \in N$, let

$$H_i = \{X_{\sigma_i} - E^{\sigma_i} X_{t_l} > 1/3\},$$

$$K_i = \{E^{\tau_i} X_{t_l} - X_{\tau_i} > 1/3\}, \qquad 1 \leq i \leq l.$$

Then

$$\mathrm{P}(H_i) < \frac{c}{2^{i+3}}, \quad \mathrm{P}(K_i) < \frac{c}{2^{i+3}}, \qquad 1 \leq i \leq l.$$

Pick $D_0 \in \mathcal{F}_{\sigma_1}, \mathrm{P}(D_0) < c/2^2$, and let

$$M_1 = B_1 \setminus (H_1 \cup D_0), \qquad D_1 = M_1 \cap (\bar{C}_1 \cup K_1).$$

Then $\mathrm{P}(M_1) > c/2, D_1 \cap D_0 = \phi, \mathrm{P}(D_1) < \mathrm{P}(K_1) + \mathrm{P}(B_1) - \mathrm{P}(C_1) < c/2^3$, and

$$\int_{M_1} X_{t_l} \geq \int_{M_1} (X_{\sigma_1} - 1/3) > 2\mathrm{P}(M_1)/3,$$

$$\int_{M_1 C_1 \bar{K}_1} X_{t_l} \leq \int_{M_1 C_1 \bar{K}_1} (X_{\tau_1} + 1/3) < \mathrm{P}(M_1)/3.$$

Hence,

$$\int_{D_1} X_{t_l} = \int_{M_1} X_{t_l} - \int_{M_1 C_1 \bar{K}_1} X_{t_l} > \mathrm{P}(M_1)/3 > c/6.$$

Assume that $(M_j, D_j), 1 \leq j \leq i - 1 < l$, have been chosen, where

$$\mathrm{P}(M_j) > c/2, \ \mathrm{P}(D_j) < \frac{c}{2^{j+2}}, \quad D_j \cap \Big[\bigcup_{k=0}^{j-1} D_k\Big] = \phi,$$

and

$$\int_{D_j} X_{t_l} > c/6, \qquad 1 \leq j \leq i - 1 < l.$$

Let

$$M_i = B_i \setminus \left[H_i \cup \left(\cup_{j=0}^{i-1} D_j\right)\right], \quad D_i = M_i \cap (\bar{C}_i \cup K_i).$$

Then

$$D_i \cap \left[\bigcup_{k=0}^{i-1} D_k\right] = \phi,$$

$$\mathrm{P}(D_i) < \mathrm{P}(K_i) + \mathrm{P}(B_i) - \mathrm{P}(C_i) < \frac{c}{2^{i+2}},$$

$$\mathrm{P}(M_i) \geq \mathrm{P}(B_i) - \mathrm{P}(H_i) - \sum_{j=0}^{i-1} \mathrm{P}(D_j) > c - \sum_{j>1} \frac{c}{2^j} = c/2,$$

and

$$\int_{D_i} X_{t_l} > \mathrm{P}(M_i)/3 > c/6.$$

Therefore,

$$\int X_{t_l}^+ \geq \sum_{i=1}^{l} \int_{D_i} X_{t_l} > lc/6.$$

and $\sup_l \int X_{t_l}^+ = \infty$, a contradiction. Hence, $s.\lim_{\tau \in T} X_\tau$ exists. Since $\liminf_t \int X_t^+ < \infty$, by Fatou's lemma, $s.\lim_{\tau \in T} X_\tau < \infty$. Now we prove that $s.\lim_{\tau \in T} X_\tau > -\infty$. Assume that the conclusion does not hold. Then there exists $h > 0$ such that

$$\mathrm{P}\left\{s.\lim_t X_t = -\infty\right\} > h.$$

Clearly, we can choose $(t_n) \subset J$ such that $s.\lim_n X_{t_n} = s.\lim_t X_t$ and $\sup_n \int X_{t_n}^+ < \infty$. For any $0 < \epsilon < h/2$, there exists $b > 0$ such that $\sup_n \int X_{t_n}^+ < \epsilon b/4$. Choose $u \in J$ such that for any $s, t \in J(u)$ and $s \geq t$,

$$\mathrm{P}\left\{X_t - E^t X_s > \epsilon\right\} < h/2. \tag{1}$$

For any fixed $t \in J(u)$, we can choose $g > 0$ such that

$$\mathrm{P}(X_t > -g) > 1 - \epsilon/4,$$

and

$$4(\epsilon + b + g) < \epsilon g^2.$$

Since $\mathrm{P}\left\{s \lim_n X_{t_n} = -\infty\right\} > h > 0$, we can choose $s \in J(t) \cap (t_n)$ such that

$$\mathrm{P}(X_s^- > g^2) > h - \epsilon/4.$$

Since $\int X_s^+ < \epsilon b/4$,

$$\mathrm{P}\{E^t X_s^+ \geq b\} < \epsilon/4.$$

Therefore,

$$\begin{aligned}
\mathrm{P}\{X_t - E^t X_s \leq \epsilon\} &\leq \epsilon/4 + \mathrm{P}\{[X_t - E^t X_s \leq \epsilon] \cap (X_t > -g)\} \\
&\leq 1 - h + \epsilon/2 + \mathrm{P}\{[E^t X_s \geq -(\epsilon + g)] \cap (X_s^- > g^2)\} \\
&\leq 1 - h + 3\epsilon/4 + \mathrm{P}\{[E^t X_s^- \leq \epsilon + b + g] \cap (X_s^- > g^2)\} \\
&\leq 1 - h + 3\epsilon/4 + \frac{1}{g^2} \int_{[E^t X_s^- \leq \epsilon + b + g]} X_s^- \\
&< 1 - h/2,
\end{aligned}$$

which contradicts (1). Hence the conclusion holds. If $(\mathcal{F}_t, J)$ satisfies the Vitali condition V, then, by Theorem 4.1 in [3] (also see [4], page 45), (X_t) converges essentially to a finite r.v. ◇

Lemma 1. *Suppose that $A \subset (s.\limsup_{\tau \in T} X_\tau > 1)$ and $P(A) > c$. Then*

$$\limsup_{\tau \in T} P(X_\tau > 1, A) \geq c. \tag{2}$$

Proof. Assume that (2) does not hold. Let $B_\tau = (X_\tau > 1, A)$. Then we can choose $t_1 \in J$ such that

$$\sup_{\tau \in T(t_1)} \mathrm{P}(B_\tau) = c_1 < c. \tag{3}$$

By the definition of $s.\limsup_\tau X_\tau$, for any event $C \subset A$ such that $\mathrm{P}(C) > 0$, we have $\limsup_{\tau \in T} \mathrm{P}(X_\tau > 1, C) > 0$. Choose $\tau_1 \in T(t_1)$ such that $\mathrm{P}(B_{\tau_1}) \geq c_1/2$ and let $C_1 = (X_{\tau_1} > 1)$. Assume that

$$\limsup_{\tau \in T} \mathrm{P}(B_\tau, \bar{C}_1) = c_2.$$

Choose $t_2 \in J(\tau_1)$ and $\tau_2 \in T(t_2)$ such that $\mathrm{P}(C_2, A \setminus C_1) \geq c_2/2$, where $C_2 = (X_{\tau_2} > 1)$. Repeating the above procedure, we get $(t_n, \tau_n, c_n, C_n), n \geq 1$, such that $t_n \in J(\tau_{n-1}), \tau_n \in T(t_n), C_n = (X_{\tau_n} > 1)$, and

$$2\mathrm{P}(C_n, A \setminus (\cup_{i=1}^{n-1} C_i)) \geq \limsup_\tau \mathrm{P}(B_\tau, A \setminus (\cup_{i=1}^{n-1} C_i)) = c_n.$$

Clearly, $c_n \downarrow 0$. Hence,

$$\limsup_\tau \mathrm{P}(B_\tau, A \setminus (\cup_{i \geq 1} C_i)) = \lim_n c_n = 0,$$

and $P(A \setminus (\cup_{i\geq 1} C_i)) = 0$. Choose $k > 1$ such that $P(A \setminus (\cup_{i>k} C_i)) < (P(A) - c_1)/2$. Let $\tau = \tau_i$ on $C_i, 1 \leq i \leq k$; $= t_{k+1}$ on $\Omega \setminus (\cup_{i=1}^k C_i)$. Then $\tau \in T(t_1)$ and

$$P(B_\tau) \geq P(A, \cup_{i=1}^k C_i) = P(A) - P(A \setminus (\cup_{i>k} C_i)) > (P(A) + c_1)/2 > c_1,$$

which contradicts (3), and (2) holds. $\diamond$

3. Proof of Theorem 2. To prove Theorem 2 we need the following lemma. When $J = N$ the lemma was proved by Talagrand [6], Theorem 6. The proof here follows the same idea in [6].

Lemma 2. *Let E be a separable Banach space, E^* the dual space of E, and D a countable norming subset of E^*, i.e., $\|x\| = \sup\{f(x), f \in D\}$ for all $x \in E$. If (X_t) is an E–valued mil(3) of class (d), and if $s.\lim_{\tau\in T} f(X_\tau) = 0$ for all $f \in D$, then $s.\lim_{\tau \in T} \|X_\tau\| = 0$.*

Proof. Assume that the conclusion does not hold. Then there exists $a > 0$ such that for any $\sigma \in T$ there exists $\tau \in T(\sigma)$, $P(\|X_\tau\| > a) > a$. Choose $(t_n) \subset J$ such that

$$\sup_{\sigma\in T(t_n)} \sup_{t \in J(\sigma)} P\{\|E^\sigma X_t - X_\sigma\| > a/4\} < \frac{a}{2^{n+2}}, \qquad n \geq 1.$$

Pick $\tau_1 \in T(t_1)$ such that $P(B_1) > a$, where $B_1 = (\|X_{\tau_1}\| > a)$. Then there exists a finite subset G_1 of D such that $P(\hat{B}_1) > 7a/8$, where

$$\hat{B}_1 = \{f(X_{\tau_1}) > a \text{ for some } f \in G_1\} \cap B_1.$$

By the assumption of the lemma there exists $\sigma_1 \in T(\tau_1)$ such that $P(C_1) < a/8$, where

$$C_1 = \{f(X_{\sigma_1}) > a/4 \text{ for some } f \in G_1\}.$$

Assume that $(\tau_k, \sigma_k, B_k, G_k, \hat{B}_k, C_k), 1 \leq k \leq i-1$, have been chosen, pick $\tau_i \in T(t_i) \cap T(\sigma_{i-1})$ such that $P(B_i) > a$, where $B_i = (\|X_{\tau_i}\| > a)$. Choose a finite subset G_i of D such that $P(\hat{B}_i) > 7a/8$, where

$$\hat{B}_i = \{f(X_{\tau_1}) > a \text{ for some } f \in G_i\} \cap B_i.$$

Then pick $\sigma_i \in T(\tau_i)$ such that $P(C_i) < \frac{a}{2^{i+2}}$, where

$$C_i = \{f(X_{\sigma_i}) > a/4 \text{ for some } f \in G_i\}.$$

Thus we get $(\tau_k, \sigma_k, B_k, G_k, \hat{B}_k, C_k), k \geq 1$. Now we can choose $(s_n) \subset J$ such that $s_n \geq \sigma_n$ and $\sup_n \int \|X_{s_n}\| < \infty$. Then for any fixed $l \in N$, pick $A_0 \in \mathcal{F}_{\tau_1}, \mathrm{P}(A_0) < a/8$, and let

$$H_1 = \{\|E^{\tau_1} X_{s_l} - X_{\tau_1}\| > a/4\},$$

$$F_1 = \{\|E^{\sigma_1} X_{s_l} - X_{\sigma_1}\| > a/4\},$$

$$M_1 = \hat{B}_1 \setminus (H_1 \cup A_0), \quad K_1 = F_1 \cup C_1.$$

Then

$$\mathrm{P}(H_1) < \frac{a}{2^{2+1}}, \quad \mathrm{P}(F_1) < \frac{a}{2^{2+1}},$$
$$\mathrm{P}(K_1) < \frac{a}{2^{1+1}}, \quad \mathrm{P}(M_1) > \frac{a}{8}.$$

Assume that the elements of G_1 are $(f_j, 1 \leq j \leq k_1)$. Let

$$S_i = M_1 \cap \{f_i(X_{\tau_1}) > a; f_j(X_{\tau_1}) \leq a, j < i\},$$

then $\bigcup_{i=1}^{k_1} S_i = M_1$, $M_1 \cap H_1 = \phi$, and

$$\int_{S_i} f_i(X_{s_l}) > \int_{S_i} [f_i(X_{\tau_1}) - a/4] > 3a\mathrm{P}(S_i)/4,$$

$$\int_{S_iK_1} f_i(X_{s_l}) < \int_{S_iK_1} [f_i(X_{\sigma_1}) + a/4] \leq a\mathrm{P}(S_i)/2.$$

Let $A_1 = M_1K_1$, then $A_0A_1 = \phi, \mathrm{P}(A_1) < a/4$, and

$$\begin{aligned} \int_{A_1} \|X_{s_l}\| &\geq \sum_{i=1}^{k_1} \int_{S_iK_1} f_i(X_{s_l}) \\ &> \sum_{i=1}^{k_1} (3a/4 - a/2)\mathrm{P}(S_i) = a\mathrm{P}(M_1)/4 > a^2/32. \end{aligned} \tag{4}$$

Assume that $A_i, 0 \leq i \leq j - 1 < l$, have been chosen, where

$$\mathrm{P}(A_i) < \frac{a}{2^{i+1}}, \quad A_i \cap A_k = \phi, \ 0 \leq i \neq k \leq j-1,$$

and

$$\int_{A_i} \|X_{s_l}\| > a^2/32.$$

Let

$$H_j = \{\|E^{\tau_j} X_{s_l} - X_{\tau_j}\| > a/4\},$$

$$F_j = \{\|E^{\sigma_j} X_{s_l} - X_{\sigma_j}\| > a/4\},$$

$$M_j = \hat{B}_j \setminus (H_j \cup (\cup_{i=0}^{j-1} A_i)), \quad K_j = F_j \cup C_j.$$

Then

$$\mathrm{P}(H_j) < \frac{a}{2^{2+j}}, \quad \mathrm{P}(F_j) < \frac{a}{2^{2+j}},$$

$$\mathrm{P}(K_j) < \frac{a}{2^{1+j}}, \quad \mathrm{P}(M_j) > \frac{a}{8}.$$

Let $A_j = M_j K_j$, then $A_j \cap (\cup_{i=0}^{j-1} A_i) = \phi, \mathrm{P}(A_j) < a/2^{j+1}$. As the proof of (4) we can get

$$\int_{A_j} \|X_{s_l}\| \geq a\mathrm{P}(M_j)/4 > a^2/32.$$

Hence,

$$\int \|X_{s_l}\| \geq \sum_{j=1}^{l} \int_{A_j} \|X_{s_l}\| > la^2/32,$$

and $\sup_l \int \|X_{s_l}\| = \infty$, a contradiction. ◇

Now we prove the theorem. We do it in two steps.

(I) Assume that E is separable. Let $D = \{f_i, i \in N\}$ be a norming subset of E^*. Clearly, for each $f_i \in D$, $(f_i(X_t), J)$ is a real-valued mil(3) of class (d). Therefore, by Theorem 1, $s.\lim_T f_i(X_\tau) = g_i$ exists. Choose $(t_n) \subset J$ and $(t_k^i) \subset J$ such that $t_k^{i+1} \in J(t_k) \cap J(t_k^i)$,

$$\sup_{\sigma \in T(t_n)} \sup_{t \in J(\sigma)} \mathrm{P}\{\|E^\sigma X_t - X_\sigma\| > 1/n\} < 1/n, \qquad n \geq 1, \tag{5}$$

and for each $i, k \geq 1$, and $\tau \in T(t_k^i)$

$$\mathrm{P}\{|f_i(X_\tau) - g_i| > 1/k\} < 1/k. \tag{6}$$

Now choose $(s_n) \subset J$ such that $s_0 = t_1, s_n \in J(t_n^n) \cap J(s_{n-1})$, and $\sup_n \int \|X_{s_n}\| < \infty$. By (5), (X_{s_n}) is an L^1-bounded, discrete parameter mil(3), and by [6], Theorem 8, there exists a Bochner integrable r.v. X, X_{s_n} converges to X almost everywhere in the norm topology. Hence, by (6),

$$f_i(X) = \lim_n f_i(X_{s_n}) = s.\lim f_i(X_\tau), \qquad i \in N.$$

Let

$$Y_t = E^t X, \qquad Z_t = X_t - Y_t.$$

Then (Y_t) is a right closed martingale, and (Z_t) is a mil(3) of class (d). Therefore, for each $i \in N$, $s.\lim_T f_i(Z_\tau)$ exists. Since $s.\lim_T Y_\tau = X$,

$$f_i(X) = s.\lim_T f_i(Y_\tau) = s.\lim_T f_i(X_\tau) - s.\lim_T f_i(Z_\tau).$$

Thus

$$s.\lim_T f_i(Z_\tau) = 0, \quad f_i \in D,$$

and, by Lemma 2, $s.\lim_T \|Z_\tau\| = 0$, $s.\lim_T X_\tau = X$.

(II) In a general case. Assume that $s.\lim_T X_\tau$ does not exists. Then there exists $a > 0$ such that for any $t \in J$ there exists $\tau, \sigma \in T(t)$,

$$\mathrm{P}(\|X_\tau - X_\sigma\| > a) > a.$$

Choose $(t_n) \subset J$, $(\tau_n^i) \subset T$, and $(s_n) \subset J$ such that (5) holds, $s_0 = t_1$, $\tau_n^i \in T(s_{n-1}), s_n \in J(\tau_n^i) \cap J(t_n), i = 1, 2$, $\sup_n \int \|X_{s_n}\| < \infty$, and

$$\mathrm{P}(\|X_{\tau_n^1} - X_{\tau_n^2}\| > a) > a. \tag{7}$$

Let

$$\bar{J} = \{\tau_n^i(\omega) : \quad \text{for some } \omega \in \Omega, n \geq 1, \text{ and } i = 1 \text{ or } 2\} \cup (s_n).$$

Since X_t being Bochner integrable is separably valued and $\bar{J}$ is a countable subset of J, $(X_t, \bar{J})$ is separably valued and is a mil(3) of class (d) via (5) and the above choises. Let $\bar{T}$ be the set of all simple stopping times with respect to $(\mathcal{F}_t, \bar{J})$. Then, by (I), $s.\lim_{\bar{T}} X_\tau$ exists. Since (τ_n^i) is a cofinal subset of $\bar{T}$, $i = 1, 2$,

$$s.\lim_n X_{\tau_n^1} = s.\lim_n X_{\tau_n^2} = s.\lim_{\bar{T}} X_\tau,$$

which contradicts (7). Hence $s.\lim_T X_\tau$ exists. If $(\mathcal{F}_t, J)$ satisfies the Vitali condition V, then, by Theorem 12.3 in [3], $e.\lim_J X_t$ exists. ◇

4. Proof of Theorem 3. If (X_t) is a mil(3) of class B, then, by Theorem 2, $s.\lim_T X_\tau$ exists, and (X_t) is a pramart ([8], Theorem 1). If (X_t) is a pramart, clearly, it is a mil(3). Now suppose that (X_t) is uniformly integrable. If (X_t) is a mil(3), then, by Theorem 2,

$s.\lim_T X_\tau$ exists. Conversely, if $s.\lim_T X_\tau = X$ exists, then $X_t \to X$ in L^1 and $E^\tau X \to X$ in L^1, and for any $\epsilon > 0$

$$\begin{aligned}
&\lim_{\tau \in T} \sup_{t \in J(\tau)} \mathrm{P}(\|E^\tau X_t - X_\tau\| > 3\epsilon) \\
&\leq \lim_{\tau \in T} \sup_{t \in J(\tau)} \{\mathrm{P}(\|E^\tau (X_t - X)\| > \epsilon) + \mathrm{P}(\|E^\tau X - X\| > \epsilon) + \mathrm{P}(\|X - X_\tau\| > \epsilon)\} \\
&\leq \lim_{\tau \in T} \sup_{t \in J(\tau)} \int \|(X_t - X)\|/\epsilon = 0,
\end{aligned}$$

(X_t) is a mil(3). ◇

Acknowledgments. The authors would like to thank Professor Sucheston for his helpful discussion and comments and to the referee for his very careful reading and valuable remarks.

REFERENCES

[1] Chatterji, S. D. (1968). Martingale convergence and the Radon–Nikodym theorem, *Math. Scand.* **22**, 21–41.

[2] Frangos, N. E. (1985). On convergence of vector valued pramarts and subpramarts, *Can. J. Math.* **37**, 260–270.

[3] Millet, A. & Sucheston, L. (1980). Convergence of classes of amarts indexed by directed sets, *Can. J. Math.* **32**, 86–125.

[4] Millet, A. & Sucheston, L. (1981). On regularity of multiparameter amarts and martingales, *Z. Wahrsch. verw. Get.* **56**, 21–45.

[5] Mucci, A. G. (1976). Another Martingale convergence theorem, *Pacific J. Math.* **64**, 539–541.

[6] Talagrand, M. (1985). Some structure results for martingales in the limit and pramarts, *Ann. Probab.* **13**, 1192–1203.

[7] Wang, Z. P. (1988). Mil-type sequences and convergence of GFT, *Acta. Math. Sinica* **31**, 372–380.

[8] Wang, Z. P. & Xue, X. H. (1988). Remarks on the pramart system, preprint.